Springer Series in Mat

Volume 195

For further volumes:
http://www.springer.com/series/856

The Springer Series in Materials Science covers the complete spectrum of materials physics, including fundamental principles, physical properties, materials theory and design. Recognizing the increasing importance of materials science in future device technologies, the book titles in this series reflect the state-of-the-art in understanding and controlling the structure and properties of all important classes of materials.

Vadim P. Veiko · Vitaly I. Konov
Editors

Fundamentals of Laser-Assisted Micro- and Nanotechnologies

 Springer

Editors
Vadim P. Veiko
Department of Laser-based Technologies
NRU ITMO
St. Petersburg
Russia

Vitaly I. Konov
Natural Sciences Center
General Physics Institute RAS
Moscow
Russia

ISSN 0933-033X
ISSN 2196-2812 (electronic)
ISBN 978-3-319-37906-7
ISBN 978-3-319-05987-7 (eBook)
DOI 10.1007/978-3-319-05987-7
Springer Cham Heidelberg New York Dordrecht London

Springer is part of Springer Science+Business Media (www.springer.com)

Preface

Over the past decades, laser microfabrication has successfully spread in micro-electronics, micromechanics, and some other areas. Nowadays, laser-induced microprocessing and, in some cases, nanoprocessing of materials constitute new and very promising application areas, such as MEMS, photonic, and fluidic devices, etc., for information, biomedical, and other technologies. In these areas, laser technologies are in the stage of transition from impressive demonstrations to practically feasible results for several reasons. First, short laser pulses of nano- and especially pico- and femtosecond duration, combined with modern focusing elements and beam scanning devices allow strong localization of laser pulse effects on the surface and in the bulk of materials. Second, the productivity of laser micro- and nanotechnologies is rapidly growing as a result of the development of reliable industrial lasers that emit intense radiation pulses with ultra-high repetition rate (up to the MHz range). The mean beam power of these lasers is now approaching the level of 1 kW and becomes comparable with that of lasers used for macro-processing of materials (welding, cutting, surface hardening, etc.).

Accordingly, various phenomena that can be induced by short intense pulses: surface and bulk modification of materials, radiation-induced chemical reactions, various types of ablation, plasma formation should be better understood for development of efficient and novel technological procedures.

Naturally, one book cannot cover all topics of interest in the field of laser micro- and nano-laser technologies. We have selected some of them—currently hot topics in our opinion—based on presentations made by a number of the world's leading experts who participated in the International Symposium, "Fundamentals of Laser-Assisted Micro- and Nanotechnologies" (FLAMN) recently. This is a serial and popular event that takes place every 3 years in St. Petersburg, Russia.

The book consists of 13 chapters divided into five parts:

  I. Laser–Matter interaction phenomena;
 II. Nanoparticles related technologies and problems;
III. Surface and thin films phenomena and applications;
 IV. Bulk micro structuring of transparent materials;
  V. Laser-Induced modification of polymers.

We expect that the reader will be able to learn important features of (i) defect formation in laser-heated surface layers, (ii) laser ablation of organic and inorganic materials, (iii) metal oxidation by laser pulses, and (iv) reversible and irreversible changes of the structure and optical properties in the bulk of a material, induced by tightly focused ultrashort laser pulses. Special emphasis is made on the application of nanoparticles as a positive factor in materials processing. Several advanced applications of laser microtechnologies are included such as laser printing and laser fabrication of three-dimensional polymer devices.

We hope that the book will be of use for researchers and engineers, as well as for young scientists and students already working or planning to work in the area of laser applications. Finally, we wish to express our appreciation to all contributors for their friendly cooperation in preparing the book. Special thanks to Dr. Koji Sugioka to whom the idea to publish this book belongs.

St. Petersburg   Vadim P. Veiko
Moscow   Vitaly I. Konov

Contents

Part II   Nanoparticles Related Technologies and Problems

Part V   Laser-Induced Modification of Polymers

Contributors

A. Barchanski Nanotechnology Department, Laser Zentrum Hannover e.V, Hannover, Germany

Nikita M. Bityurin Institute of Applied Physics, Nizhny Novgorod, Russian Federation

Jodie E. Bradby Electronic Materials Engineering, Research School of Physics and Engineering, The Australian National University, Canberra, Australia

Boris N. Chichkov Nanotechnology Department, Laser Zentrum Hannover e.V, Hannover, Germany

Nathalie Destouches Laboratory Hubert Curien, University of Lyon-University Jean Monnet, UMR 5516 CNRS, Saint-Étienne, France

A. B. Evlyukhin Nanotechnology Department, Laser Zentrum Hannover e.V, Hannover, Germany

Eugene G. Gamaly Research School of Physics and Engineering, Laser Physics Centre, The Australian National University, Canberra, Australia

Evgeny L. Gurevich Laser Applications Technology, Department of Mechanical Engineering, Ruhr-University Bochum, Bochum, Germany

Bianca Haberl Electronic Materials Engineering, Research School of Physics and Engineering, The Australian National University, Canberra, Australia

Frank Hubenthal Institut für Physik and Center for Interdisciplinary Nanostructure Science and Technology, Universität Kassel, Kassel, Germany

Saulius Juodkazis Faculty of Engineering and Industrial Sciences, Centre for Micro-Photonics, Swinburne University of Technology, Hawthorn, VIC, Australia; Department of Semiconductor Physics, Vilnius University, Vilnius, Lithuania

Eaman T. Karim Department of Materials Science and Engineering, University of Virginia, Charlottesville, VA, USA

Sergey M. Klimentov A. M. Prokhorov General Physics Institute of Russian Academy of Sciences, Moscow, Russia

Alexander V. Kolobov Nanoelectronics Research Institute, National Institute of Advanced Industrial Science and Technology, Tsukuba, Ibaraki, Japan

V. V. Kononenko A. M. Prokhorov General Physics Institute of RAS, Moscow, Russia

Vitaly I. Konov A. M. Prokhorov General Physics Institute of Russian Academy of Sciences, Moscow, Russia

A. Koroleva Nanotechnology Department, Laser Zentrum Hannover e.V, Hannover, Germany

Boris S. Luk'yanchuk Data Storage Institute, Agency for Science, Technology and Research, Singapore, Singapore

Mangirdas Malinauskas Physics Faculty, Department of Quantum Electronics, Vilnius University, Vilnius, Lithuania

Andreas Ostendorf Laser Applications Technology, Department of Mechanical Engineering, Ruhr-University Bochum, Bochum, Germany

Alexander G. Poleshchuk Institute of Automation and Electrometry Siberian Branch of Russian Academy of Science, Novosibirsk, Russian Federation

Ludovic Rapp Research School of Physics and Engineering, Laser Physics Centre, The Australian National University, Canberra, Australia

C. Reinhardt Nanotechnology Department, Laser Zentrum Hannover e.V, Hannover, Germany

Andrei V. Rode Research School of Physics and Engineering, Laser Physics Centre, The Australian National University, Canberra, Australia

C. L. Sajti Nanotechnology Department, Laser Zentrum Hannover e.V, Hannover, Germany

Xiao Shizhou Laser Applications Technology, Department of Mechanical Engineering, Ruhr-University Bochum, Bochum, Germany

Junji Tominaga Nanoelectronics Research Institute, National Institute of Advanced Industrial Science and Technology, Tsukuba, Ibaraki, Japan

Michael I. Tribelsky Faculty of Physics, Lomonosov Moscow State University, Moscow, Russia; Moscow State Institute of Radioengineering Electronics and Automation, Technical University MIREA, Moscow, Russia

Tigran Vartanyan Center of Information Optical Technologies, St. Petersburg National Research University of Information Technologies, Mechanics and Optics, St. Petersburg, Russia

Vadim P. Veiko Mechanics and Optics Chair of Laser Technologies and Applied Ecology, St. Petersburg National Research University of Information Technologies, St. Petersburg, Russian Federation

Jim S. Williams Electronic Materials Engineering, Research School of Physics and Engineering, The Australian National University, Canberra, Australia

Chengping Wu Department of Materials Science and Engineering, University of Virginia, Charlottesville, VA, USA

Irina N. Zavestovskaya P. N. Lebedev Physical Institute of RAS, Moscow, Russia; National Research Nuclear University MEPhI, Moscow, Russia

Leonid V. Zhigilei Department of Materials Science and Engineering, University of Virginia, Charlottesville, VA, USA

U. Zywietz Nanotechnology Department, Laser Zentrum Hannover e.V, Hannover, Germany

Part I
Laser–Matter Interaction Phenomena

Chapter 1
Ultrafast Laser Induced Confined Microexplosion: A New Route to Form Super-Dense Material Phases

Ludovic Rapp, Bianca Haberl, Jodie E. Bradby, Eugene G. Gamaly, Jim S. Williams and Andrei V. Rode

Abstract Intense ultrafast laser pulses tightly focused in the bulk of transparent material interact with matter in the condition where the conservation of mass is fulfilled. A strong shock wave generated in the interaction region expands into the surrounding cold material and compresses it, which may result in the formation of new states of matter. Here we show that the extreme conditions produced in the ultrafast laser driven micro-explosions can serve as a novel microscopic laboratory for high pressure and temperature studies well beyond the pressure levels achieved in a diamond anvil cell.

L. Rapp · A. V. Rode (✉) · E. G. Gamaly
Laser Physics Centre, Research School of Physics and Engineering, The Australian National University, Oliphant Building #60, Canberra ACT 0200, Australia
e-mail: avr111@physics.anu.edu.au

L. Rapp
e-mail: ludovic.rapp@anu.edu.au

E. G. Gamaly
e-mail: eugene.gamaly@anu.edu.au

B. Haberl · J. E. Bradby · J. S. Williams
Electronic Materials Engineering, Research School of Physics and Engineering, The Australian National University, Canberra ACT 0200, Australia
e-mail: bianca.haberl@anu.edu.au

J. E. Bradby
e-mail: jodie.bradby@anu.edu.au

J. S. Williams
e-mail: jodie.bradby@anu.edu.au

V. P. Veiko and V. I. Konov (eds.), *Fundamentals of Laser-Assisted Micro- and Nanotechnologies*, Springer Series in Materials Science 195, DOI: 10.1007/978-3-319-05987-7_1, © Springer International Publishing Switzerland 2014

1.1 Introduction

Ultrafast laser pulse at a µJ energy level tightly focused inside the bulk of a transparent solid can easily generate the MJ/cm^3 energy density, higher than the Young modulus of any solid, within a focal volume less than a cubic micron. The pressure of the order of several TPa (1 TPa = 10 Mbar) inside a focal volume leads to formation of a cavity (void) surrounded by a shell of compressed material. These two features of the phenomenon delineate two areas in the high-pressure material studies and their potential applications. The first area relates to the formation of different 3D structures, like photonic crystals, waveguides, and gratings etc. making use of multiple voids, separated or interconnected. For these studies the most important part is the process of void formation by a rarefaction wave following the shock wave. In order to produce a void one has to generate a pressure in excess of the strength (the modulus) of a material. The second area of research relates to the studies of material transformations under high pressure-temperature conditions, which are possible to create in tabletop laboratory experiments using powerful ultrafast laser pulses. The interaction of a laser with matter at the intensity above the ionisation and ablation threshold proceeds in a similar way for all the materials. The material converts into plasma in a few femtoseconds at the very beginning of the pulse, changing the interaction to the laser-plasma mode, increasing the absorption coefficient and reducing the absorption length, which ensures fast energy release in a very small volume. A strong shock wave generated in the interaction region propagates into the surrounding cold material. The shock wave propagation is accompanied by the compression of the solid material at the wave front and decompression behind it, leading to the formation of a void inside the material. The laser and shock wave affected material is in the shell that surrounds the void. This shell is the major object for studies of new phases and new material formation in strongly non-equilibrium conditions of confined microexplosion. Single pulse action thereby allows a formation of various three-dimensional structures inside a transparent solid in a controllable and predictable way.

First notion that the extreme conditions produced in the ultrafast laser driven confined micro-explosion may serve as a novel microscopic laboratory for high pressure and temperature studies, well beyond the levels achieved in diamond anvil cell, was presented by Glezer and Mazur in 1997 [1]. Recently it was experimentally demonstrated that it is possible to create super-high pressure and temperature conditions in table-top laboratory experiments with ultra-short laser pulses focused inside transparent material to the level significantly above the threshold for optical breakdown [2–4]. The laser energy absorbed in a sub-micron volume confined inside a bulk of pristine solid is fully converted into the internal energy. Therefore high energy density, several times higher than the strength of any material, can be achieved with ~100 fs, 1 µJ laser pulses focused down to a 1 µm^3 volume inside the solid.

Let us first to underline the differences between the intense laser-matter interaction at the surface of a solid and the case when laser-matter interaction is confined deep inside a solid by comparing the pressure created at the absorption region at the same intensity and total absorbed energy. At the intensity well over the ionisation and ablation thresholds any material converts into plasma in a few fs time. Therefore the interaction proceeds most of the time in laser-plasma interaction mode. In these conditions the pressure at the ablated plume-solid interface (in laser-surface interaction) constitutes from the sum of thermal pressure of plasma next to the boundary plus the pressure from the recoil momentum of expanding plasma. Significant part of absorbed energy is spent on the expansion and heating of the ablated part of a solid. Therefore the ablation pressure in the case of surface interaction depends on the absorbed intensity by the power law $P_{abl} \propto I_{abs}^m$; $m < 1$. Alternatively, there is no expansion loss in confined interaction. Hence the maximum pressure is proportional to the absorbed intensity $P_{conf} \propto I_{abs}$ and it is almost twice larger than in the surface interaction mode.

Full description of the laser-matter interaction process and laser-induced material modification from the first principles embraces the self-consistent set of equations that includes the Maxwell's equations for the laser field coupling with matter, complemented with the equations describing the evolution of energy distribution functions for electrons and phonons (ions) and the ionisation state. A resolution of such a system of equations is a formidable task even for modern supercomputers. Therefore, the thorough analytical aprouch is needed. We split below this complicated problem into a sequence of simpler interconnected problems: the absorption of laser light, the ionisation and energy transfer from electrons to ions, the heat conduction, and hydrodynamic expansion.

1.2 Energy Density in Confined Ultra-Short Laser Interaction with Solids

The sequence of processes in the ultrashort laser induced microexplosion schematically presented in Fig. 1.1. In the following section we consider the major mechanisms of laser absorption, ionisation, and shock wave formation in confined geometry.

1.2.1 Absorbed Energy Density

Let us first define the range of laser and focussing parameters necessary for obtaining high pressure inside the interaction region. A 100 nJ laser pulse of duration $t_p \leq 100$ fs with the average intensity $I > 10^{14}$ W/cm^2 focussed into the area $S_{foc} \propto \lambda^2$ delivers the energy density > 10 J/cm^2, well above the ionisation and ablation thresholds for any material [5]. The focal volume has a complicated three-dimensional structure. As a first approximation (that is also useful for scaling purposes) the focal volume is the focal area multiplied by the absorption length. The absorbed laser energy per unit time and per unit volume during the pulse reads:

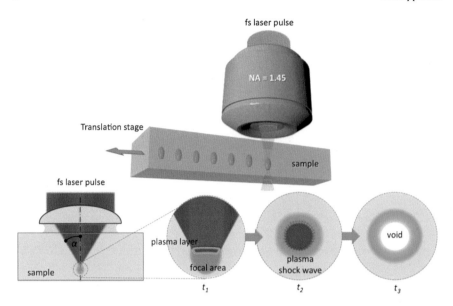

Fig. 1.1 Schematic representation of the experiments and the processes in fs-laser induced micro-explosion inside transparent dielectric. Three blown-up focal areas are shown in time sequences $t_1 < t_2 < t_3$. The first time slot t_1 during the pulse $\sim$170 fs shows the formation of the energy absorbing plasma layer inside the focal area; the second interval $t_2 \sim$ 1–100 ps shows hydrodynamic expansion of plasma and formation of a shock wave; the third $t_3 \sim$ 100–1 ns shows rarefaction wave, and formation of a void surrounded by a densified shell due to quenching

$$\frac{dE_{abs}}{dt} = \frac{2A}{l_{abs}} I\,(r, z, t) \tag{1.1}$$

l_{abs} is the electric field absorption depth $l_{abs} = c/\omega k$. We assume that the electric field exponentially decays inside a focal volume, $E = E_0 \exp\{-x/l_{abs}\}$ as it does in the skin layer; A is the absorption coefficient defined by the Fresnel formula [6] as the following:

$$A = \frac{4n}{(n+1)^2 + k^2} = \frac{2\varepsilon''}{\left|1 + \varepsilon^{1/2}\right|^2 k}. \tag{1.2}$$

The duration of a typical short pulse of $\sim$100 fs is shorter than the electron-phonon and electron-ion collision times. Therefore the electron energy distribution during the pulse has a delta function like shape peaked near the energy that can be estimated from the general formula of Joule heating (1.1) under the assumption that the spatial intensity distribution inside a solid, and material parameters are time independent. We denote the energy per single electron by ε_e (it should not be confused with the dielectric function that is always without a subscript here). Then the electron energy density change reads:

$$\frac{d\,(n_e \varepsilon_e)}{dt} = \frac{2A}{l_{abs}} I(t)\,. \tag{1.3}$$

We show later that the ionisation degree at $I > 10^{14}$ W/cm^2 is high, $Z > 1$, the number density of electrons is large, and electrons heat capacity can be taken as that for ideal gas. Thus from the above one can make a rough estimate of the electron temperature to the end of the pulse:

$$T_e \approx \frac{2A}{1.5 k_B n_e l_{abs}} I(t)\, t. \tag{1.4}$$

The electron temperature rises to tens of electron volts at the very beginning of the pulse. Fast ionisation of a solid occurs that affects absorption coefficient and absorption length. Thus, the next step is to introduce the model where the optical properties are dependent on the changing electron density and electron energy.

1.2.2 Ionisation Processes

Optical breakdown of dielectrics and optical damage produced by an intense laser beam has been extensively studied over several decades. Analytical estimates of the breakdown threshold, ionisation rates and transient number density of electrons created in the absorption region allows one to obtain the general picture of the processes in qualitative and quantitative agreement with computer simulations.

1.2.2.1 Ionisation Thresholds

It is generally accepted that the breakdown occurs when the number density of electrons in the conduction band reaches the critical density expressed through the frequency of the incident light by the familiar relation, $n_c = m_e \omega^2 / 4\pi e^2$. Thus, laser parameters, (intensity, wavelength, pulse duration) and material parameters (band-gap width and electron-phonon effective collision rate) at the breakdown threshold are combined by condition, $n_e = n_c$.

The ionisation threshold for the majority of transparent solids lies at intensities in between $(10^{13}-10^{14})$ W/cm^2 ($\lambda \sim 1\,\mu$m) with a strong non-linear dependence on intensity. The conduction-band electrons gain energy in an intense short pulse much faster than they transfer energy to the lattice. Therefore the actual structural damage (breaking inter-atomic bonds) occurs after the electron-to-lattice energy transfer, usually after the end of the pulse. It was determined that in fused silica the ionisation threshold was reached to the end of 100 fs pulse at 1064 nm at the intensity 1.2×10^{13} W/cm^2 [7]. Similar breakdown thresholds in a range of $(2.8 \pm 1) \times 10^{13}$ W/cm^2 were measured in interaction of 120 fs, 620 nm laser with glass, MgF$_2$, sapphire, and the fused silica [8]. This behaviour is to be expected, since all transparent dielectrics share the same general properties of slow thermal diffusion, fast electron-phonon scattering and similar ionisation rates. The breakdown threshold fluence (J/cm^2) is an appropriate parameter for characterization conditions at different pulse duration.

It is found that the threshold fluence varies slowly if pulse duration is below 100 fs. For example, for the most studied case of fused silica the following threshold fluences were determined: ~2 J/cm^2 at 1053 nm; ~300 fs and ~1 J/cm^2 at 526 nm; ~200 fs [9]; 1.2 J/cm^2 (620 nm; ~120 fs) [8]; 2.25 J/cm^2 at 780 nm; ~220 fs [10, 11]; 3 J/cm^2 at 800 nm; 10–100 fs [12].

1.2.2.2 Ionisation Rates: Avalanche Ionisation

In interaction of lasers in a visible range with wide band gap dielectrics the direct photon absorption by electrons in a valence band is rather small. However, a few seed electrons can always be found in the conduction band. These electrons oscillate in the laser electromagnetic field and can be gradually accelerated to the energy in excess of the band-gap. Electrons with $\varepsilon_e > \Delta_{gap}$ collide with electrons in the valence band and can transfer them a sufficient energy to excite into the conduction band. Thus the number of free electrons increases, which provokes the effect of avalanche ionisation. The probability of such event per unit time (ionisation rate) can be estimated as follows:

$$w_{imp} \approx \frac{1}{\Delta_{gap}} \frac{d\varepsilon_e}{dt} = \frac{2\varepsilon_{osc}}{\Delta_{gap}} \frac{\omega^2 v_{eff}}{\left(v_{eff}^2 + \omega^2\right)}. \tag{1.5}$$

Electron is accelerated continuously in this classical approach. The oscillation energy is proportional to the laser intensity and to the square of the laser wavelength. At relatively low temperature corresponding to low intensities below the ablation threshold the effective collision rate, v_{eff}, equals to the electron-phonon momentum exchange rate $v_{eff} = v_{e-ph}$. The electron-phonon momentum exchange rate increases in proportion to the temperature. For example, the electron-phonon momentum exchange rate in SiO$_2$ is of $v_{e-ph} = 5 \times 10^{14}$ s^{-1} [7] and it is lower of the laser frequency for visible light, $\omega \geq 10^{15}$ s^{-1}. Then the ionisation rate from (1.5) grows in proportion to the square of the laser wavelength in correspondence with the Monte-Carlo solutions to the Boltzmann kinetic equation for electrons [7]. With further increase in temperature and due to the ionisation, the effective collision rate becomes equal to the electron-ion momentum exchange rate, and reaches maximum approximately at the plasma frequency ($\sim 10^{16}$ s^{-1}) [5, 13]. At this stage the wavelength dependence of the ionisation rate almost disappears due to $\omega < v_{e-i} \approx \omega_{pe}$, as it follows from (1.3) in agreement with rigorous calculations of [7]. In the beginning of the ionisation process, when $\omega > v_{e-ph}$, the ionisation rate is $w_{imp} \approx 2v_{e-ph} \sim 10^{14}$–$10^{15}$ s^{-1}. When the collision rate reaches its maximum, $\omega < v_{e-i} \approx \omega_{pe}$, the ionisation rate equals to $w_{imp} \approx 2\omega^2/v_{e-i} \approx 2\omega^2/\omega_{pe} \sim 5 \times 10^{14}$ s^{-1}.

1.2.2.3 Ionisation Rates: Multi-Photon Ionisation

Multiphoton ionisation has no intensity threshold and hence its contribution can be important even at relatively low intensity. Multi-photon ionisation creates the initial (seed) electron density, n_0, which then grows by the avalanche process. It proved to be a reasonable estimate of the ionisation probability (per atom, per second) in the multi-photon form [14]:

$$w_{mpi} \approx \omega n_{ph}^{3/2} \left(\frac{\varepsilon_{osc}}{2\Delta_{gap}} \right)^{n_{ph}}; \qquad (1.6)$$

here $n_{ph} = \Delta_{gap}/\hbar\omega$ is the number of photons necessary for electron to be transferred from the valence to the conductivity band. The multi-photon ionisation process is important at low intensities as it generate the initial number of seed electrons, though small number, they are further multiplied by the avalanche process. The multi-photon ionisation rate dominates, $w_{mpi} > w_{imp}$, for any relationship between the frequency of the incident light and the effective collision frequency in conditions when $\varepsilon_{osc} > \Delta_{gap}$. However, even at high intensity the contribution of avalanche process is crucially important: at $w_{mpi} \sim w_{imp}$ the seed electrons are generated by multi-photon effect whilst final growth is due to the avalanche ionisation. Such an inter-play of two mechanisms has been demonstrated with the direct numerical solution of kinetic Fokker–Planck (1.14). Under the condition $\varepsilon_{osc} = \Delta_{gap}$, $\hbar\omega = 1.55$ eV, $n_{ph} = \Delta_{gap}/\hbar\omega \sim 6.4$, and $\omega = 2.356 \times 10^{15}\,\mathrm{s}^{-1}$, the multi-photon rate comprises $w_{mpi} \sim 5.95 \times 10^{15}\,\mathrm{s}^{-1}$. The ionisation time estimates as $t_{ion} \approx w_{mpi}^{-1}$. Thus, the critical density of electrons (the ionisation threshold) is reached in a few femtoseconds in the beginning of a 100-fs pulse. After that the interaction proceeds in a laser-plasma interaction mode.

1.2.2.4 Ionisation State During the Laser Pulse

In order to estimate the electron number density generated by the ionisation during the laser pulse the recombination processes should be taken into account. Electron recombination proceeds in dense plasma mainly by three-body Coulomb collisions with one of the electrons acting as a third body [15]. The cross section for the Coulomb collision reads $\sigma_{e-i} \approx \pi \left(e^2/\varepsilon_{el} \right)^2 Z^2$, while the probability for a third body (electron) presence in the vicinity of colliding particles is proportional to the cube of the Coulomb impact distance, $p_{3b} \propto r_{Coul}^3 = \left(e^2/\varepsilon_{el} \right)^3$. The growth rate of the electron number density is the balance of ionisation and recombination terms [4]:

$$\frac{dn_e}{dt} \approx n_e w_{ion} - \beta_e n_i n_e^2. \qquad (1.7)$$

Here the ionisation rate is $w_{ion} = \max\{w_{mpi}, w_{imp}\} \sim 10^{15}\,\text{s}^{-1}$, and the recombination rate is $\beta_e n_i n_e$, where the coefficient β_e is expressed as the following [15]:

$$\beta_e = 8.75 \times 10^{-27} \ln \Lambda Z^2 / \varepsilon_{el}^{9/2}. \tag{1.8}$$

We assumed that $n_e = Zn_i$, the electron energy ε_{el} is in eV; $\ln \Lambda$ is the Coulomb logarithm. One can see that ionisation time, $t_{ion} \approx w_{ion}^{-1}$, and recombination time $t_{rec} \approx (\beta_e n_i n_e)^{-1}$, are of the same order of magnitude, $\sim 10^{-15}\,\text{s}$, and both are much shorter then the pulse duration. This is a clear indication of the ionisation equilibrium, and that the multiple ionisations $Z > 1$ take place. Therefore, the electron number density at the end of the pulse can be estimated in a stationary approximation as the follows: $n_e^2 \approx w_{ion}/Z\beta_e$. Taking $w_{ion} \sim 10^{15}\,\text{s}^{-1}$; $\varepsilon_e \sim 30\,\text{eV}$; $Z = 5$; $\ln\Lambda \sim 2$, one obtains, that number density of electrons at the end of the pulse becomes comparable to the atomic number density $n_e \sim 10^{23}\,\text{cm}^{-3}$.

1.2.3 Increase in the Absorbed Energy Density Due to Modification of Optical Properties

We demonstrated above that the swift ionisation during the first femtoseconds in the beginning of the pulse produces the electron number density comparable to the critical density for the incident laser light, $n_e = n_c$. The free-electron number density grows up and becomes comparable to the ion density to the end of the pulse. Respectively, the electron-ion collision rate reaches its maximum that equals approximately to the plasma frequency in the dense non-ideal plasma. With further increase of electron temperature the electron-ion exchange rates decrease due to domination of the Coulomb collisions. The optical properties of this transient plasma are described by the Drude-like dielectric function; they are changing in accord with the change in electron density and temperature. Let us estimate the absorption coefficient and absorption length in the beginning of the laser pulse and at the end of the pulse. The dielectric function and refractive index in conditions, $\nu_{e-i} \approx \omega_{pe} \gg \omega$, are estimated as the following:

$$\varepsilon_{re} \approx \frac{\omega^2}{\omega_{pe}^2}; \ \varepsilon_{im} \approx \frac{\omega_{pe}}{\omega}; \ n \approx k = \left(\frac{\varepsilon_{im}}{2}\right)^{1/2}. \tag{1.9}$$

For example, after the optical breakdown of fused silica glass by 800 nm laser at high laser intensity ($\omega = 4.7 \times 10^{15}\,\text{s}^{-1}$; $\omega_{pe} = 1.45 \times 10^{16}\,\text{s}^{-1}$) the real and imaginary parts of refractive index are $n \sim k = 1.18$ thus giving the absorption length of $l_s = 54\,\text{nm}$, and the absorption coefficient $A = 0.77$ [4]. Therefore, the optical breakdown and further ionisation and heating converts silica into a metal-like plasma medium reducing the energy deposition volume by up to two orders of magnitude when compared with the focal volume, and correspondingly massively increasing the absorbed energy density and consequently the maximum pressure in the absorption region. For the interaction parameters presented above ($I = 10^{14}$ W/cm^2; $A = 0.77$; $l_s = 54$ nm; $t_p = 150$ fs) the pressure corresponding to the absorbed energy density equals to 4.4 TPa, ten times higher than the Young modulus of sapphire, one of the hardest of dielectrics. The general approach presented above is applicable for estimating parameters of any wide band gap dielectric affected by high intensity short pulse laser.

1.2.4 Energy Transfer From Electrons to Ions: Relaxation Processes After the Pulse

The hydrodynamic motion starts after the electrons transfer the absorbed energy to the ions. The following processes are responsible for the energy transfer from electrons to ions: recombination; electron-to-ion energy transfer in the Coulomb collisions; ion acceleration in the field of charge separation; electronic heat conduction. Below we compare the characteristic times of different relaxation processes.

1.2.4.1 Impact Ionisation and Recombination

The electron temperature in the energy units at the end of the pulse is much higher then the ionisation potential. Therefore, the ionisation by the electron impact continues after the pulse end. The evolution of the electron number density can be calculated in the frame of the familiar approach [15]:

$$\frac{dn_e}{dt} \approx \alpha_e n_e n_a - \beta_e n_i n_e^2; \tag{1.10}$$

here $\alpha_e = \sigma_e v_e n_e (I/T_e + 2) \exp(-I/T_e) \left[\text{cm}^3/\text{s}\right]$ is the impact ionisation rate and β_e is the recombination rate connected to α_e by the principle of detailed balance. One can see that for parameters of experiments in question ($\sigma_e \sim 2 \times 10^{-16}$ cm^2; $\varepsilon_e \sim 30$ eV) the time for establishing the ionisation equilibrium is very short $\tau_{eq} \approx (\alpha_e n_e)^{-1} \sim 10^{-16}$ s. Thus the average charge of multiple ionised ions can be estimated from the equilibrium conditions using Saha equations. Losses for ionisation lead to temporary decrease in the electron temperature and in the total pressure [4]. However the fast recombination results in the increase in the ionic pressure.

1.2.4.2 Electron-to-Ion Energy Transfer by the Coulomb Collisions

The Coulomb forces dominate the interactions between the charged particles in the dense plasma created by the end of the pulse. The parameter that characterizes the plasma state is the number of particles in the Debye sphere, $N_D =$

$1.7 \times 10^9 \left(T_e^3/n_e \right)^{1/2}$ [16]. Plasma is in ideal state when $N_D \gg 1$. In the plasma with parameters estimated above for the fused silica ($Z = 5$, ln $\Lambda = 1.7$; $n_e = 3 \times 10^{23}\,\mathrm{cm}^{-3}$; $\varepsilon_e = 50\,\mathrm{eV}$) N_D is of the order of unity, that is a clear signature of the non-ideal conditions. The maximum value for the electron-ion momentum exchange rate in non-ideal plasma approximately equals to the plasma frequency, $\nu_{ei} \approx \omega_{pe} \sim 3 \times 10^{16}\,\mathrm{s}^{-1}$ [13, 17, 18]. Hence electrons in ionised fused silica transfer the energy to ions over a time $t_{ei}^{en} \approx \left(\nu_{ei}^{en} \right)^{-1} \approx (M_i/m_e\nu_{ei}) \sim (1-2)\,\mathrm{ps}$.

1.2.4.3 Ion Acceleration by the Gradient of the Electron Pressure

Let us estimate the time for the energy transfer from electrons to ions under the action of electronic pressure gradient when ions are initially cold. The Newton equation for ions reads:

$$\frac{\partial M_i n_i u_i}{\partial t} \approx -\frac{\partial P_e}{\partial x}$$

The kinetic velocity of ions then estimates as follows:

$$u_i \approx \frac{P_e}{M_i n_i \Delta x} t \qquad (1.11)$$

The time for energy transfer from electrons to the ions is defined by condition that the ions kinetic energy compares to that of the electrons, $M_i u_i^2/2 \sim \varepsilon_e$. With the help of (1.11) one obtains the energy transfer time ($Zn_i \approx n_e$):

$$t_{el-st} \sim \frac{\Delta x}{Z} \left(\frac{\varepsilon_e}{2M_i} \right)^{-1/2} ; \qquad (1.12)$$

here $\Delta x \approx l_{abs} = 54\,\mathrm{nm}$ is the characteristic space scale. Then taking the time for the electron-to-ion energy transfer by the action of the electrostatic field of charge separation equals to $t_{el-st} \sim 1\,\mathrm{ps}$.

1.2.4.4 Electronic Heat Conduction

Energy transfer by non-linear electronic heat conduction starts immediately after the energy absorption. Therefore heat wave propagates outside of the heated area before the shock wave emerges. The thermal diffusion coefficient is defined conventionally as the following, $D = l_e\nu_e/3 = \nu_e^2/3\nu_{ei}$, where l_e, ν_e and n_{ei} are the electron mean free path, the electron velocity and the momentum transfer rate respectively. The characteristic cooling time is conventionally defined as $t_{cool} = l_s^2/D$. For the conditions of experiments [2] $n_{ei} \sim \omega_{pe} \sim 3 \times 10^{16}\,\mathrm{s}^{-1}$; $\varepsilon_e = 50\,\mathrm{eV}$, and the cooling time is $t_{cool} = 3\omega_{pe}m_e l_{abs}^2/2\varepsilon_e = 14.9\,\mathrm{ps}$.

Summing up the results of the energy deposition in confined microexplosion we shall note that the major processes responsible for the electron-to-ion energy transfer in the dense plasma created by the tight focussing inside a bulk solid are different, and much shorter, when compared to those in the plume created by laser ablation. The ions acceleration by the gradient of the electronic pressure and the electron-to-ion energy transfer by the Coulomb collisions both comprise ~ 1 ps. The thermal ionization and recombination are in equilibrium, this permits the description of the plasma state by the Saha equations. The electronic non-linear heat conduction becomes important in the first ~ 15 ps after the pulse, and dominates the return to the ambient conditions.

1.3 Shock Wave Propagation and Void Formation

It was shown above that a focal volume as small as $0.2\,\mu m^3$ can be illuminated by focusing 800 nm laser pulses in the bulk of fused silica glass ($n = 1.453$) with a microscope objective with $NA = 1.35$ [2, 3]. The original focal volume shrinks to much smaller energy deposition region due to fast decrease in the absorption length, approximately five times less than the averaged focal radius. Modified shape of the absorption region has a complicated shape, which is practically unknown. Therefore, it is reasonable to assume for the further calculations that the absorption volume is a sphere of a smaller radius than the focal volume. One can see that 100 nJ of laser energy focussed in the volume of $0.2\,\mu m^3$ create the energy density of $5 \times 10^5\,J/cm^3$ equivalent to the pressure of 0.5 TPa (5 Mbar). However this energy absorbs in much smaller volume thereby generating a pressure in excess of $P = 10$ TPa. All absorbed energy is confined in the electron component at the end of the 150-fs pulse.

1.3.1 Shock Wave Generation and Propagation

The hydrodynamic motion of atoms and ions starts when the electrons have transferred their energy to ions. This process is completed in a few picoseconds time, however one should note that the energy transfer time by the Coulomb collisions increases in proportion to the electron temperature. So, in solid-state density plasma formed in confined microexplosion the higher the absorbed energy, the longer the time for hydrodynamic motion to start. The pressure in a range of several TPa builds up after electron-ion energy equilibration; this pressure considerably exceeds the Young modulus for majority of materials. For example, the Young modulus for sapphire equals to 0.3–0.4 TPa, and that for silica is ~ 0.07 TPa. The high pressure generates the shock wave propagating from the energy absorption region into the surrounding cold material. The bulk modulus of the cold material equals to the cold pressure, P_c. This cold counter pressure finally decelerates and stops the shock wave. Because the shock driving pressure significantly exceeds the cold pressure, $P \gg P_c$, the strong shock emerges compressing a solid to the limit allowed by the equation of state of a solid, which does not depend on the magnitude of the driving pressure. The maximum density of a perfect gas with the adiabatic constant γ is as the follows:

$$\rho = \frac{\gamma + 1}{\gamma - 1} \rho_{0_e}. \tag{1.13}$$

The adiabatic constant for a cold solid is conventionally estimates as $\gamma \sim 3$ [15]. Therefore maximum density after the shock front is expected to be $\rho_{max} = 2\rho_0$. The compression ratio gradually decreases to unity along the shock propagation, deceleration and transformation into a sound wave. Note that the temperature in the compressed solid behind the shock front in the limit of $P \gg P_c$ grows in proportion to the driving pressure:

$$T = \left(\frac{\gamma - 1}{\gamma + 1} \right) \frac{P}{P_c} T_0 \tag{1.14}$$

Thus material is compressed and heated behind the shock wave front. Hence, the conditions for transformation to another phase might be created and this phase might be preserved after unloading to the normal pressure. The final state of matter may possess different properties from those in the initial state.

1.3.2 Shock Wave Dissipation

The shock wave propagating in a cold material loses its energy due to dissipation, e.g. due to the work done against the internal pressure (Young modulus) that resists material compression. The distance at which the shock wave effectively stops defines the shock-affected area. At the stopping point the shock wave converts into a sound wave, which propagates further into the material without inducing any permanent changes to a solid. The distance where the shock wave stops can be estimated from the condition that the internal energy in the whole volume enclosed by the shock front is comparable to the absorbed energy: $4\pi P_0 r_{stop}^3 / 3 \approx E_{abs}$ [15]. The stopping distance obtained from this condition reads:

$$r_{stop} \approx \left(\frac{3 E_{abs}}{4\pi P_0} \right)^{1/3}. \tag{1.15}$$

In other words, at this point the pressure behind the shock front equals to the internal pressure of the cold material. One can reasonably suggest that the sharp boundary observed between the amorphous (laser-affected) and crystalline (pristine) sapphire in the experiments [2, 3] corresponds to the distance where the shock wave effectively stopped. The sound wave continues to propagate at $r > r_{stop}$ apparently not affecting the properties of material. For 100 nJ of absorbed energy and sapphire, taking $P_c = 0.4$ TPa for sapphire, one gets from (1.15) $r_{stop} = 180$ nm, which is in qualitative agreement with the experimental values.

1.3.3 Rarefaction Wave: Formation of Void

The experimentally observed formation of void, which is a hollow or low-density region, surrounded by a shell of the laser-affected material, can be qualitatively understood from the following simple reasoning. Let us consider for simplicity spherically symmetric explosion. The strong spherical shock wave starts to propagate outside the centre of symmetry of the absorbed energy region, compressing the material. At the same time, a rarefaction wave propagates to the centre of symmetry decreasing the density in the area of the energy deposition. This problem qualitatively resembles the familiar hydrodynamic phenomenon of strong point explosion ($P \gg P_0$) in homogeneous atmosphere with counter pressure taken into account. It is characteristic of a strong spherical explosion that material density decreases rapidly in space and time behind the shock front in direction to the centre of symmetry. Practically, the entire mass of material, initially uniformly distributed in the energy deposition region inside a sphere of radius $r \sim l_{abs}$, is concentrated within a thin shell near the shock front some time after the explosion. The temperature increases and density decreases towards the centre of symmetry, while the pressure is almost constant along the radius [15]. This picture is qualitatively similar to that observed in the experiments [4] as a result of fs-laser pulses tightly focussed inside sapphire, silica glass and polystyrene. A void surrounded by a shell of laser-modified material was formed at the focal spot. Hence, following the strong point explosion model, one can suggest that that the whole heated mass in the energy deposition region was expelled out of the centre of symmetry and was frozen after shock wave unloading in the form of a shell surrounding the void.

One can apply the mass conservation law for estimate of the density of compressed material from the void size measured in the experiment. Indeed, the mass conservation relates the size of the void to compression of the surrounding shell. We assume that in conditions of confinement no mass losses could occur. One can use the void size and size of the laser-affected material from the experiments and deduce the compression of the surrounding material. The void formation inside a solid only possible if the mass initially contained in the volume of the void was pushed out and compressed. Thus after the micro-explosion the whole mass initially confined in a volume with of radius r_{stop} resides in a layer in between r_{stop} and r_v, which has a density $\rho = \delta\rho_0$; $\delta > 1$:

$$\frac{4\pi}{3} r_{stop}^3 \rho_0 = \frac{4\pi}{3} \left(r_{stop}^3 - r_{void}^3 \right) \rho. \tag{1.16}$$

Now, the compression ratio can be expressed through the experimentally measured void radius, r_{void}, and the radius of laser-affected zone, r_{stop}, as follows:

$$r_{void} = r_{stop} \left(1 - \delta^{-1} \right)^{1/3}. \tag{1.17}$$

It was typically observed that $r_{void} \sim 0.5 r_{stop}$ in experiments of [4]. Applying (1.16), (1.17) one obtains that compressed material in a shell has a density 1.14 times higher than that of crystalline sapphire. Note that the void size was measured at the room temperature long after the interaction.

1.4 Density and Temperature in the Shock-Wave and Heat-Wave Affected Solid

1.4.1 Two Characteristic Areas in Confined Microexplosion

There are two distinctive regions in the area affected by laser action, which is confined inside a cold solid. First is the area where the laser energy is absorbed. Second area relates to the zone where shock wave propagates outside the absorption zone, compresses and heats the initially cold crystal and then decelerates into the sound wave, which apparently does not affect the pristine crystal. In the first area the crystal is heated to the temperature of tens of eV ($\sim 10^5$ K). It is swiftly ionised at the density close to that of a cold solid. In picosecond time the energy is transferred to ions and shock wave emerges. Conservatively, the heating rate estimates approximately as 10 eV/ps $\sim 10^{17}$ K/s. Then this material undergoes fast compression under the action of the micro-explosion and its density may increase to the maximum of ~ 2 times the solid density. The energy dissipation in the shock wave and by the heat conduction takes nanoseconds. In the second zone heated and compressed exclusively by the shock wave the maximum temperature reaches several thousands Kelvin, approximately 10 times less than in the first zone. Heating and cooling rates in this zone are of the order of $\sim 10^{14}$ K/s.

Phase transformations in quartz, silica and glasses induced by strong shock waves have been studied for decades, see [15, 19] and references therein. However all these studies were performed in one-dimensional (plane) open geometry when unloading into air was always present. The pressure ranges for different phase transitions to occur under shock wave loading and unloading have been established experimentally and understood theoretically [19]. Quartz and silica converts to dense phase of stishovite (mass density 4.29 g/cm^3) in the pressure range between 15–46 GPa. The stishovite phase exists up to a pressure 77–110 GPa. Silica and stishovite melts at pressure >110 GPa that is in excess of the shear modulus for liquid silica ~ 10 GPa. Dense phases usually transform into a low density phases (2.29–2.14 g/cm^3) when the pressure releases back to the ambient level. Numerous observations exist of amorphisation upon compression and decompression. An amorphous phase of silica denser than the initial state sometimes forms when unloading occurs from 15–46 GPa. Analysis of experiments shows that the pressure release and the reverse phase transition follows an isentropic path.

In the studies of shock compression and decompression under the action of shock waves induced by explosives or kJ-level ns lasers, the loading and release time scales are in the order of $\sim$1–10 ns [20–23]. The heating rate in the shock wave experiments is 10^3 K/ns $= 10^{12}$ K/s, 5 orders of magnitude slower than in the confined micro-explosion.

In contrast, the peak pressure at the front of shock wave, driven by the laser in confined geometry, reaches the level of TPa, that is, 100 times in excess of pressure value necessary to induce structural phase changes and melting. Therefore, the region where the melting may occur is located very close to that where the energy is deposited. The zones where structural changes and amorphisation might occur are located further away. Super-cooling of the transient dense phases may occur if the quenching time is sufficiently short. Short heating and cooling time along with the small size of the area where the phase transition takes place can affect the rate of the direct and reverse phase transitions. In fact, phase transitions in these space and time scales have been studied very little.

The refractive index changes in a range of 0.05–0.45 along with protrusions surrounding the central void that were denser than silica were observed as a result of laser-induced micro-explosion in a bulk of silica [24]. This is the evidence of formation of a denser phase during the fast laser compression and quenching; however, little is known of the exact nature of the phase. Thus, we can conclude that a probable state of a laser-affected glass between the void and the shock stopping edge may contain amorphous or micro-crystallite material denser than the pristine structure and with a larger refractive index than the initial glass [24].

1.4.2 Upper Limit for the Pressure Achievable in Confined Interaction

The micro-explosion can be considered as a confined one if the shock wave affected zone is separated from the outer boundary of a crystal by the layer of pristine crystal m-times thicker than the size of this zone. On the other side the thickness of this layer should be equal to the distance at which the laser beam propagates without self-focussing $L_{s-f} (W/W_c)$ [25]:

$$L_{s-f} = \frac{2\pi n_0 r_0^2}{\lambda} \left(\frac{W_0}{W_{cr}} - 1 \right)^{-1/2} \tag{1.18}$$

W is the laser power, W_c is the critical power for self-focussing:

$$W_{cr} = \frac{\lambda^2}{2\pi\, n_0\, n_2} \tag{1.19}$$

This condition expresses as the following:

$$L_{s-f} = m r_{stop} \tag{1.20}$$

The absorbed energy, $E_{abs} = AE_{las}$, expressed through the laser power $W = E_{las}/t_p$ (E_{las}, t_p are respectively energy per pulse and pulse duration), and the radius of shock wave affected zone are connected by the equation:

$$r_{stop} \approx \left(\frac{3A W t_p}{4\pi P_{cold}} \right)^{1/3}.$$ (1.21)

The condition of (1.20) with (1.21) for the self-focusing length inserted then turns to the equation for the maximum laser power at which micro-explosion remains confined and self-focussing does not affect the crystal between the laser affected zone and outer boundary:

$$\frac{2n_0\pi r_0^2}{m\lambda} \left(\frac{4\pi P_{cold}}{3AW_c t_p} \right)^{1/3} = \left(\frac{W}{W_c} \right)^{1/3} \left(\frac{W}{W_c} - 1 \right)^{1/2}.$$ (1.22)

The left-hand side in (1.22) can be calculated if the laser pulse duration and focal spot area are both known. Taking, for example, sapphire ($n_0 = 1.75$); $W_c = 1.94$ MW; $\lambda = 800$ nm; $4\pi P_{cold}/3 = 1.67$ MJ; $t_p = 100$ fs; $\pi r_0^2 = 0.2 \,\mu m^2$; $m = 3$, one gets 0.6 in the LHS of (1.22). Thus the maximum laser power allowed in these conditions equals to $\sim 1.3W_c = 2.5$ MW or 250 nJ of the energy in 100 fs laser pulse. For conditions considered above the maximum pressure that can be achieved in absorption volume confined inside the transparent crystal might be up to 27 TPa, approximately three times higher that was achieved in the experiments [2].

1.4.3 Ionisation Wave Propagation Towards the Laser Beam

There is an additional effect in the focal zone which can influence the size of the volume absorbing the laser energy at the laser fluence above the optical breakdown threshold F_{thr}, namely, the motion of ionisation front with critical density in the direction towards the laser beam propagation. It was first discovered in studies of optical breakdown in gases [14]. Indeed, the intense beam with the total energy well above the ionisation threshold reaches the threshold value at the beginning of the pulse. Laser energy increases and the beam cross-section where the laser fluence is equal to the threshold value, the ionisation front, starts to move in the opposite to the beam direction. The beam is focussed to the focal spot area, $S_f = \pi r_f^2$. The spatial shape of the beam path is a truncated cone, the intensity at any time to be independent of the transverse coordinates (see Fig. 1.2).

The ionisation time, i.e. the time required for generating the number of electrons to reach the critical density in the conduction band, is defined by (1.1). Therefore, the threshold fluence is achieved at the beam cross-section with a radius increasing as the following:

$$r(z, t) = r_f + z(t) \, tg\alpha$$ (1.23)

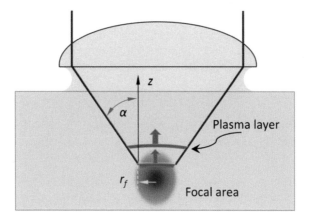

Fig. 1.2 The path of the converging focused pulse as a truncated cone. The motion of the ionisation front toward the beam propagation is indicated by the *red arrows*

where, z is the distance from the focal spot, which is usually a circle with radius r_f. Then at any moment t during the pulse the relation holds:

$$\frac{E_{las}(t)}{\pi r^2(z,t)} = F_{thr} \qquad (1.24)$$

We introduce the dimensionless parameter, $f = E_{las}(t_p)/\pi r_f^2 F_{thr} = F_{las}/F_{thr}$, as the ratio of the maximum fluence to the threshold fluence. Then the ionisation front moved the distance $z(t_p)$:

$$z(t_p) = \frac{r_f}{tg\alpha}\left(f^{1/2} - 1\right). \qquad (1.25)$$

Correspondingly, the ionisation time can be evaluated as:

$$t_{ion} = t_p\left[1 - \left(1 - \frac{1}{f}\right)^{1/2}\right]. \qquad (1.26)$$

One can see that if the total fluence equals to that for the threshold, $f = 1$, the n_{cr} is reached only at the end of the pulse [9, 26, 27], the ionisation time equals to the pulse duration and thus there is no movement of the ionisation front. These simple geometrical considerations are in qualitative agreement with the experiments in sapphire and silica. Indeed, the voids measured in sapphire and in silica in the references [2–4] are slightly elongated; the (1.25) gives $z_m = 0.67r_f$ and $z_m = 0.45r_f$ for sapphire and silica respectively.

Summing up the results of this section, the effects of the blue shift of the pulse spectrum and the intensity dependence of the group velocity are small and rather positive for achieving high absorbed energy density. The negative effect of the ion-isation front motion at the laser energy well above the ionisation threshold leads to a large decrease in the absorbed energy density. The negative effect of defocusing needs further careful studies with the solution of Maxwell equations.

1.4.4 Modelling of Macroscopic Explosions by Micro-Explosion

The micro-explosion process can be described solely in the frames of the ideal hydro-dynamics if the heat conduction and other dissipative processes, all characterized by specific length scales, could be ignored. The hydrodynamic equations contain five variables: the pressure, P, the velocity, v, the density, ρ, the distance, r, and the time, t. The last three of them are independent parameters, and the other two can be expressed through the previous three. The micro-explosion can be fully character-ized by the following independent parameters: the radius of the energy deposition zone, R_0, the total absorbed energy, E_0, and the initial density ρ_0. Then the initial pressure, $P_0 = E_0/R_0^3$, and the initial velocity, $v_0 = (P_0/\rho_0)^{1/2}$ are combinations of the independent parameters. One can neglect the energy deposition time and time for the energy transfer from electrons to ions (picosecond) in comparison to hydrody-namic time of a few nanoseconds. Then, the hydrodynamic equations can be reduced to the set of the ordinary equations with one variable [15], $\xi = r/v_0 t$, describing any hydrodynamic phenomena with the same initial pressure and density (velocity), but with the characteristic distance and time scales changed in the same proportion. When the energy of explosion increases, the space, R_0, and time scales are increased accordingly to $R_0 = (E_0/p_0)^{1/3}$; $t_0 = R_0/v_0$. The similarity laws of hydrodynam-ics in micro- and macroscopic explosions suggest that micro-explosion in sapphire $(E_0 = 10^{-7}\,\text{J}; \rho_0 \sim 4\,\text{g/cm}^3; R_0 = 1.5 \times 10^{-5}\,\text{cm}; t_0 = 5.5 \times 10^{-12}\,\text{s})$ is a reduced copy of macroscopic explosion that produces the same pressure at the same initial density but with the energy deposition area size and time scales changed in accor-dance with the above formulae. For example, the energy of 10^{14} J (that is equivalent to 25,000 tons of high explosive or one nuclear bomb) released in a volume of 4 cubic meters $(R_0 = 1.59\,\text{m})$ during the time of $20\,\mu\text{s}$ exerts the same pressure of 12.5 TPa as the laser-induced micro-explosion in sapphire does. Thus, exactly the same physical phenomena occur at the scale 10^7 times different in space and in time, and 10^{21} times different in energy. Therefore, all major hydrodynamic aspects of powerful macroscopic explosion can be reproduced in the laboratory tabletop exper-iments with ultra-short laser pulses.

1.5 Formation of Void at Si/SiO$_2$ Interface

The experiments were conducted with 170 fs, 790 nm, laser pulses from a MXR-2001 CLARK laser system. Pulses with up to 2.5 µJ per pulse were focused using an optical microscope (Olympus IX70) equipped with an oil-immersion ×150 objective (NA = 1.45) The focal spots were measured using a knife-edge technique with a sharp edge of a Si(100) wafer etched at 54.74° to the surface along the (111) direction and mounted on a nano-positioning stage. The focal spots were measured to have a radius of 0.368 µm at the full-width at half-maximum level. The experiments were conducted with laser pulses at 1 kHz repetition rate in a sample moved at a rate 2 mm s^{-1} to guarantee a single shot per spot regime, so that each of the shots was located 2 µm apart.

Silicon is not transparent for 790 nm, the wavelength that we used for the experiments. While the laser induced micro-explosion method was previously considered as suitable only for transparent materials, we propose to expand it into the unexplored domain of non-transparent materials. To confine laser-matter interaction inside a material the distance ought to be far enough from the crystal surface in order the interaction region can be considered as confined. The interaction of intense laser radiation with matter when the beam is tightly focused inside a transparent material is radically different from the case of focusing the beam onto a surface. In the laser-surface interaction the temperature has maximum at the outermost atomic surface layer. If the absorbed energy density in the surface layer is in excess of ablation threshold, the atomic bonding breaks and the ablated atoms leave the surface. In tightly focused interaction mode the focal zone with high energy density is confined inside a bulk of a cold and dense solid. The laser-affected material remains at the focal area inside the pristine material.

By growing a transparent oxide on the sample, laser induced confined micro-explosion can be applied to virtually any opaque material. The confinement conditions were formed in silicon wafers covered by 10-µm thick layer of dioxide of silicon (SiO$_2$). SiO$_2$ is transparent to the laser irradiation and allow tightly focusing femtosecond laser pulses on the buried surface of the Si crystal. The thickness of the SiO$_2$ layer was not so deep for developing large spherical aberrations with high-NA focusing optics and at the same time guarantee the absence of optical breakdown at the surface.

Silicon crystal was exposed to strong shock wave induced by fs-laser micro-explosion in confined geometry [18]. Figure 1.1 shows a schematic representation of the process and the realization of array of voids using focusing fs-pulses at the Si/SiO$_2$ interface. After processing the array have been investigated by optical microscopy. Figure 1.3 presents an optical microscope image of a top-view of the sample of the laser-affected areas at the Si/SiO$_2$ interface. The dark dots in the picture are the voids. The voids were spaced from 2 µm. The voids are observed through the SiO$_2$ layer.

To analyse the dimensions of the voids and of the shock wave modified areas, we opened the sample using a focused ion beam (FIB) technique. The sample was gold coated by sputtering (5 nm). The layer of SiO$_2$ was removed, as shown on Fig. 1.4a, and then the sample was characterized with scanning electron microscopy (SEM).

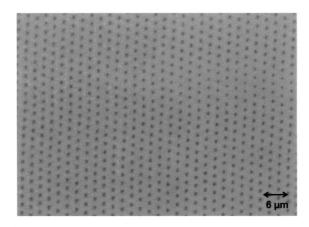

Fig. 1.3 Optical microscope image of arrays of void made at Si/SiO₂ interface viewed through the SiO₂ layer produced fs-pulse laser micro-explosion. The spacing between the void is 2 μm

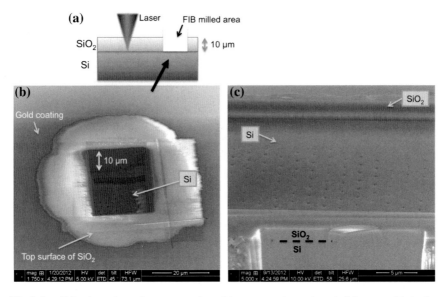

Fig. 1.4 **a** Side-view schematic representation of the processed sample and of the area milled using the focused ion beam (*FIB*). **b** SEM image of the hole made in the SiO₂ layer by FIB to reach the surface of the Si layer, following the schematic in (**a**). **c** SEM image of the side-view of the milling by FIB of the SiO₂ layer to reach the surface of the Si layer

Figure 1.4b presents a top-view SEM image of the sample where the 10 μm thick SiO₂ was removed where we can see the Si surface and arrays of voids. Voids were observed under the surface in the region where the fs-laser was focused. Figure 1.4c shows the Si surface and the arrays of voids; on the side-view we can observe the Si/SiO₂ boundary.

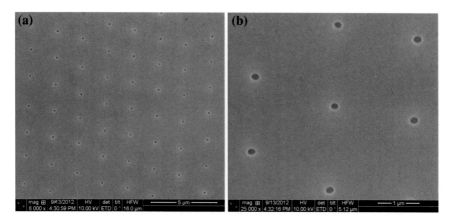

Fig. 1.5 Laser produced voids on a Si-surface buried 10 μm beneath the SiO$_2$ surface of the oxidised wafer, the SiO$_2$ was removed using FIB milling, the spacing between the voids is 2 μm: **a** SEM image of FIB-opened section showing an array of voids produced by 170-fs, 800 nm, 300 nJ single laser pulses focused 2 μm apart with ×150 objective; **b** Close-view of (**a**)

Each of the regions irradiated by a single laser pulse at the fluence above ∼1 J/cm^2 contains a void located at the focal spot. Figure 1.5 presents arrays of voids at the Si surface produced by 300 nJ single laser pulses focused 2 μm apart with ×150 objective. On the top-view, all the voids had a circular shape and the diameter was 250 nm.

On the area where the SiO$_2$ layer was not completely removed, cross-section has been obtained using FIB through a void and characterized with SEM. Figure 1.6 shows a side-view of a void produced with an energy of 700 nJ. The maximal horizontal length of the void in the SiO$_2$ region is 720 nm, in Si 550 nm. The vertical size (including Si and SiO$_2$) is 1.25 μm. A shock-wave-modified Si surrounded the voids. The thickness of the boundary between the transparent oxidised layer and crystalline Si where the laser radiation is focused is of the order of only 2 nm, it can be clearly seen in electron microscope. Material should be removed from the energy deposition region in order to form a void involving that a denser shell surrounded the void. Therefore the observation of a large void is unequivocal evidence of creation the pressure well in excess the Young modulus of both materials, $Y_{SiO_2} \sim 75$ GPa for SiO$_2$ and $Y_{Si} \sim 165$ GPa for Si.

The characterization of the laser-affected area, the dense shell surrounding the void and the shock wave affected area by transmission electron microscopy and by Raman micro-spectroscopy will be discuss elsewhere.

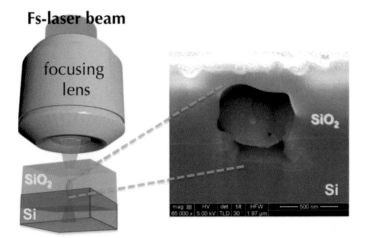

Fig. 1.6 Schematic representation of the laser induced micro-explosion at the Si/SiO2 interface with the associated side-view SEM image of a void produced at the energy of 700 nJ

1.6 Summary

Let us summarise the main conclusions on ultrafast laser-induced material modifications in confined geometry:

- In the conditions close but below the optical breakdown threshold the femtosecond laser pulse creates optically detectable changes in the refractive index. The modifications in refractive index are short-lived and transient. The short-lived modification occurs due to excitation of electrons of all constituent atoms. Permanent modification occurs in the doped sites due to the field of spontaneous polarisation.
- Femtosecond laser pulse tightly focussed by high-NA optics leads to absorbed energy density in excess of the strength of any existing material. A void surrounded by a compressed shell is formed as a result of the confined micro-explosion in the focal volume.
- Warm Dense Matter at the pressure exceeding TPa and the temperature more then 100,000 K is created in the table-top experiments, mimicking the conditions in the cores of stars and planets.
- Confined micro-explosion studies open several broad avenues for research, such as formation of three-dimensional structures for applications in photonics, studies of new materials formation, and imitation the inter-planetary conditions at the laboratory tabletop.

It was demonstrated that tight focusing of a conventional tabletop laser inside the bulk of a transparent solid creates pressure exceeding the strength of any material, and the shock wave compresses a solid, which afterwards remains confined inside a crystal. High pressure and temperature are necessary to produce super-dense, super-hard and super-strong phases or materials, which may possess other unusual properties, such as ionic conductivity. In nature, such conditions are created in the cores of planets and stars. Extreme pressures were recreated by strong explosions, by diamond anvil cell presses and with powerful ns-lasers. All these methods were cumbersome and expensive. By contrast, ultra-short lasers create extreme pressure and temperature along with record high heating and cooling rates by focusing 100–200 nanoJoules of conventional femtosecond laser pulse into a sub-micron volume confined inside a solid [28]. Recently, the crystalline phase of aluminium, bcc-Al has been discovered in ultra-fast laser-induced micro-explosions [29]. These results open the possibilities of formation of new high-pressure phases and prospects of modelling in the laboratory the conditions in the cores of planets and macro-explosions. The first results might be considered a proof-of-principles step. However, it is obvious that, with this simple and inexpensive method for creation of extreme pressure/temperature, the focus in research is shifted to post-mortem diagnosis of shock-compressed material. Micro-Raman, x-ray and electron diffraction, and AFM and STM studies with resolution on the sub-micron level are needed for identification of the new phases. Another challenge is to develop a pump-probe technique with time resolution capable if *in situ* observation of shock-front propagation inside a crystal.

Summing up, the prospects for the fundamental study of ultra-fast laser-matter interaction and its technological applications look extremely encouraging. As the technology becomes smaller, less expensive, more robust, less power-hungry and more energy-efficient, it allows the increased exploitation of ultrafast phenomena, ultimately entering our everyday lives.

Acknowledgments This research was supported under Australian Research Council's Discovery Project funding scheme (project number DP120102980). Partial support to this work by Air Force Office of Scientific Research, USA (FA9550-12-1-0482) is gratefully acknowledged.

References

1. E. Glezer, E. Mazur, Appl. Phys. Lett. **71**, 882–884 (1997)
2. S. Juodkazis, H. Misawa, E. Gamaly, B. Luther-Davies, L. Hallo, P. Nicolai, V.T. Tikhonchuk, Phys. Rev. Lett. **96**, 166101 (2006)
3. S. Juodkazis, H. Misawa, T. Hashimoto, E. Gamaly, B. Luther-Davies, Appl. Phys. Lett. **88**, 1 (2006)
4. E.G. Gamaly, E. G., S. Juodkazis, H. Misawa, B. Luther-Davies L. Hallo, P. Nicolai, V.T. Tikhonchuk, Phys. Rev. B **73**, 214101 (2006)
5. E.G. Gamaly, A.V. Rode, B. Luther-Davies, V.T. Tikhonchuk, Phys. Plasmas **9**, 949–957 (2002)
6. L.D. Landau, E.M. Lifshitz, L.P. Pitaevskii, *Electrodynamics of Continuous Media* (Pergamon Press, Oxford, 1984)
7. D. Arnold, E. Cartier, Phys. Rev. B **46**, 15102–15115 (1992)
8. K. Sokolowski-Tinten, K. J. Bialkowski, A. Cavalieri, M. Boing, H. Schuler, D. von der Linde, in *High-Power Laser Ablation, Proceedings SPIE*, vol. 3343, ed. by C. Phipps, Part 1, 46–57 (1998)
9. B.C. Stuart, M.D. Feit, S. Herman, A.M. Rubenchik, B.W. Shore, M.D. Perry, J. Opt. Soc. Am. B **13**, 459–468 (1996)
10. W. Kautek, J. Krüger, M. Lenzner, S. Sartania, Ch. Spielmann, F. Krausz, Appl. Phys. Lett. **69**, 3146 (1996)

11. M. Lenzner, J. Kruger, S. Sartania, Z. Cheng, Ch. Spielmann, G. Mourou, W. Kautek, F. Krausz, Phys. Rev. Lett. **80**, 4076–4079 (1998)

12. An-Chun Tien, S. Backus, H. Kapteyn, M. Murname, G. Mourou, Phys. Rev. Lett. **82**, 3883–3886 (1999)

13. K. Eidmann, J. Meyer-ter-Vehn, T. Schlegel, S. Huller, Phys. Rev. E **62**, 1202–1214 (2000)

14. YuP Raizer, *Laser-Induced Discharge Phenomena* (Consultant Bureau, New York, 1978)

15. Ya. B. Zel'dovich, Yu. P. Raizer, *Physics of Shock Waves and High-Temperature Hydrodynamic Phenomena* (Dover, New York, 2002)

16. W.L. Kruer, *The Physics of Laser Plasma Interactions* (Addison-Wesley, New-York, 1988)

17. E.G. Gamaly, Phys. Rep. **508**, 91–243 (2011)

18. E. G. Gamaly, L. Rapp, V. Roppo, S. Juodkazis, A. V. Rode, New J. Phys. **15**, 025018 (2013)

19. Sheng-Nian Luo, T. J. Arens, P. D. Asimov, J. Geophys. Res. **108**, 2421 (2003)

20. S. Brygoo, E. Henry, P. Loubeyre, J. Eggert, M. Koenig, B. Loupias, A. Benuzzi-Mounaix, M.R. Le Gloahec, Nat. Mater. **6**, 274–277 (2007)

21. D.G. Hicks, P.M. Celliers, G.W. Collins, J.H. Eggert, S.J. Moon, Phys. Rev. Lett. **91**, 035502 (2003)

22. D. C. Swift, J. A. Hawreliak, D. Braun, A. Kritcher, S. Glenzer, G. Collins, S. D. Rothman, D. Chapman and S. Rose, *in Gigabar material properties experiments on NIF and Omega. Shock Compression of Condense Matter - 2011, AIP Conf. Proc.*, vol. 1426, 477–480 (2012)

23. R.F. Trunin, Physics-Uspekhi **37**, 1123–1146 (1994)

24. E.N. Glezer, M. Milosavjevic, L. Huang, R.J. Finlay, T.-H. Her, J.P. Callan, E. Masur, Opt. Lett. **21**, 2023–2026 (1996)

25. S.A. Akhmanov, V.A. Vyspoukh, A.S. Chirkin, *Optics of Femtosecond Laser Pulses* (Nauka, Moscow, 1988)

26. V.V. Temnov, V. V. K. Sokolowski-Tinten, P. Zhou, A. El-Khamhawy, D. von der Linde, Phys. Rev. Lett. **97**, 237403 (2006)

27. B.C. Stuart, M.D. Feit, A.M. Rubenchick, B.W. Shore, M.D. Perry, Phys. Rev. Lett. **74**, 2248–2251 (1995)

28. E.G. Gamaly, A. Vailionis, V. Mizeikis, W. Yang, A.V. Rode, S. Juodkazis, High Energy Density Phys. **8**, 13–17 (2012)

29. A. Vailionis, E.G. Gamaly, V. Mizeikis, W. Yang, A.V. Rode, S. Juodkazis, Nat. Commun. **2**, 445 (2011)

Chapter 2
Molecular Dynamics Simulations of Laser-Materials Interactions: General and Material-Specific Mechanisms of Material Removal and Generation of Crystal Defects

Eaman T. Karim, Chengping Wu and Leonid V. Zhigilei

Abstract Molecular dynamics simulations of laser-materials interactions are capable of providing detailed information on the complex processes induced by the fast laser energy deposition and can help in the advancement of laser-driven applications. This chapter provides a brief overview of recent progress in the atomic- and molecular-level modeling of laser-materials interactions and presents several examples of the application of atomistic simulations for investigation of laser melting and resolidification, generation of crystal defects, photomechanical spallation, and ablation of metals and molecular targets. A particular focus of the analysis of the computational results is on revealing the general and material-specific phenomena in laser-materials interactions and on making connections to experimental observations.

2.1 Introduction

Rapid expansion of the area of practical applications of short pulse laser processing (see, e.g., Chaps. 4, 5, 7, and 9 of this book) has been motivating growing interest in the fundamental mechanisms of laser-materials interactions. Computer modeling is playing an important role in the development of the theoretical understanding of laser-induced processes and the advancement of laser applications. The need for computer modeling is amplified by the complexity of the material response to the rapid laser energy deposition, which includes transient modification of the material properties by strong electronic excitation, fast non-equilibrium structural and phase transformations occurring under conditions of extreme overheating/undercooling and ultrahigh deformation rates, generation of crystal defects, photomechanical fracture and spallation, vaporization and explosive boiling of strongly overheated surface region,

E. T. Karim · C. Wu · L. V. Zhigilei (✉)
Department of Materials Science and Engineering, University of Virginia,
395 McCormick Road, P.O. Box 400745, Charlottesville, VA 22904-4745, USA
e-mail: lz2n@virginia.edu

V. P. Veiko and V. I. Konov (eds.), *Fundamentals of Laser-Assisted Micro- and Nanotechnologies*, Springer Series in Materials Science 195, DOI: 10.1007/978-3-319-05987-7_2, © Springer International Publishing Switzerland 2014

ionization and plasma formation. Computational description of this diverse range of processes is challenging and requires a combination of different computational approaches, ranging from quantum mechanics based (*ab initio*) electronic structure calculations [1–6], to classical molecular dynamics (MD) simulations [7–44], and to continuum-level kinetic and hydrodynamic modeling [45–54].

The continuum models, in particular, have been demonstrated to be capable of computationally efficient treatment of laser-induced processes at experimental time and length-scales and have been actively used for optimization of irradiation conditions in laser processing applications. Several examples of continuum-level simulations of the phase transformations occurring in laser processing of metal targets are provided in Chap. 3 of this book. The predictive power of continuum modeling, however, is limited by the need for *a priory* knowledge of all the processes that may take place during the simulations. The highly non-equilibrium nature of the processes induced by short pulse laser irradiation challenges the basic assumptions of the continuum models that are commonly designed based on equilibrium material behavior and properties.

An alternative computational approach, free of assumptions on the nature of laser-induced processes and capable of providing atomic-level insights into rapid structural and phase transformations, is presented by the classical MD simulation technique. The MD technique is based on the numerical integration of the classical equations of motion for all atoms or molecules in the system. The interatomic interaction is described by a potential energy function that defines the equilibrium structure and thermodynamic properties of the material. The main strength of the MD method is that it does not require any assumptions about the processes taking place in the systems that are investigated. This characteristic of the MD technique presents a significant advantage over the continuum-level methods where all relevant processes have to be known and described mathematically before the simulations can be performed.

The main limitations of MD method are the relatively small time- and length-scales accessible for atomic-level simulations that, even with the use of high-performance parallel computers, are typically limited to tens of nanoseconds and hundreds of nanometers. The effect of the severe limitations on the time- and length-scales of the simulations can be partially alleviated through an appropriate choice of boundary conditions capable of mimicking the interaction of the simulated small part of the system with the surrounding material [34, 55] or through design of coarse-grained mesoscopic computational approaches aimed at extending the time- and length-scales of the simulations [26, 34, 56].

The incorporation of a realistic description of laser coupling to the optically active states in the irradiated material and relaxation of the photo-excited states is another pre-requisite for application of MD technique to simulation of laser-materials interactions. A number of material-specific models have been developed for computational description of the laser excitation within the general framework of classical MD technique. For metals, a combined atomistic-continuum model that couples the classical MD method with a continuum-level description of the laser excitation and subsequent relaxation of the conduction-band electrons based on two-temperature model (TTM) [57] has been developed and applied for investigation of laser melting [9–15], generation of crystal defects [16–18], photomechanical spallation and ablation [9, 12, 15, 21, 23, 24, 32, 37, 42, 44]. A number of computational approaches have also been suggested for MD simulations of laser interactions with Si, including the models based on local treatment of individual excitation events accounting for bond weakening, ionization and electron–ion recombination [27, 58], as well as stochastic treatment of carrier diffusion and scattering [35, 38]. For insulators, a continuum description of the laser coupling and generation of free electrons has been combined with MD modified to include the energy transfer from the excited electrons to ions [59, 60] and to account for local changes in interatomic interactions due to the ionization [59].

For molecular systems, a coarse-grained "breathing sphere" model accounting for the finite rate of the vibrational relaxation of photo-excited molecules has been developed [26, 34] and actively used in investigations of laser desorption, ablation, and spallation of one-component molecular targets [19, 21, 26, 29, 33, 34] and polymer solutions [31, 40, 43, 61, 62]. Recent extensions of the model include incorporation of ionization mechanisms that enables investigation of processes that control the yield of ions in matrix-assisted laser desorption/ionization (MALDI) mass spectrometry technique [63–65], addition of semi-quantitative representation of photochemical reactions [30, 34, 41, 66] and integration of a mesoscopic model for carbon nanotubes [67, 68] into the framework of the breathing sphere model for investigation of the ejection and deposition of nanotube-based films and coatings in matrix-assisted pulsed laser evaporation (MAPLE) technique [43].

The results of MD simulations of laser interactions with various materials have provided a wealth of information on the structural and phase transformations responsible for material modification and/or removal (ablation) in laser processing applications, e.g., see recent reviews [69–71]. While some of the laser-induced processes are found to be highly sensitive to the structure and thermodynamic properties of the target material, some other computational predictions appear to be surprisingly similar for systems as different as molecular solids and metals. In this chapter, we use the results of recent simulations of laser interactions with metal targets of different crystal structure, namely face-centered cubic (fcc) Ag, Ni, Al and body-centered cubic (bcc) Cr, and an amorphous molecular system to discuss the general and material-specific characteristics of laser-induced structural modification, spallation and ablation of irradiated targets. The processes responsible for the transitions between different regimes of material response to the laser irradiation are discussed first and related to the laser fluence dependence of the total ablation yield and the yield of individual (vapor-phase) atoms or molecules. The generation of crystal defects as a result of laser melting and resolidification, photomechanical spallation and phase explosion of overheated surface regions of irradiated targets are then discussed based on the simulation results obtained for different target materials and a broad range of irradiation conditions.

2.2 Physical Regimes of Laser-Material Interactions

In this section we use the results of TTM-MD simulations of a bulk Cr target irradiated by 200 fs laser pulses to discuss three distinct regimes of material response to laser irradiation: surface melting and resolidification, photomechanical spallation, and phase explosion regimes. We then compare the results obtained for Cr with the ones for Ni and molecular targets and discuss the differences and similarities in these results. The interatomic interactions are described by the embedded-atom method (EAM) potentials for Cr [16] and Ni [72], whereas the molecular target is simulated with the breathing sphere model [26, 34].

The conditions leading to the transitions between the different regimes can be established based on the analysis of the evolution of temperature and pressure in the surface regions of the Cr target shown in the form of contour plots in Fig. 2.1. The absorbed laser fluence of 85 mJ/cm^2 used in the simulation illustrated in Fig. 2.1a is above the threshold for surface melting, $\sim$50 mJ/cm^2 [16] but below the spallation threshold of $\sim$95 mJ/cm^2 [15]. The temperature plot in Fig. 2.1a shows that the electronic excitation by the 200 fs laser pulse and the energy transfer to the lattice due to the electron-phonon coupling result in a fast lattice heating. The maximum temperature reached by 5.5 ps at a depth of 10–20 nm below the surface exceeds 3200 K, which is 37% above the equilibrium melting temperature of the EAM Cr material, $T_m = 2332$ K [15]. At this level of superheating, the surface region undergoes a fast homogeneous melting that proceeds in a form of the fast collapse of the crystal lattice and does not involve the formation of well-defined liquid nuclei [9–11, 14]. The fast homogeneous melting of $\sim$20 nm surface region is followed by an additional slow propagation of the melting front deeper into the target, with the maximum depth of the melted region reaching 24 nm by the time of 40 ps. The short time of the homogeneous melting under the conditions of strong superheating observed in the simulation is consistent with the results of time-resolved electron diffraction experiments [73, 74], where the melting time of several picoseconds is reported for thin Al and Au films irradiated by femtosecond laser pulses.

The temperature near the liquid-crystal interface drops below the equilibrium melting temperature by the time of 50 ps and the melting turns into epitaxial recrystallization of the melted region. The strong temperature gradient created by the laser excitation results in the fast electronic heat conduction to the bulk of the target and leads to a rapid cooling of the surface region, with the initial rate of cooling of the melted region exceeding 5×10^{12} K/s. The velocity of the solidification front increases with increasing undercooling below the melting temperature and reaches the maximum value of about 80 m/s by the time when the melting front reaches the surface of the target and the temperature at the liquid-crystal interface drops down to about $0.8 T_m$. The values of the velocity of solidification front observed in the simulations are comparable to the ones estimated from pump-probe measurements performed for Ag films [75].

The rapid lattice heating during the first picoseconds after the laser pulse takes place under conditions of the inertial stress confinement [19, 21] and results in the buildup of compressive stresses in the surface region of the irradiated target. The generation of strong compressive stresses can be seen in all pressure plots shown in Fig. 2.1. The relaxation of the compressive stresses in the presence of the free surface of the target results in the generation of a bimodal stress wave consisting of a compressive component followed by a tensile one. To simulate the propagation of the stress wave from the surface into the bulk of the target, a pressure-transmitting boundary condition [34, 55] is applied at the bottom of the MD part of the TTM-MD model. The strength of the compressive and tensile stresses generated by the laser pulses increases with increasing laser fluence and, in a simulation performed at a fluence of $106 \, mJ/cm^2$ (Fig. 2.1b), the tensile stresses exceed the dynamic strength of the melted material causing spallation or separation of a melted layer from the target. The mechanisms of spallation, which proceeds through the nucleation, growth, coalescence, and percolation of multiple voids in a surface region of the target, are discussed in more detail in Sect. 2.4.

Further increase of the laser fluence above the spallation threshold results in the separation and ejection of multiple layers/droplets from the target and, above $\sim 275 \, mJ/cm^2$, leads to the transition from the regime of photomechanical spallation to the regime of phase explosion. As discussed below, this transition occurs when a surface region of the irradiated target reaches and exceeds the threshold temperature at which the strongly overheated melted layer becomes thermodynamically unstable and undergoes an explosive decomposition into a mixture of vapor and liquid droplets, as shown, e.g., in Fig. 2.1c. A brief discussion of the material ejection in the phase explosion regime is provided and illustrated by snapshots from a large-scale TTM-MD simulation in Sect. 2.5. The transition from the spallation to the phase explosion regime also affects the characteristics of the pressure wave generated by the laser irradiation, Fig. 2.1. While the compressive component of the pressure wave continues to increase linearly with increasing laser fluence, the tensile component starts to decrease in the spallation and phase explosion regimes. This decrease has been attributed [12, 21] to the reduced ability of the strongly overheated part of the target to support the transient tensile stresses, as well as to the compressive ablation recoil pressure that partially cancels the tensile component of the wave.

The transitions between the regimes of surface melting, photomechanical spallation, and phase explosion can also be identified from the fluence dependence of the total ablation yield and the yield of vapor-phase atoms or molecules shown in Fig. 2.2 for the Cr target as well as for Ni and amorphous molecular targets. Despite the apparent differences between the properties of the target materials and irradiation conditions (pulse durations are 200 fs for Cr, 1 ps for Ni, and 15 ps for the molecular system), a unifying feature is the condition of stress confinement [21] that is satisfied in all three series of simulations [12, 15, 19]. Under the condition of stress confinement, the characteristic time of the laser heating of the absorbing material is shorter than the time required for the mechanical relaxation (expansion) of the heated volume, causing the generation of the stress waves (such as the ones in Fig. 2.1) and, at sufficiently high laser intensities, driving the spallation of surface layer(s).

At low laser fluences, below the spallation threshold, the material ejection from the transiently melted metal surfaces is limited to thermal desorption of just several

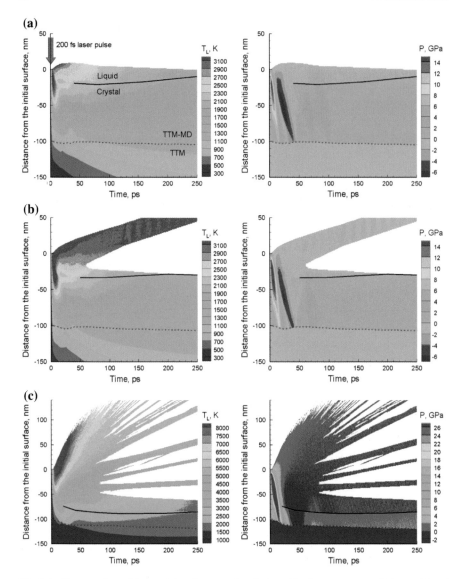

Fig. 2.1 Contour plots of the spatial and temporal evolution of lattice temperature (*left panels*) and pressure (*right panels*) in TTM-MD simulations of a bulk Cr target irradiated with a 200 fs laser pulse at absorbed fluences of 85 mJ/cm^2 (**a**), 106 mJ/cm^2 (**b**), and 298 mJ/cm^2 (**c**). The laser pulse is directed along the Y axes, from the top of the contour plots. The *black solid lines* separate the melted and crystalline regions of the target. The *red dashed lines* separate the continuum (TTM) and atomistic (TTM-MD) parts of the computational system. The results are adopted from [15]

atoms (e.g., 3 atoms evaporate from Cr target in a simulation illustrated in Fig. 2.1a). For the molecular target, the evaporation of molecules below the spallation threshold

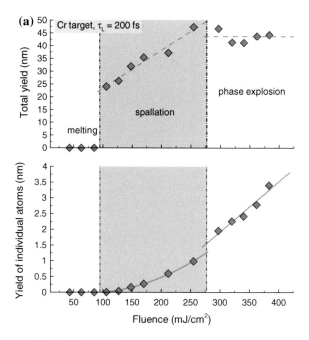

Fig. 2.2 Total ablation yield and yield of individual (vapor-phase) atoms or molecules as functions of the absorbed laser fluence predicted in TTM-MD simulations of Cr targets irradiated with 200 fs laser pulses (**a**), TTM-MD simulations of Ni targets irradiated with 1 ps laser pulses (**b**), and coarse-grained MD simulations of molecular targets irradiated with 15 ps laser pulses (**c**). The values of the total yield and yield of individual atoms/molecules are expressed in units of depth in the initial target (a layer of this depth in the initial target has the number of atoms/molecules equal to those ejected from the target). The *vertical dash-dotted lines* mark the approximate values of the threshold fluences for onset of photomechanical spallation of surface layers(s) and explosive decomposition of the surface region into vapor and liquid droplets (phase explosion). The range of fluences where the photomechanical spallation is the dominant mechanism of material ejection is highlighted by *gray color*. The *red dashed* and *blue solid curves* show the predictions of the ablation and desorption models discussed in [19, 34] in (**c**) and are guides to the eye in (**a**) and (**b**). The results for the Cr, Ni, and molecular targets are adopted from [15], [12], and [19], respectively

is more active, but still remains at a level of sub-monolayer desorption per laser pulse. In all three systems, the transition to the spallation regime manifests itself in a step increase in the total ablation yield. The increase is from 0.3 nm surface layer at 2.8 mJ/cm^2 to 13.6 nm layer at 3.1 mJ/cm^2 for molecular target (Fig. 2.2c) and from essentially zero to more than 20 nm layers for metal targets, Fig. 2.2a, b. At the same time, no abrupt changes in the yield of vapor-phase atoms or molecules are observed at the spallation threshold, thus highlighting the mechanical rather than the thermodynamic nature of the driving forces responsible for the spallation onset. As the laser fluence increases above the spallation threshold, the temperature of the surface increases and the ability of the surface region to support tensile stresses diminishes, leading to the disintegration of the overheated top melted region of the

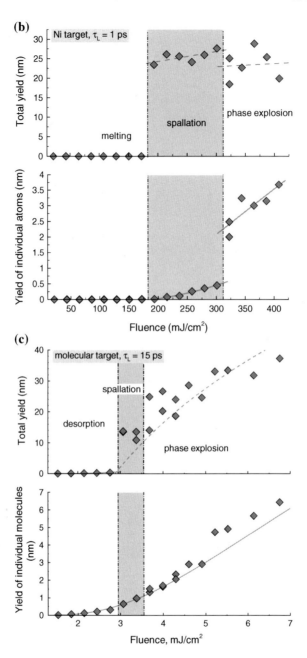

Fig. 2.2 continued

target into multiple droplets. The increase in the laser fluence in the spallation regime leads to a substantial raise of the total ablation yield for Cr, but causes only marginal or no increase of the yields for Ni and molecular targets. The number of the vapor-phase atoms or molecules in the ejected plume increases with fluence but remains below 2 % of the total yield for metal targets (Fig. 2.2a, b) and below 8 % for the molecular target (Fig. 2.2c) in the spallation regime.

Further increase of the laser fluence brings surface regions of the irradiated targets to temperatures that reach and exceed the threshold temperatures for the phase explosion [76–78] and leads to abrupt changes in the characteristics of the ablation process. The threshold temperatures for the onset of the phase explosion, T^*, have been determined in constant-pressure MD simulations of a slow heating of a metastable liquid, with zero-pressure values found to be $T^* \approx 6000$ K for the EAM Cr [15], $T^* \approx 9000$ K for the EAM Ni [12], and $T^* \approx 1060$ K for the model molecular system [40]. A surface region of the target overheated above T^* (e.g., the top region of Cr target in Fig. 2.1c) undergoes a rapid decomposition into a mixture of vapor and liquid droplets. Thus, the transition to the regime of phase explosion corresponds to the change in the dominant mechanism responsible for the material ejection from the photomechanical spallation driven by the relaxation of the laser-induced stresses to the phase explosion driven by the explosive release of the vapor. The values of the threshold fluences for the transition from spallation to the phase explosion regime also correspond to the ablation thresholds in simulations performed with longer laser pulses (e.g., 50 ps for Ni [12] and 150 ps for the molecular targets [19, 33, 34, 40]), when the condition of stress confinement is not satisfied and the spallation regime of material ejection is not activated. In these cases, direct transitions from surface evaporation to the phase explosion are observed in the simulations.

The transition to the phase explosion regime does not result in an increase in the total amount of the ejected material, Fig. 2.2. Quite the reverse, in the case of metals the total yield decreases somewhat as the fluence increases above the threshold for the phase explosion (Fig. 2.2a, b). The decrease of the total yield can be explained by two factors: (1) a higher energy cost of the decomposition of the surface region of the target into a mixture of vapor and small liquid droplets as compared to the ejection of larger droplets in the spallation regime, and (2) redeposition of some of the droplets ejected at the end of the ablation process back to the target due to the vapor pressure from the upper part of the plume. Indeed, deceleration and redeposition of some of the large droplets is observed in simulations performed for Cr and Ni at laser fluences above the thresholds for the phase explosion [12, 15]. For the molecular system, the increase in the total ablation yield in the phase explosion regime roughly follows predictions of a model that assumes the ejection of all material down to the depth in the target where a certain critical energy density is reached [19]. The prediction of this model is shown by the dashed line in Fig. 2.2c.

For metal targets, the transition to the phase explosion regime is also signified by an increase in the fraction of vapor-phase atoms in the ablation plume, from ~2 % of the total yield right below the threshold for the phase explosion to 4–7 % above the threshold for Cr and from 2 % to more than 10 % of the total yield for Ni. This increase in the fraction of the vapor-phase atoms upon the transition from the spallation to the phase explosion regimes can be related to the results of plume imaging experiments [79], where the maximum ejection of nanoparticles in laser ablation of Ni targets is observed at low fluences (possible spallation), whereas the degree of the plume atomization increases at higher fluences (possible phase explosion regime). The transition from spallation to phase explosion has also been related [12, 24] to the results of pump-probe experiments [39, 80], where the observation of optical interference patterns (Newton rings) can be explained by the spallation of a thin liquid layer from the irradiated target [21, 81] and the disappearance of the interference fringes in the central part of the laser spot [39, 82] can be related to the transition to the phase explosion regime.

2.3 Generation of Crystal Defects Below the Spallation Threshold

The discussion of the irradiation regime of melting and resolidification provided in the previous section may leave an impression of complete recovery of the initial state of the metal targets irradiated below the spallation thresholds. Detailed structural analysis of the targets that experienced the rapid melting and resolidification, however, reveals the presence of a high density of crystal defects, which may have important implications on physical, chemical, and mechanical properties of the surface layer. Two examples of defect configurations generated in surface regions of Cr and Ag targets irradiated at laser fluences that result in transient melting of about 20 nm thick layers of the targets are shown in Fig. 2.3. The simulation for Cr is the one illustrated by the contour plots shown in Fig. 2.1a and the mechanisms and kinetics of melting and resolidification in this simulation are discussed in the previous section. The simulation for Ag is performed with 100 fs laser pulse and an absorbed fluence of 70 mJ/cm^2. The interatomic interaction in Ag is described by EAM potential with parameters given in [83].

To provide a clear view of the crystal defects, the atoms that retain the original bcc or fcc local structure are blanked in the snapshots, while the remaining atoms are colored according to their potential energies in Fig. 2.3a and according to their local structure environment in Fig. 2.3b. With this visualization method, the vacancies, which are the most abundant defects introduced by the laser irradiation in these simulations, appear as clusters of atoms that surround the lattice sites with missing atoms. The distributions of vacancy concentration plotted in Fig. 2.4 indicate that most of the vacancies are located in the regions of the target that have experienced the transient melting and resolidification. To identify the mechanisms of the vacancy formation, several series of simulations of the resolidification process have been performed under well-controlled temperature and pressure conditions. These simulations reveal that most of the vacancies are generated at the rapidly advancing solidification front and are stabilized by the fast cooling of the surface region. Essentially, the vacancies are generated as "errors" made in the process of building the crystal structure at the solidification front, which, under conditions of strong

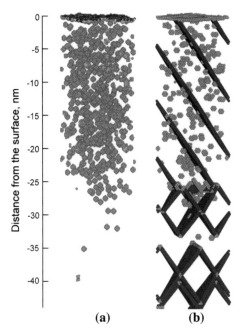

Fig. 2.3 Snapshots of the defect structures generated in surface regions of bulk bcc Cr (**a**) and fcc Ag (**b**) targets irradiated by femtosecond laser pulses at laser fluencies close to the thresholds for surface melting. The snapshots are from TTM–MD simulations performed with relatively small (8×8 nm^2) lateral sizes of the computational cells. The laser pulse durations and absorbed fluences are 200 fs and 85 mJ/cm^2 for Cr (**a**) and 100 fs and 70 mJ/cm^2 for Ag (**b**). The snapshots are taken at the end of the resolidification process, at 400 ps in (**a**) and 600 ps in (**b**). The atomic configurations are quenched for 1 ps to reduce thermal noise in atomic positions and energies. The atoms are colored according to their potential energies in (**a**) and local structure environment in (**b**). The atoms that belong to local configurations with the original bcc (**a**) or fcc (**b**) structure are blanked to expose crystal defects. Each *blue ball* (a compact cluster of atoms) in the snapshots corresponds to a vacancy, a small cross at the bottom of (**a**) corresponds to an interstitial in a <110>-dumbbell configuration, and *dark blue planes* in (**b**) correspond to stacking faults with displacement vectors a/6 <112>. The results for Cr are described in [15, 16]

undercooling, moves too fast to allow for atomic rearrangements needed to correct these "errors".

While the vacancy concentrations observed in both simulations are very high (about two orders of magnitude higher than the equilibrium vacancy concentrations at the melting temperatures of Cr and Ag), the vacancy concentration in bcc Cr is about twice higher than the one in fcc Ag. Given that the energies of vacancy formation expressed in units of the thermal energy at the melting temperature, $E_v^f / k_B T_m$, are similar in the two metals (10.35 for EAM Cr and 9.88 for EAM Ag), the difference in the vacancy concentrations is likely to reflect the differences in atomic-level mechanisms responsible for the growth of the close-packed fcc and more open bcc crystal lattices.

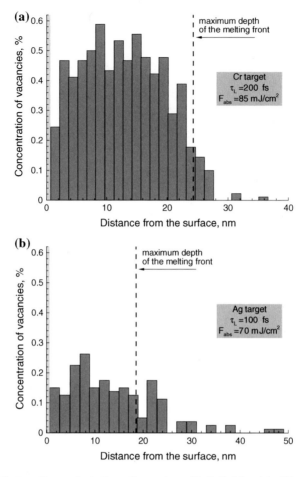

Fig. 2.4 Distribution of vacancies in the surface regions of bulk Cr (**a**) and Ag (**b**) targets irradiated by femtosecond laser pulses at laser fluencies close to the thresholds for surface melting. Snapshots of atomic configurations used in the analysis are shown in Fig. 2.3. Each bar in the histograms is the result of averaging over ten individual (001) atomic planes. The *dashed lines* mark the depths of the regions that experienced transient melting and resolidification in response to the laser irradiation

   The generation of strong supersaturation of vacancies in the surface regions of the irradiated targets may have important practical implications, including the formation of nanovoids and degradation of the mechanical properties of the surface region of the target in the multi-pulse irradiation regime. The generation of vacancies in this case may be related to experimental observations of the incubation effect, when the laser fluence threshold for ablation/damage decreases with increasing number of laser pulses applied to the same area, e.g. [84–86]. The high density of vacancies generated in the surface regions of irradiated targets may also play an important role in the redistribution of impurities or mixing/alloying in multi-component or composite targets.

In addition to vacancies, the snapshot shown in Fig. 2.3b reveals the presence of multiple stacking fault planes with displacement vectors of $a/6 <112>$, where a is the fcc lattice constant. The stacking faults located below the region that experienced the transient melting are left behind by the partial dislocations emitted from the melting front at a time when the tensile component of the laser-induced stress wave passes through the melting front. In the single crystal fcc target with {001} orientation of the irradiated surface, the partial dislocations can be activated on four different {111} slip planes. Interactions between the dislocations propagating along the different slip planes can result in the formation of immobile dislocation segments and stable dislocation configurations [71], thus leading to the hardening of the laser-treated surface. In contrast to a relatively ductile fcc Ag material, the higher resistance of bcc crystals to the movement of dislocations results in the absence of laser-generated dislocations in the bcc Cr target (Fig. 2.3a).

2.4 Evolution of Voids in Photomechanical Spallation

As discussed in Sect. 2.2 and illustrated in Fig. 2.1b, the relaxation of laser-induced stresses in the surface region of the irradiated target can generate tensile stresses that are sufficiently strong to induce cavitation and fragmentation in the melted surface region, leading to the ejection (or photomechanical spallation) of liquid layer(s) or droplets. In this section we consider the microscopic mechanisms of the spallation process and compare computational predictions obtained in simulations performed for two very different targets, an amorphous molecular solid and Ag (001) single crystal.

The visual picture of the evolution of voids in a surface region of a Ag target irradiated by a 100 fs laser pulse at a fluence of 85 mJ/cm^2, just below the spallation threshold, is shown in Fig. 2.5, where the top-view snapshots of the void evolution are shown for a relatively large-scale TTM-MD simulation performed for a system with $98.7 \times 98.7 \times 150$ nm^3 dimensions of the atomistic part of the computational domain (84.2 million atoms). The large (by atomistic modeling standards) lateral size of the computational cell provides a clear view of the nucleation, growth and coarsening of multiple voids at a depth of $\sim$40–60 nm under the irradiated surface and ensures that the initial evolution of voids is not affected by the periodic boundary conditions applied in the lateral directions. The appearance of voids coincides with passage of the tensile component of the stress wave through the melted surface region of the target. The voids grow as the surface layer accelerated by the initial relaxation of the laser-induced stresses moves away from the bulk of the target. In the simulation illustrated in Fig. 2.5, the outward motion of the surface layer slows down with time and reverses the direction of its motion at about 800 ps. At this time the growth of the voids turns into recession. At the same time, the fast cooling of the surface region creates conditions for fast epitaxial resolidification of the melted region, with the

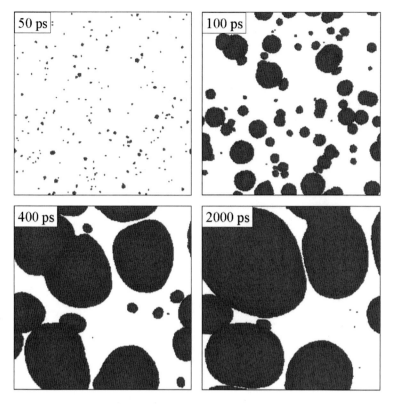

Fig. 2.5 The *top-view* snapshots of the evolution of voids (*empty space*) in the sub-surface region of a Ag (001) target irradiated by a 100 fs laser pulse at an absorbed fluence of 85 mJ/cm². The nucleation of voids in response to the tensile stresses associated with laser-generated stress wave (snapshot taken at 50 ps) is followed by void growth/coalescence (100 and 400 ps) and capture by the solidification front (2,000 ps)

velocity of the solidification front increasing up to 90 m/s. The solidification front crosses the region of the void evolution during the time from 500 to 1,500 ps and prevents the collapse of the voids. As a result, the voids shown in Fig. 2.5 for 2,000 ps are completely surrounded by the crystalline material and remain stable upon further cooling of the surface. At higher fluences (e.g., at 90 mJ/cm² for the Ag target), the expansion of voids leads to the eventual percolation of the growing empty regions and separation of liquid layer(s) or large droplets from the target.

A similar sequence of void nucleation, growth, coarsening, coalescence and percolation has been observed in earlier smaller-scale simulations performed for metals [9, 21–23, 87] and molecular systems [19, 21], as well as in recent large-scale simulations of laser spallation and ablation of Al targets [24]. In particular, an animated sequence of snapshots from a coarse-grained MD simulation of laser spallation of a molecular target posted at this web site [88] shows a picture of the void evolution that is visually similar to the one in Fig. 2.5. To quantify the evolution of voids in laser spallation, the void size distributions are shown in Fig. 2.6a for different times during the simulation of incomplete laser spallation of Ag target discussed above and illustrated in Fig. 2.5. The distributions are fitted to power law dependences with exponents that are increasing with time, Fig. 2.6b. This increase of the power law exponent is reflecting the void coarsening and coalescence, when the size and the number of large voids are growing at the expense of quickly decreasing population of small voids. The void size distributions plotted in Fig. 2.6a are very similar not only to the distributions observed earlier in a simulation of laser spallation of a Ni film [87] and a bulk Al target [24], but also to the ones shown in Fig. 2.6c for a molecular target [21]. The time dependences of the power law exponents predicted for the two very different targets, Fig. 2.6b, d, are also very similar, except for the fact that the exponent saturates in Fig. 2.6b due to the capture of the voids by the solidification front but continues to growth in Fig. 2.6d reflecting the growth and eventual percolation of voids in the spallation of the molecular target.

The prediction of the capture of voids by the solidification front can be related to the recently reported experimental observation of surface swelling, or "frustrated ablation," in Al targets irradiated by 100 fs laser pulses [89]. Similarly to the experiments, the voids captured by the solidification front increase the volume of the surface region, leading to an effective "swelling" of the irradiated target by about 17 nm. The larger thermal conductivity and smaller melting depth near the spallation threshold in Ag, as compared to Al, makes it possible to observe this interesting phenomenon with smaller computational systems and shorter simulation times.

2.5 The Visual Picture of Phase Explosion

The transition from the spallation to the phase explosion regime of material ejection discussed in Sect. 2.2 does not result in an increase in the total amount of the ejected material (Fig. 2.2) but can still be clearly identified from changes in the composition of the ablation plume (larger fraction of vapor and small liquid droplets) and the dynamics of the plume expansion (faster expansion of the front part of the plume). These changes are the reflection of the change in the dominant driving force responsible for the material ejection from the relaxation of laser-induced stresses in the spallation regime to the explosive decomposition of material overheated up to the limit of its thermodynamic stability [12, 15, 40, 76–78] in the phase explosion regime.

The discussion of the thermodynamic conditions leading to the onset of the phase explosion provided in Sect. 2.2 and illustrated by Fig. 2.2c can be complemented by a visual picture of the material ejection in the phase explosion regime shown in Fig. 2.7. The snapshots in Fig. 2.7 are from a large-scale TTM-MD simulation of laser ablation of a bulk Al target irradiated by a 100 fs laser pulse at an absorbed fluence of $200 \, mJ/cm^2$ [24]. The interatomic interaction in Al is described by EAM potential with parameters given in [90]. In the first snapshot, shown for 50 ps after the laser pulse, one can see a fine "Swiss cheese"-like structure of liquid cells enclosing

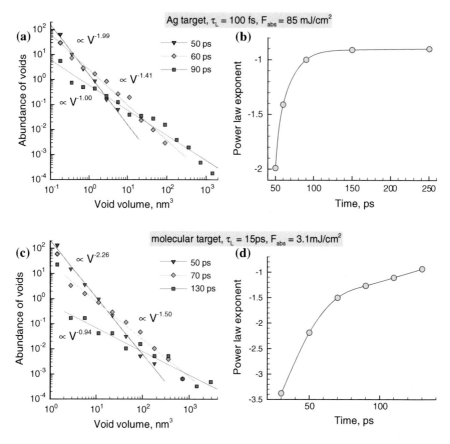

Fig. 2.6 Void abundance distributions as a function of void volume predicted for different times in simulations of short pulse laser irradiation of a single crystal Ag (001) target (**a**) and an amorphous molecular target (**c**). The *lines* in (**a**) and (**c**) are power law fits of the data points with the exponents indicated in the figures. Time dependences of the power-law exponents are shown in (**b**) and (**d**). The values of laser pulse duration, τ_L, and absorbed fluence, F_{abs}, are listed in the figure. The irradiation conditions in both simulations correspond to the regime of stress confinement. Laser fluence is just below the spallation threshold in the simulation of the Ag target and about 7 % above the spallation threshold in the simulation of the molecular target. The results for the molecular target (**c**, **d**) are adopted from [21]

dense hot vapor forming in a relatively broad surface region of the irradiated target. This cellular structure is generated by a rapid (explosive) release of vapor in the melted metal overheated above the threshold temperature for phase explosion, T^*. The expansion of the cellular structure leads to the coarsening of the liquid and vapor regions and results in the formation of a foamy structure of interconnected liquid regions surrounded by vapor, see snapshot shown for 150 ps in Fig. 2.7. Deeper into the target, the relaxation of the initial compressive pressure and the pulling force from the expanding foamy structure leads to the cavitation of the liquid and formation

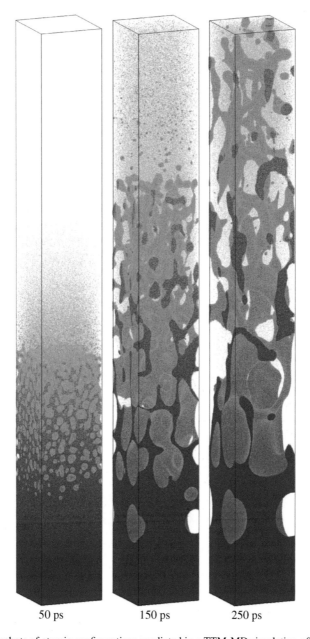

50 ps   150 ps   250 ps

Fig. 2.7 Snapshots of atomic configurations predicted in a TTM-MD simulation of laser ablation of a bulk Al target irradiated by a 100 fs laser pulse at an absorbed fluence of $200 \, \text{mJ}/\text{cm}^2$ [24]. Atoms are colored by their instantaneous potential energy. The initial dimensions of the atomistic (MD) part of the computational system are $94.3 \times 94.3 \times 300 \, \text{nm}^3$. The irradiation regime in this simulation corresponds to the phase explosion and ablation of a surface region of the target

of coarse liquid structures extending in the direction of the plume expansion (see snapshots for 150 and 250 ps in Fig. 2.7).

The eventual decomposition of the complex hierarchical foamy structure into individual liquid droplets, clusters and vapor-phase atoms leads to the formation of a multi-component ablation plume moving away from the target. An important implication of the hierarchical ablation process illustrated in Fig. 2.7 and observed in earlier MD simulations of molecular systems [33, 91] is the effect of spatial segregation of clusters and droplets of different sizes in the ablation plume. The vapor-phase atoms/molecules and small clusters are predominantly present in the front part of the expanding plume, the medium size clusters are localized in the middle of the expanding plume, and the large liquid droplets formed at the final stage of the plume development tend to be slower and are closer to the original surface. The cluster segregation effect predicted in the simulations can be related to the results of plume imaging experiments [79, 92–95], where splitting of the plume into a fast component with optical emission characteristic for neutral atoms and a slow component with blackbody-like emission associated with presence of hot clusters is observed.

2.6 Conclusions

A comparative analysis of the results of MD simulations of laser-materials interactions performed for various target materials reveals the general and material-specific characteristics of laser-induced structural modification, spallation and ablation. The driving forces and microscopic mechanisms of laser spallation and ablation are found to be similar for a broad class of target materials. In particular, the same physical conditions are governing the onset of photomechanical spallation and the transition between the spallation and phase explosion regimes in material systems as different as metals and amorphous molecular systems. Moreover, some of the quantitative characteristics of laser-induced processes, such as the evolution of the void size distributions in spallation or the degree of segregation of clusters and liquid droplets of different sizes in ablation plumes generated in the phase explosion regime, are also surprisingly similar for different materials, reflecting the common mechanical and thermodynamic origins of the underlying processes. The material-specific predictions of the simulations include the microstructural modification of the irradiated surface, with the type and density of crystal defects generated in the melting and resolidification regime being sensitive to the structure of the target material, as well as the quantitative characteristics of the yield versus fluence dependence, which are found to be sensitive to the thermodynamic characteristics of the target material.

Given the fast advancement of the computing technology and the development of new computational models for increasingly realistic MD simulations of laser interactions with metals, semiconductors, dielectrics and organic materials, it is reasonable to expect that atomic- and molecular-level modelling will remain at the forefront of computational investigation of laser-materials interactions. The ability to simulate increasingly larger systems is likely to open up opportunities for investigation of laser processing of multi-phase and nanocomposite systems and for establishing direct connections to experimental observations.

Acknowledgments Financial support for this work was provided by the National Science Foundation (NSF) through Grants DMR-0907247 and CMMI-1301298, Electro Scientific Industries, Inc., and the Air Force Office of Scientific Research through Grant FA9550-10-1-0541. Computational support was provided by the Oak Ridge Leadership Computing Facility (project MAT048) and NSF through the Extreme Science and Engineering Discovery Environment (project TG-DMR110090).

References

1. P.L. Silvestrelli, A. Alavi, M. Parrinello, D. Frenkel, Ab initio molecular dynamics simulation of laser melting of silicon. Phys. Rev. Lett. **77**, 3149–3152 (1996)
2. T. Dumitrica, A. Burzo, Y. Dou, R.E. Allen, Response of Si and InSb to ultrafast laser pulses. Phys. Status Solidi B **241**, 2331–2342 (2004)
3. V. Recoules, J. Clérouin, G. Zérah, P.M. Anglade, S. Mazevet, Effect of intense laser irradiation on the lattice stability of semiconductors and metals. Phys. Rev. Lett. **96**, 055503 (2006)
4. Z. Lin, L.V. Zhigilei, V. Celli, Electron-phonon coupling and electron heat capacity of metals under conditions of strong electron-phonon nonequilibrium. Phys. Rev. B **77**, 075133 (2008)
5. H.O. Jeschke, M.S. Diakhate, M.E. Garcia, Molecular dynamics simulations of laser-induced damage of nanostructures and solids. Appl. Phys. A **96**, 33–42 (2009)
6. Z. Lin, R.E. Allen, Ultrafast equilibration of excited electrons in dynamic simulations. J. Phys. Condens. Matter **21**, 485503 (2009)
7. C.F. Richardson, P. Clancy, Picosecond laser processing of copper and gold: a computer simulation study. Mol. Sim. **7**, 335–355 (1991)
8. X. Wang, X. Xu, Molecular dynamics simulation of heat transfer and phase change during laser material interaction. J. Heat Transf. **124**, 265–274 (2002)
9. D.S. Ivanov, L.V. Zhigilei, Combined atomistic-continuum modeling of short pulse laser melting and disintegration of metal films. Phys. Rev. B **68**, 064114 (2003)
10. D.S. Ivanov, L.V. Zhigilei, The effect of pressure relaxation on the mechanisms of short pulse laser melting. Phys. Rev. Lett. **91**, 105701 (2003)
11. Z. Lin, L.V. Zhigilei, Time-resolved diffraction profiles and atomic dynamics in short pulse laser induced structural transformations: molecular dynamics study. Phys. Rev. B **73**, 184113 (2006)
12. L.V. Zhigilei, Z. Lin, D.S. Ivanov, Atomistic modeling of short pulse laser ablation of metals: Connections between melting, spallation, and phase explosion. J. Phys. Chem. C **113**, 11892–11906 (2009)
13. D.A. Thomas, Z. Lin, L.V. Zhigilei, E.L. Gurevich, S. Kittel, R. Hergenröder, Atomistic modeling of femtosecond laser-induced melting and atomic mixing in Au film - Cu substrate system. Appl. Surf. Sci. **255**, 9605–9612 (2009)
14. Z. Lin, E.M. Bringa, E. Leveugle, L.V. Zhigilei, Molecular dynamics simulation of laser melting of nanocrystalline Au. J. Phys. Chem. C **114**, 5686–5699 (2010)
15. E.T. Karim, Z. Lin, L.V. Zhigilei, Molecular dynamics study of femtosecond laser interactions with Cr targets. AIP Conf. Proc. **1464**, 280–293 (2012)
16. Z. Lin, R.A. Johnson, L.V. Zhigilei, Computational study of the generation of crystal defects in a bcc metal target irradiated by short laser pulses. Phys. Rev. B **77**, 214108 (2008)
17. D.S. Ivanov, Z. Lin, B. Rethfeld, G.M. O'Connor, Th.J. Glynn, L.V. Zhigilei, Nanocrystalline structure of nanobump generated by localized photo-excitation of metal film. J. Appl. Phys. **107**, 013519 (2010)

18. C. Wu, D.A. Thomas, Z. Lin, L.V. Zhigilei, Runaway lattice-mismatched interface in an atomistic simulation of femtosecond laser irradiation of Ag film—Cu substrate system. Appl. Phys. A **104**, 781–792 (2011)

19. L.V. Zhigilei, B.J. Garrison, Microscopic mechanisms of laser ablation of organic solids in the thermal and stress confinement irradiation regimes. J. Appl. Phys. **88**, 1281–1298 (2000)

20. S.I. Anisimov, V.V. Zhakhovskii, N.A. Inogamov, K. Nishihara, A.M. Oparin, Yu.V. Petrov, Destruction of a solid film under the action of ultrashort laser pulse. Pis'ma Zh. Eksp. Teor. Fiz. **77**, 731 (JETP Lett. **77**, 606–610 (2003))

21. E. Leveugle, D.S. Ivanov, L.V. Zhigilei, Photomechanical spallation of molecular and metal targets: molecular dynamics study. Appl. Phys. A **79**, 1643–1655 (2004)

22. A.K. Upadhyay, H.M. Urbassek, Melting and fragmentation of ultra-thin metal films due to ultrafast laser irradiation: a molecular-dynamics study. J. Phys. D **38**, 2933–2941 (2005)

23. B.J. Demaske, V.V. Zhakhovsky, N.A. Inogamov, I.I. Oleynik, Ablation and spallation of gold films irradiated by ultrashort laser pulses. Phys. Rev. B **82**, 064113 (2010)

24. C. Wu, L.V. Zhigilei, Microscopic mechanisms of laser spallation and ablation of metal targets from large-scale molecular dynamics simulations. Appl. Phys. A **114**, 11–32 (2014)

25. E. Ohmura, I. Fukumoto, Molecular dynamics simulation on laser ablation of fcc metal. Int. J. Jpn. Soc. Precis. Eng. **30**, 128–133 (1996)

26. L.V. Zhigilei, P.B.S. Kodali, B.J. Garrison, Molecular dynamics model for laser ablation of organic solids. J. Phys. Chem. B **101**, 2028–2037 (1997)

27. R.F.W. Herrmann, J. Gerlach, E.E.B. Campbell, Ultrashort pulse laser ablation of silicon: an MD simulation study. Appl. Phys. A **66**, 35–42 (1998)

28. X. Wu, M. Sadeghi, A. Vertes, Molecular dynamics of matrix-assisted laser desorption of leucine enkephalin guest molecules from nicotinic acid host crystal. J. Phys. Chem. B **102**, 4770–4778 (1998)

29. L.V. Zhigilei, P.B.S. Kodali, B.J. Garrison, A microscopic view of laser ablation. J. Phys. Chem. B **102**, 2845–2853 (1998)

30. Y.G. Yingling, L.V. Zhigilei, B.J. Garrison, The role of photochemical fragmentation in laser ablation: a molecular dynamics study. J. Photochem. Photobiol. A **145**, 173–181 (2001)

31. T.E. Itina, L.V. Zhigilei, B.J. Garrison, Microscopic mechanisms of matrix assisted laser desorption of analyte molecules: insights from molecular dynamics simulation. J. Phys. Chem. B **106**, 303–310 (2002)

32. C. Schäfer, H.M. Urbassek, L.V. Zhigilei, Metal ablation by picosecond laser pulses: A hybrid simulation. Phys. Rev. B **66**, 115404 (2002)

33. L.V. Zhigilei, Dynamics of the plume formation and parameters of the ejected clusters in short-pulse laser ablation. Appl. Phys. A **76**, 339–350 (2003)

34. L.V. Zhigilei, E. Leveugle, B.J. Garrison, Y.G. Yingling, M.I. Zeifman, Computer simulations of laser ablation of molecular substrates. Chem. Rev. **103**, 321–348 (2003)

35. P. Lorazo, L.J. Lewis, M. Meunier, Short-pulse laser ablation of solids: from phase explosion to fragmentation. Phys. Rev. Lett. **91**, 225502 (2003)

36. N.N. Nedialkov, P.A. Atanasov, S.E. Imamova, A. Ruf, P. Berger, F. Dausinger, Dynamics of the ejected material in ultra-short laser ablation of metals. Appl. Phys. A **79**, 1121–1125 (2004)

37. C. Cheng, X. Xu, Mechanisms of decomposition of metal during femtosecond laser ablation. Phys. Rev. B **72**, 165415 (2005)

38. P. Lorazo, L.J. Lewis, M. Meunier, Thermodynamic pathways to melting, ablation, and solidification in absorbing solids under pulsed laser irradiation. Phys. Rev. B **73**, 134108 (2006)

39. M.B. Agranat, S.I. Anisimov, S.I. Ashitkov, V.V. Zhakhovskii, N.A. Inogamov, K. Nishihara, Yu.V. Petrov, V.E. Fortov, V.A. Khokhlov, Dynamics of plume and crater formation after action of femtosecond laser pulse. Appl. Surf. Sci. **253**, 6276–6282 (2007)

40. E. Leveugle, L.V. Zhigilei, Molecular dynamics simulation study of the ejection and transport of polymer molecules in matrix-assisted pulsed laser evaporation. J. Appl. Phys. **102**, 074914 (2007)

41. M. Prasad, P.F. Conforti, B.J. Garrison, On the role of chemical reactions in initiating ultraviolet ablation in poly (methyl methacrylate). J. Appl. Phys. **101**, 103113 (2007)

42. M. Gill-Comeau, L.J. Lewis, Ultrashort-pulse laser ablation of nanocrystalline aluminum. Phys. Rev. B **84**, 224110 (2011)
43. L.V. Zhigilei, A.N. Volkov, E. Leveugle, M. Tabetah, The effect of the target structure and composition on the ejection and transport of polymer molecules and carbon nanotubes in matrix-assisted pulsed laser evaporation. Appl. Phys. A **105**, 529–546 (2011)
44. X. Li, L. Jiang, Size distribution control of metal nanoparticles using femtosecond laser pulse train: a molecular dynamics simulation. Appl. Phys. A **109**, 367–376 (2012)
45. R.K. Singh, J. Narayan, Pulsed-laser evaporation technique for deposition of thin films: Physics and theoretical model. Phys. Rev. B **41**, 8843–8858 (1990)
46. A. Peterlongo, A. Miotello, R. Kelly, Laser-pulse sputtering of aluminum: Vaporization, boiling, superheating, and gas-dynamic effects. Phys. Rev. E **50**, 4716–4727 (1994)
47. J.R. Ho, C.P. Grigoropoulos, J.A.C. Humphrey, Computational study of heat transfer and gas dynamics in the pulsed laser evaporation of metals. J. Appl. Phys. **78**, 4696–4709 (1995)
48. X. Xu, G. Chen, K.H. Song, Experimental and numerical investigation of heat transfer and phase change phenomena during excimer laser interaction with nickel. Int. J. Heat Mass Transf. **42**, 1371–1382 (1999)
49. O.A. Bulgakova, N.M. Bulgakova, V.P. Zhukov, A model of nanosecond laser ablation of compound semiconductors accounting for non-congruent vaporization. Appl. Phys. A **101**, 53–59 (2010)
50. K. Eidmann, J. Meyer-ter-Vehn, T. Schlegel, S. Huller, Hydrodynamic simulation of subpicosecond laser interaction with solid-density matter. Phys. Rev. E **62**, 1202–1214 (2000)
51. J.P. Colombier, P. Combis, F. Bonneau, R. Le Harzic, E. Audouard, Hydrodynamic simulations of metal ablation by femtosecond laser irradiation. Phys. Rev. B **71**, 165406 (2005)
52. A.N. Volkov, L.V. Zhigilei, Hydrodynamic multi-phase model for simulation of laser-induced non-equilibrium phase transformations. J. Phys. Conf. Ser. **59**, 640–645 (2007)
53. M.E. Povarnitsyn, T.E. Itina, K.V. Khishchenko, P.R. Levashov, Suppression of ablation in femtosecond double-pulse experiments. Phys. Rev. Lett. **103**, 195002 (2009)
54. M.E. Povarnitsyn, T.E. Itina, P.R. Levashov, K.V. Khishchenko, Mechanisms of nanoparticle formation by ultra-short laser ablation of metals in liquid environment. Phys. Chem. Chem. Phys. **15**, 3108–3114 (2013)
55. C. Schäfer, H.M. Urbassek, L.V. Zhigilei, B.J. Garrison, Pressure-transmitting boundary conditions for molecular dynamics simulations. Comp. Mater. Sci. **24**, 421–429 (2002)
56. L.V. Zhigilei, A.N. Volkov, A.M. Dongare, in *Encyclopedia of Nanotechnology*, ed. by B. Bhushan (Springer, Heidelberg, 2012), Part 4, pp. 470–480
57. S.I. Anisimov, B.L. Kapeliovich, T.L. Perel'man, Electron emission from metal surfaces exposed to ultrashort laser pulses. Sov. Phys. JETP **39**, 375–377 (1974)
58. R. Holenstein, S.E. Kirkwood, R. Fedosejevs, Y.Y. Tsui, Simulation of femtosecond laser ablation of silicon. Proc. SPIE **5579**, 688–695 (2004)
59. Y. Wang, X. Xu, L. Zheng, Molecular dynamics simulation of ultrafast laser ablation of fused silica film. Appl. Phys. A **92**, 849–852 (2008)
60. Y. Cherednikov, N.A. Inogamov, H.M. Urbassek, Atomistic modeling of ultrashort-pulse ultraviolet laser ablation of a thin Lif film. J. Opt. Soc. Am. B **28**, 1817–1824 (2011)
61. E. Leveugle, L.V. Zhigilei, A. Sellinger, J.M. Fitz-Gerald, Computational and experimental study of the cluster size distribution in MAPLE. Appl. Surf. Sci. **253**, 6456–6460 (2007)
62. A. Sellinger, E. Leveugle, J.M. Fitz-Gerald, L.V. Zhigilei, Generation of surface features in films deposited by matrix-assisted pulsed laser evaporation: the effects of the stress confinement and droplet landing velocity. Appl. Phys. A **92**, 821–829 (2008)
63. R. Knochenmuss, L.V. Zhigilei, Molecular dynamics model of ultraviolet matrix-assisted laser desorption/ionization including ionization processes. J. Phys. Chem. B **109**, 22947–22957 (2005)
64. R. Knochenmuss, L.V. Zhigilei, Molecular dynamics simulations of MALDI: laser fluence and pulse width dependence of plume characteristics and consequences for matrix and analyte ionization. J. Mass Spectrom. **45**, 333–346 (2010)

65. R. Knochenmuss, L.V. Zhigilei, What determines MALDI ion yields? A molecular dynamics study of ion loss mechanisms. Anal. Bioanal. Chem. **402**, 2511–2519 (2012)
66. Y.G. Yingling, B.J. Garrison, Coarse-grained chemical reaction model. J. Phys. Chem. B **108**, 1815–1821 (2004)
67. L.V. Zhigilei, C. Wei, D. Srivastava, Mesoscopic model for dynamic simulations of carbon nanotubes. Phys. Rev. B **71**, 165417 (2005)
68. A.N. Volkov, L.V. Zhigilei, Mesoscopic interaction potential for carbon nanotubes of arbitrary length and orientation. J. Phys. Chem. C **114**, 5513–5531 (2010)
69. L. V. Zhigilei, Z. Lin, D.S. Ivanov, E. Leveugle, W.H. Duff, D. Thomas, C. Sevilla, S. J. Guy, Atomic/molecular-level simulations of laser-materials interactions. in *Laser-Surface Interactions for New Materials Production: Tailoring Structure and Properties*, ed. by A. Miotello, P.M. Ossi. Springer Series in Materials Science, vol. 130.(Springer, New York, 2010), pp. 43–79
70. L.V. Zhigilei, E. Leveugle, D.S. Ivanov, Z. Lin, A.N. Volkov, Molecular dynamics simulations of short pulse laser ablation: Mechanisms of material ejection and particle generation. in *Nanosized Material Synthesis by Action of High-Power Energy Fluxes on Matter* (Siberian Branch of the Russian Academy of Sciences, Novosibirsk, 2010), pp. 147–220 (in Russian)
71. C. Wu, E. T. Karim, A. N. Volkov, and L. V. Zhigilei, Atomic movies of laser-induced structural and phase transformations from molecular dynamics simulations. in *Lasers in Materials Science*, ed. by M. Castillejo, P.M. Ossi, L.V. Zhigilei. Springer Series in Materials Science, vol. 191. (Springer, New York, 2014), pp. 67–100
72. X.W. Zhou, H.N.G. Wadley, R.A. Johnson, D.J. Larson, N. Tabat, A. Cerezo, A.K. Petford-Long, G.D.W. Smith, P.H. Clifton, R.L. Martens, T.F. Kelly, Atomic scale structure of sputtered metal multilayers. Acta Mater. **49**, 4005–4015 (2001)
73. B.J. Siwick, J.R. Dwyer, R.E. Jordan, R.J.D. Miller, An atomic-level view of melting using femtosecond electron diffraction. Science **302**, 1382–1385 (2003)
74. J.R. Dwyer, R.E. Jordan, C.T. Hebeisen, M. Harb, R. Ernstorfer, T. Dartigalongue, R.J.D. Miller, Femtosecond electron diffraction: an atomic perspective of condensed phase dynamics. J. Mod. Opt. **54**, 905–922 (2007)
75. W.-L. Chan, R.S. Averback, D.G. Cahill, Y. Ashkenazy, Solidification velocities in deeply undercooled silver. Phys. Rev. Lett. **102**, 095701 (2009)
76. B.J. Garrison, T.E. Itina, L.V. Zhigilei, The limit of overheating and the threshold behavior in laser ablation. Phys. Rev. E **68**, 041501 (2003)
77. A. Miotello, R. Kelly, Laser-induced phase explosion: new physical problems when a condensed phase approaches the thermodynamic critical temperature. Appl. Phys. A **69**, S67–S73 (1999)
78. N.M. Bulgakova, A.V. Bulgakov, Pulsed laser ablation of solids: transition from normal vaporization to phase explosion. Appl. Phys. A **73**, 199–208 (2001)
79. S. Amoruso, R. Bruzzese, C. Pagano, X. Wang, Features of plasma plume evolution and material removal efficiency during femtosecond laser ablation of nickel in high vacuum. Appl. Phys. A **89**, 1017–1024 (2007)
80. K. Sokolowski-Tinten, J. Bialkowski, A. Cavalleri, D. von der Linde, A. Oparin, J. Meyer-ter-Vehn, S.I. Anisimov, Transient states of matter during short pulse laser ablation. Phys. Rev. Lett. **81**, 224–227 (1998)
81. N.A. Inogamov, Y.V. Petrov, S.I. Anisimov, A.M. Oparin, N.V. Shaposhnikov, D. von der Linde, J. Meyer-ter-Vehn, Expansion of matter heated by an ultrashort laser pulse. JETP Lett. **69**, 310–316 (1999)
82. A.A. Ionin, S.I. Kudryashov, L.V. Seleznev, D.V. Sinitsyn, Dynamics of the spallative ablation of a GaAs surface irradiated by femtosecond laser pulses. JETP Lett. **94**, 753–758 (2011)
83. S.M. Foiles, M.I. Baskes, M.S. Daw, Embedded-atom-method functions for the fcc metals Cu, Ag, Au, Ni, Pd, Pt, and their alloys. Phys. Rev. B **33**, 7983–7991 (1986)
84. P.T. Mannion, J. Magee, E. Coyne, G.M. O'Connor, T.J. Glynn, The effect of damage accumulation behaviour on ablation thresholds and damage morphology in ultrafast laser micromachining of common metals in air. Appl. Surf. Sci. **233**, 275–287 (2004)

85. S.E. Kirkwood, A.C. Van Popta, Y.Y. Tsui, R. Fedosejevs, Single and multiple shot near-infrared femtosecond laser pulse ablation thresholds of copper. Appl. Phys. A **81**, 729–735

86. G. Raciukaitis, M. Brikas, P. Gecys, M. Gedvilas, Accumulation effects in laser ablation of metals with high-repetition-rate lasers. Proc. SPIE **7005**, 70052L (2008)

87. L.V. Zhigilei, D.S. Ivanov, E. Leveugle, B. Sadigh, E.M. Bringa, Computer modeling of laser melting and spallation of metal targets, in *High-Power Laser Ablation V*, ed. by C.R. Phipps. Proc. SPIE **5448**, 505–519 (2004)

88. Animated sequences of snapshots from a MD simulation of laser spallation of a molecular target, http://www.faculty.virginia.edu/CompMat/spallation/animations/

89. J.-M. Savolainen, M.S. Christensen, P. Balling, Material swelling as the first step in the ablation of metals by ultrashort laser pulses. Phys. Rev. B **84**, 193410 (2011)

90. Y. Mishin, D. Farkas, M.J. Mehl, D.A. Papaconstantopoulos, Interatomic potentials for monoatomic metals from experimental data and *ab initio* calculations. Phys. Rev. B **59**, 3393–3407 (1999)

91. L.V. Zhigilei, Computational model for multiscale simulation of laser ablation, ed. by V.V. Bulatov, F. Cleri, L. Colombo, L.J. Lewis, N. Mousseau. *Advances in Materials Theory and Modeling-Bridging Over Multiple-Length and Time Scales* Mat. Res. Soc. Symp. Proc. **677**, AA2.1.1–AA2.1.11 (2001)

92. S. Noël, J. Hermann, T. Itina, Investigation of nanoparticle generation during femtosecond laser ablation of metals. Appl. Surf. Sci. **253**, 6310–6315 (2007)

93. T.E. Itina, K. Gouriet, L.V. Zhigilei, S. Noël, J. Hermann, M. Sentis, Mechanisms of small clusters production by short and ultra-short pulse laser ablation. Appl. Surf. Sci. **253**, 7656–7661 (2007)

94. O. Albert, S. Roger, Y. Glinec, J.C. Loulergue, J. Etchepare, C. Boulmer-Leborgne, J. Perriere, E. Millon, Time-resolved spectroscopy measurements of a titanium plasma induced by nanosecond and femtosecond lasers. Appl. Phys. A **76**, 319–323 (2003)

95. N. Jegenyes, J. Etchepare, B. Reynier, D. Scuderi, A. Dos-Santos, Z. Tóth, Time-resolved dynamics analysis of nanoparticles applying dual femtosecond laser pulses. Appl. Phys. A **91**, 385–392 (2008)

Chapter 3
Laser Nanocrystallization of Metals

Irina N. Zavestovskaya

Abstract The results of experimental and theoretical studies of surface micro- and nanostructuring of metals and other materials irradiated directly by short and ultra-short laser pulses are reviewed. Special attention is paid to direct laser action involving melting of the material (with or without ablation), followed by ultrafast surface solidification, which is an effective approach to producing surface nanostructures.

3.1 Introduction

Laser micro- and nanostructuring of materials is important in many scientific, techno-logical and medical applications [1–5]. Nanostructures resulting from laser material processing have unique properties and often cannot be produced by other, nonlaser techniques.

The production of nanoscale structures on the surface of metals improves their physical and mechanical properties, enhances the biocompatibility of implants with body issues, etc. For example, the strength of structural materials increases with decreasing grain size, with no loss in plasticity [6]. Nanostructured materials strongly differ in optical properties from unmodified bulk materials, acquiring properties of metamaterials. This can be used to increase the absorptance of materials by laser processing [7], to produce black and coloured films on the surface of various met-als [8], and to fabricate "black silicon" structures [5] for solar-cell manufacturing. Nanostructured surfaces find other various applications: in selective nanocatalysis, microelectronics for information recording, process for creating optical memory based on phase transitions in metal nanoparticles [9].

I. N. Zavestovskaya (✉)
P. N. Lebedev Physical Institute of RAS, 53 Leninskiy prospekt, Moscow 119991, Russia

National Research Nuclea University MEPhI, Moscow, Russia
e-mail: INZavestovskaya@mephi.ru

V. P. Veiko and V. I. Konov (eds.), *Fundamentals of Laser-Assisted Micro- and Nanotechnologies*, Springer Series in Materials Science 195, DOI: 10.1007/978-3-319-05987-7_3, © Springer International Publishing Switzerland 2014

The development of laser nanostructuring technologies has been stimulated by recent advances in laser engineering, which have made it possible to generate short and ultrashort laser pulses. The femtosecond laser technique, which turned to be an effective instrument in laser applications [10–13], is actively updated, and its price reduced. In addition, in recent years the solid diode-pumped lasers are more and more widely used in laser technology. Such lasers produce the pulses of nanosecond duration. The advantages of lasers are the small size, good quality of the beam, and high efficiency [14, 15].

3.1.1 Laser Matter Interaction

Issues pertaining to laser—matter interactions have been addressed in many reports ([16–18] and references therein). Intense, detailed research into the physics of laser—matter interactions was probably triggered by the idea put forward by Basov and Krokhin in 1964 that lasers could be used to achieve controlled fusion [19]. The fundamental works by Afanasiev and Krokhin [20–22] presented a physical picture of processes initiated in solids by nanosecond laser pulses of various intensities. Later, using theoretical analysis and numerical modelling, Afanasiev et al. [23–29] explored various aspects of metal and polymer ablation with ultra-short (pico- and femtosecond) laser pulses.

The energy of a laser pulse is first absorbed in the affected zone and then transferred to the bulk of the material and its lattice through heat conduction processes due to electron–phonon coupling. Mechanisms of laser radiation absorption by electrons may be both linear and nonlinear [30, 31]. Nonlinear light absorption at high incident laser intensities has been extensively discussed in literature because it determines the effect of ultrashort laser exposure on transparent materials such as wide-gap semiconductors and dielectrics [32–34]. Ultrashort laser pulses rapidly heat the electronic system to temperatures in the order of tens of thousands of kelvins, whereas the lattice remains at room temperature. Energy transfer from electrons to lattice phonons takes several picoseconds. The peculiarities of the kinetics of electron thermolization, the energy transfer to the lattice, and the times of those processes have been discussed in a number of publications [18, 35]. The heated region then acts as a source of a thermal wave which propagates to the bulk of the material (heat conduction processes in metals exposed to femtosecond pulses under nonequilibrium conditions were examined in [36, 37]).

One can ensure any time-temperature regime of the heated region by dosing thermal loads (the laser radiation density at the material surface). The intensity range in laser technology is very wide (from 10^4 to 10^{13} W/cm^2) and is determined by the parameters of the material and technological process. And, correspondingly, the range of laser pulse durations used in various laser technologies is also wide: from milliseconds to femtoseconds. The choice of the regime determines the type of processing.

Sufficiently high incident laser intensities may cause structural and phase changes, such as polymorphous transformations, melting, vaporization, possibly accompanied by laser plasma generation; and ablation, involving the ejection of surface atoms or particles of the material [16–18, 38–46].

Laser-induced material ejection from an exposed surface region may lead to the formation of various micron- and nano-meter-scale periodic structures in the affected zone [47–49]. Moreover, laser-induced surface material removal leads to the formation of nanoclusters [5, 12, 50–52], which may then redeposit on a substrate, leading to the formation of nanostructured surface [53]. Nanocluster formation is possible both in the ablation regime, where laser plasma effects are minimal, and with direct participation of a laser plasma. Cluster formation in a liquid may yield a suspension of nanoparticles widely used in biomedical applications [5, 52].

3.1.2 Direct Surface Material Nanostructuring

In recent years, a great deal of attention is paid to a possibility of so-called direct surface nanostructuring with ultrashort laser pulses, where structures are formed at specific laser radiation parameters [2–5, 11, 51]. Direct surface nanostructuring means that nanoscale surface structures are produced without redeposition of ejected particles onto specially prepared substrates, without any screening masks, etc. Nanostructuring is only caused by a laser beam incident directly on the surface of bulk material or film. One can use repetitive-pulse exposures. The size of forming nanostructures can be controlled by varying the number of pulses [49]. A distinctive feature of direct laser nanostructuring is that the processes involved are inherent in laser ablation, which involves melting and/or material ejection in the form of atoms, followed by redeposition of the atoms and/or clustering in the vapour phase. In the case of melting, surface nanostructuring is governed by the surface cooling rate after the laser pulse.

Direct laser micro- and nanostructuring of material surfaces is receiving increasing attention since this method is simple and cheap [51, 54]. It ensures both locality of the action and the possibility of large-area processing using programmable beam scanning. An important practical issue is then the ability to control and reproduce laser nanostructuring processes.

Laser nanostructuring with short and ultrashort pulses is realized under different regimes of material laser ablation, as well as with the use of subthreshold regimes which ensure intense melting of the material.

Ultrashort laser pulses may cause ablation without melting the material. In surface processing with femtosecond pulses, the ablated material may have high energy (exceptionally in the vapor phase), and only the material in the zone under laser irradiation is heated and ejected. Here the irradiated area undergoes morphological changes and nanostructuring [51–53].

If there is a liquid material, it rapidly solidifies by virtue of the heat transfer to the interior of the sample after the laser pulse. Due to the very high cooling rate (10^7 K/s or above), the size of the forming crystallites may be comparable to interatomic distances [55–58]. This causes structural changes in the irradiated zone, including effective nanosreucturing of the solidified material. If the cooling rate exceeds the crystallization rate, an amorphous layer may form [59, 60].

In optimizing laser nanostructuring conditions, an important point is to optimize the size of the resulting structures. The minimum possible size of the structures (tens of nanometers) produced by direct laser ablation of the surface is possible in two cases: (1) intense laser ablation due to cluster redeposition in and outside the ablation crater and (2) ultrashort laser pulses which ensure very rapid heating, melting and solidification of metal surfaces. Laser structuring without ablation is possible when the pulse energy density is slightly below the ablation threshold but exceeds the melting threshold. Use is typically made of short and ultrashort (nano-, pico- and femtosecond) laser pulses with near-threshold intensities, which cause local metal melting with no significant ablation.

In recent years, controlled direct surface nanostructuring with laser pulses under such conditions has been the subject of intense research. Attempts were made to systematically study laser nanostructuring mechanisms and optimize processing conditions (intensity, duration, number and repetition rate of laser pulses) with consideration for the thermodynamic characteristics, mechanical properties and surface quality of the target material [7, 8, 11, 12].

This chapter reviews the results of experimental and theoretical studies of micro- and nanomodification processes induced on the surface of metals and other materials by short and ultrashort laser pulses. Particular attention is paid to the possibility of direct laser nanostructuring through melting of the material (with or without ablation), followed by ultra-rapid solidification of the molten surface. A model has been presented for melting of a nanoporous material by means of nanosecond laser pulses. The focus of attention has been the results on studying the kinetics of ultra-rapid solidification of the molten surface, and determination of the conditions which ensure forming of surface nano-scale structures.

3.2 Peculiarities of Laser Nano- and Microstructuring of Metal Surfaces: Experimental Results

3.2.1 Properties of Nanocrystalline Materials

Direct laser micro- and nanostructuring (modification) of materials surfaces involves both producing a fine-crystalline structure (which is also referred to in literature as "nanocrystalline structure"), in order to improve the mechanical properties and corrosion resistance of the material, and creating arrays of micro- and nanoparticles (nanoclusters) on its surface, in order to improve its optical, electrical and other properties, e.g. with the aim of enhancing its compatibility with biological tissues. Nanostructures may strongly differ in texture and properties from the bulk material they have been produced on.

Fig. 3.1 Strength and plasticity variation at the transition
from a coarse-grained structure to the nanocrystalline one

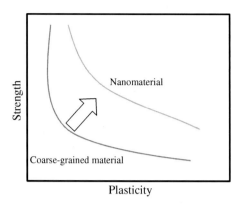

Structural details of nanomaterials (grain size, fraction of grain boundaries, structural perfection) have a significant effect on their properties and depend on the preparation procedure [6]. One should note that this refers primarily to the mechanical
strength of nanomaterials. At a large grain size, the increase in strength and hardness
with decreasing grain size is due to the formation of additional grain boundaries,
which prevent dislocation motion.

It is known that crystallites are dislocation-free when their size is below a certain
critical value (for example, the diameters of iron and nickel dislocation-free particles
are 23 nm and 140 nm, respectively [61]). The high strength of nanograined materials
is due to the low dislocation density and the difficulties in the formation of new
dislocations.

The microhardness of nanocrystalline materials is a factor of 2–7 higher than that
of their coarse-grained analogues [62]. An important point is that they have sufficiently high plasticity [63]. In addition, grain size reduction improves the corrosion
resistance. Superplacticity and other characteristics of the material are improved.
Figure 3.1 illustrates the strength and plasticity variations at the transition from a
coarse-grained to a nanocrystalline structure.

Note that surface nanostructuring influences the optical properties of metals, and,
in particular, it increases their absorptance. In [7, 50] the authors demonstrated an
increase in the absorptance of gold and copper. Selecting an appropriate combination of nano-, micro- and macrostructures on a metal surface, one can increase the
absorptance of copper to 85 %, and that of gold to 100 %. The ablation crater on
gold was shown to be surrounded by a black halo. The black halo is composed of
spherical aggregates, whose size decreases with increasing distance from the crater.
Spherical redeposited aggregates consist of spherical nanoparticles.

3.2.2 Laser Glassing

The formation of a fine-crystalline or amorphous surface structure under the action
of laser pulses is referred to as "laser glassing" [55–57]. The first experiments aimed

at surface modification of various metals and alloys in order to produce an ultrafine-crystalline or amorphous structure by laser pulses of millisecond up to nanosecond duration were performed more than three decades ago. The experiments revealed many of the main trends of laser vitrification (see e.g. [64–66]). Micro- and nanos-tructured materials were shown to have substantially better physical and, especially, mechanical properties.

First of all, the fabrication of amorphous and nanocrystalline structures by pulsed laser irradiation of metals and various alloys was shown to be conceptually feasible. The formation of fine-crystalline and amorphous structures was observed under the action of laser pulses from milliseconds to nanoseconds.

Amorphous layers of the order of 150 nm in thickness on pure Al were obtained using ∼15 ns ruby laser pulses with an energy density of ∼3.5 J/cm^2 [65]. The layers contained nanometer-scale (∼10 nm) crystalline inclusions. Note that alloy amorphization under the laser action takes place in a narrow range of irradiation parameters. A 10–15 % difference in the irradiation energy turns back to the process of crystallization.

There is *a critical laser cooling rate* at which crystallization can be suppressed to give a non-crystalline (amorphous) structure. The critical cooling rate depends not only on laser irradiation parameters (laser pulse intensity and duration) but also on the material properties (thermodynamic characteristics, alloy composition, metal purity and surface condition). The experimentally determined cooling rate needed to completely suppress crystallization in pure metals under laser irradiation made up about 10^9–10^{10} K/s. The reason for this is that any crystal at the boundary of the irradiated region is a potential seed.

Several approaches have been proposed for reducing the critical cooling rate for amorphization: the use in different Fe alloys of various additions of so-called glass formers (B, Si, P and C); a mixture of Ni-Nb and Cu-Zr powders; the substrates with high thermal conductivity (Ni and Cu) used as heat sinks, etc. This has made it possible to reduce the critical cooling rate to 10^5 K/s and to ensure the obtaining of amorphous phases [66].

3.2.3 The Role of Laser Pulse Duration

A number of reports analyzed the effect of laser pulse duration on materials process-ing conditions and compared processing conditions at various laser pulse durations (see, e.g. [47, 57]). In [47] a variety of metals (Ni, Al, Cu, Cr, Au) and semiconduc-tors (Si, Ge) were irradiated with UV (248 nm) laser pulses, and the morphologies of the irradiated regions of different materials at pulse durations 100 fs, 500 fs, 5 ps and 50 ps were compared.

It was shown that there exist *critical laser pulse durations* for metals and semi-conductors. Below critical pulse durations the irradiated materials differed very little in morphology. For the metals this is true at pulse durations under 5 ps, and for

800 nm, 100 fs, 10 kJ/m²   248 nm, 30 ns, 30 kJ/m²

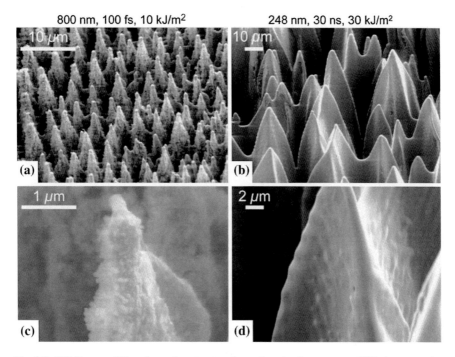

Fig. 3.2 SEM image of Si surface microstructure formed, under the presence of SF_6, by processing of femtosecond (100 fs) (**a**), (**c**) and nanosecond (30 ns) (**b**), (**d**) laser pulses. The sample is shown at 45° angle to the normal

semiconductors—under 200 fs. In [57] the authors compared the optical properties of microstructures produced on silicon by nanosecond and femtosecond laser pulses.

The forming penguin-like structures produced by 30 ns and 100 fs laser pulses differed in size by a factor of 5, and the spikes were 40 and 8 μm in height and 20 and 4 μm in width, respectively (Fig. 3.2). Even though the structures differed markedly in dimensions, morphology and crystallinity, the materials were very similar in optoelectronic properties and chemical composition. Thus, the morphology and physical properties of the layers produced by pulsed laser irradiation are governed not only by the characteristics of the material but also by the laser pulse energy density and duration.

3.2.4 Metal Surface Nanostructuring with Femtosecond Laser Pulses

A detailed study into the morphology of direct surface nanomodification of various metals under femtosecond Ti—sapphire laser pulses (800 nm wavelength) is presented in a number of papers, e.g. [7, 8, 51].

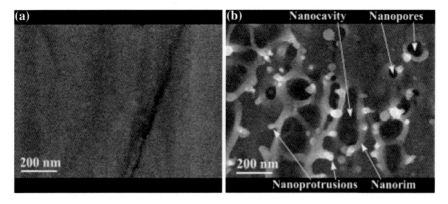

Fig. 3.3 a Copper sample surface before laser irradiation. **b** Nanostructures produced on copper surface under femtosecond laser ablation, F = 0.35 J/cm^2 and N = 1

Nanostructuring of gold films at incident laser intensities that ensured melting of the metal in the affected zone was studied in [11]. The laser pulse duration was 30 fs, and the incident intensity was just below the ablation threshold but above the melting threshold. For noble metals (e.g. gold), the lifetime of the melt pool was shown to be sufficient for the formation of various surface textures. In particular, conditions were selected for producing 500–600 nm bumps with a nanojet of 100–200 nm diameter in their central part.

Further attempts to study the morphology of laser-processed gold, copper and titanium surfaces were made by the authors in [8, 49, 50]. The experiments have shown that the size and shape of surface nanostructures depended not only on the incident laser intensity and number of pulses but also on the properties of the material and the initial structure of the processed surface.

It was shown that adjusting the incident intensity and number of pulses one is able to produce nano-, micro- and macrostructures and various combinations of these on metal surfaces. There were *optimal laser processing conditions* for each type of structure, in particular for pure surface nanostructuring, i.e., for producing only nanometer-sized surface structures. The nanostructures in three metals (Au, Ti and Cu) were similar in size and shape. Using 65 fs laser pulses and pulse energy densities just above the damage threshold (the ablation threshold for Au and Ti is $F_{th} = 0.067$ J/cm^2, and that for Cu—$F_{th} = 0.084$ J/cm^2), various nanostructures were produced by a small number of pulses (1–10).

These typically had the form of porous structures (round nanopores) 40–100 nm in diameter, randomly arranged nanoprotrusions 20–70 nm in diameter and 20–80 nm in length, nanocraters and various structures (nanorings, and elevations) 20–100 nm in size around the nanocraters.

Figure 3.3b [50] shows typical nanostructures produced on Cu surface by a single 65 fs laser pulse with an energy density $F = 0.35$ J/cm^2. The minimum size of the forming crystalline nanostructure was 20 nm. Studies of the morphology and texture of surface structures demonstrate that nanostructuring is due to randomly arranged

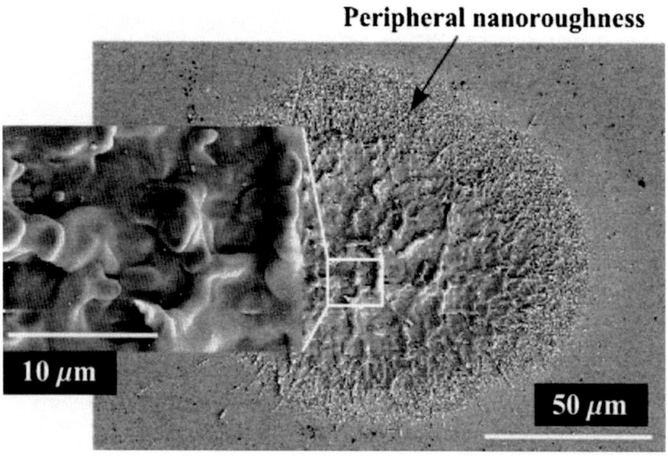

Peripheral nanoroughness

Fig. 3.4 SEM image of a copper surface after exposure to two laser pulses ($F = 9.6 \, \text{J/cm}^2$). The central part of the irradiated zone contains only microstructures; nanostructures are observed in the peripheral parts of the ablation zone. *Inset* mictostructure details in the central part of the irradiated zone

nanometer-sized melt pools. The shape and size of the forming nanostructures depend both on the dynamics of melt motion in nanoregions in the zone being irradiated and on the cooling kinetics.

Multiple irradiations produce ensembles of nanostructures up to 500 nm in size. The process is accompanied by the formation of microstructures, which prevail at a number of pulses $N = 1000$. Similar results can be obtained at higher energy densities.

Figure 3.4 [50] presents a scanning electron microscopy (SEM) image of a copper surface after exposure to two laser pulses with an energy density $F = 9.6 \, \text{J/cm}^2$. The central part of the irradiated zone contains only microstructures. At the same time, in the peripheral parts of the ablation zone, where the laser energy density is substantially lower, the formation of the melt with further crystallization is possible. Nanometre-scale structures are observed in this region.

Also of interest is laser ablation at a high energy density and a small number of laser pulses [49], which causes complete melting of the region under irradiation. Subsequent ultrarapid melt solidification leads to the formation of a smooth surface covered with redeposited metal nanoparticles down to $\sim$10 nm in size.

Figure 3.5 [49] illustrates the surface topography of titanium after irradiation by femtosecond laser pulses with an energy density of 2.9 J/cm^2. Similar structures resulted from picosecond laser irradiation [67], but the number of nanostructures produced by a femtosecond pulse considerably exceeded that in the case of a picosecond pulse.

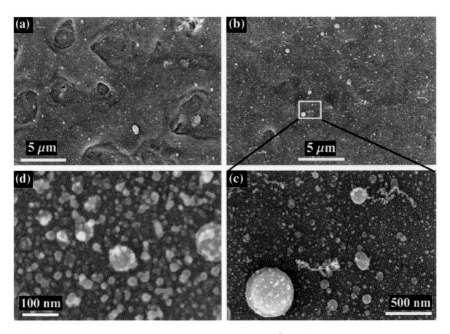

Fig. 3.5 Titanium surface after femtosecond (F = 2.9 J/cm^2) laser exposures: **a** smooth surface with microinhomogeneities after exposure to one laser pulse; **b** smooth surface with nanostructures after two pulses; **c** P. N. Lebedev Physical Institute of RAS higher magnification image of the area outlined in (**b**), showing surface nanostructures; **d** nanotopography of a smooth surface after exposure to four laser pulses (one can see spherical nanostructures down to ∼10 nm in size)

3.2.5 Micro- and Nano-Structuring with Nanosecond Laser Pulses

Despite of effectiveness of femtosecond laser nanostructuring and nanostructuring with the use of picosecond pulses, there is also practical interest in employing cheaper laser systems generating nanosecond pulses.

In [68] the authors irradiated different materials by 25-ns F$_2$-laser pulses of wavelength 157 nm. The results demonstrated that nanostructures can be produced on solid surfaces by multiple exposures that ensure ablation in the centre of the beam spot and surface melting at its periphery. An important factor in nanostructuring is inhomogeneity of the surface profile, which may cause inhomogeneity of the ablation process. Ablation first takes place at grain boundaries. After a sufficiently large number of pulses, nanostructuring is determined by laser-induced surface instability. Therefore, optimizing the number of laser pulses, one can control the size of the resulting surface nanostructures, in particular, minimize it.

In [14, 15] the authors reported the experiments on laser modification of metal and semiconductor surfaces with the use of a diode-pumped laser. The use of a compact and effective diode-pumped laser with good beam quality in a single mode regime

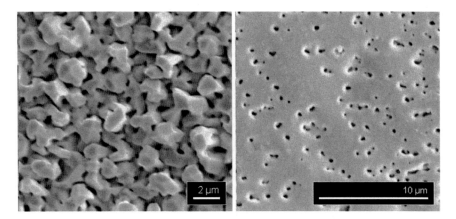

Fig. 3.6 SEM images of a 2.7 μm thick indium film before (*left*) and after (*right*) laser exposure (diode-pumped YAG:Nd laser)

has allowed one to examine laser modification of porous material—indium solder aimed at improving the effectiveness of heat removal from the crystal of high-power laser diodes [14]. The indium films produced by the magnetron evaporation method had been used as a solder at mounting of high-power semiconductor laser crystals. The use of the indium films is advantageous due to such attractive properties as low melting temperature (156–161 °C, by different data), high plasticity, availability, and others. The pulse energy density was 0.1 J/cm^2 and the pulse duration was 6.5 ns and initial thickness of the indium film, 2.7 μm. Figure 3.6 [14] illustrates SEM images of the indium solder structure before and after irradiation. As seen from Fig. 3.6, the microstructure of untreated and treated films is far from being perfect; one can observe the imperfections at the surface and in the film volume after the treatment.

The indium film is seen to amorphise (laser glassing). Typical size of the observed flaws (crystalline inclusions) ranged from 100 nm to 1 μm. Film treatment had made the porosity and roughness of the surface lower, and led to the damage of oxide films and surface purification from contamination. After the modification, the efficiency of the heatsink from a semiconductor laser crystal had increased because the area of the solder interaction with a metallic film on a laser crystal had become greater. Besides, the heat-conductivity of a soldering layer had increased due to a decrease in the number of the pores. Modifying the structure of the solder film immediately before the laser mounting process ensured a significant increase in diode laser output power [15].

Thus, the obtained results on the structure of materials exposed to short and ultra-short laser pulses point to the formation of a fine structure with a grain size from several nanometers to several micrometers. The dimensions and shape of nanostructures depend on the initial surface condition and the number of laser pulses. Nanostructures thus produced can be used in technological applications (e.g. to control the optical properties of materials, enhance the compatibility of implants with biological tissues, etc.). Laser ablation by femtosecond pulses can be used to manufacture coatings such as black gold films. A necessary condition for modification processes to take place is the presence of a melt pool in the zone under irradiation. Note that the advantages of particular laser processing conditions with a given laser pulse duration depend on the type of material, laser processing conditions and processing procedure.

To optimize laser nanostructuring one should know the processing parameters such as the heating and cooling time, the rate and the volumes of liquid and crystalline phases, to determine the average cryslallite size. All the above parameters depend on laser pulse parameters and thermodynamic properties of the material.

3.3 Theoretical Modeling of Laser-Induced Nanocrystallization Processes

3.3.1 Melting Processes

To construct a model of nanostructuring by short and ultrashort laser pulses, one should study in detail the mechanisms of laser absorption, nonequilibrium processes induced by rapid laser deposition and structural and phase transformations in laser-processed surface layers of materials. All these processes are closely interrelated. A distinctive feature of these processes is that they may occur within extremely small volumes, down to several hundred microns in size, in just 10^{-12}–10^{-13} s. As a consequence there are high heating and cooling rates and large temperature gradients. Melting processes have been analysed in detail for both cw laser radiation and various laser pulse durations (from milli- to femtoseconds) [69–73].

An issue that is often addressed in literature is whether the mechanism of melting induced by ultrashort pulses is thermal or nonthermal [17, 58]. The mechanism of phase transformations is nonthermal when the structural changes involved occur more rapidly than the absorbed energy transfer from electrons to the lattice. The phase transitions that develop after heat transfer from electrons to the lattice are ultrarapid thermal phase transitions.

There is an experimental evidence [74] (exemplified by the effect of a femtosecond laser pulse on aluminum) that the mechanism of metal melting by short and ultrashort laser pulses is thermal at pulse durations down to 40 fs, and is due to lattice heating to the melting point and takes in the order of several picoseconds (1.5–2 ps for Al). (The new phase nucleates through a homogeneous mechanism). The heating rate may reach 10^{14} K/s. The phase transition time is independent of the laser energy, in contrast to nonthermal melting mechanisms.

In [72] the authors reported a computer simulation study of the melting of 50-nm-thick Ni and Au films exposed to laser pulses 200 fs to 150 ps in duration. The simulated melting time was 2 ps. The ablation, melting and vaporization of nickel, copper, and quartz were studied using molecular dynamics simulation [73]. It should be noted that the melting and solidification times of materials exposed to femtosecond

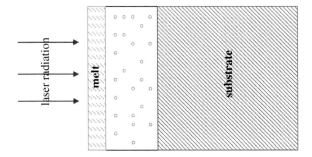

Fig. 3.7 The scheme of the process of laser processing of the porous metallic film

pulse have been variously reported to be: 2–20 ps for melting time, and 50–120 ps— for crystallization one.

The surface structure of a material exposed to a laser pulse is governed by both the melt flow dynamics and solidification kinetics. Fundamental metal solidification processes during ultrarapid cooling have not yet been studied in sufficient detail. This is particularly true of the kinetics and dynamics of the melting and solidification behavior of metals irradiated by ultrashort laser pulses.

3.3.2 Laser Processing of Porous Metal Films

The use of a pulsed laser radiation is, nowadays, one of the methods for the production and treatment of so-called nanoporous materials [75–78]. Much attention has been paid of late to studying the nanoporous materials, whose pores do not exceed 100 nm in dimension [79]. This interest is generated by unique structural and surface characteristics of nanoporous materials and wide possibilities of their application [79–82]. The processes of melting and evaporation occurring under the laser action on the nanoporous materials have unique characteristics and had been studied insufficiently thoroughly.

The model for laser modification of porous metallic films called as "the instant collapse of the pores" was considered in [81, 82]. Figure 3.7 illustrates the scheme of the process. A porous material deposited on a metallic substrate is exposed to the action of a pulsed laser radiation. The pores represent the spherical isolated cavities whose radius is much less than the thickness of the material h.

If the temperature of some region of a material reaches the melting temperature T_m, the pores should be under pressure within the volume of a viscous melt. They should be instantly collapsed due to surface tension, and the melted material around a pore should move the latter to the center. There occurs a simultaneous transition of "the solid state body-melt" and "the porous body-solid material". In the given approximation the time of the pore collapse is around 1 ns, under the above laser radiation parameters, which is much less than all the characteristics times of the heat problem [81].

The process of melting of the metallic-substrate porous material by means of the laser pulses is described by a system of thermal-conductivity equations with taking into account the depending thermophysical values of porous metal on the temperature and the material porosity of the system [79]. The material porosity is defined as a ratio of the total volume of emptiness to the body volume $\Phi = V_P/V$, where V_P is the total volume of the pores; V, the material volume. The heat capacity versus porosity ratio may be defined as follows [79]: $C = C_o(1 - \Phi)$, where C_o is the material heat capacity at $\Phi = 0$. The dependence of latent heat of melting on the porosity is analogous to that of the heat capacity: $\Delta H = \Delta H_O(1 - \Phi)$. Here ΔH_O is the melting latent heat at $\Phi = 0$. The heat capacity coefficient was finded using the Maxwell model [79] which considers the heat capacity of a porous medium with isolated pores distributed in a solid phase:

$$\lambda = \lambda_0 \frac{2(1 - \Phi)}{2 + \Phi} \tag{3.1}$$

A numerical calculation has been made in [81] for a 2.7 μm-thick porous indium film mounted on a copper substrate whose thickness is much greater than the film thickness (see experimental data, described in 2.5 [14]). The numerical solution with an allowance for indium porosity gives the melting depth value z_m about 1.5 μm. With no account of the material porosity the melting depth turns to be 1.1 μm. From this follows that taking into account the material porosity results in a 36 % increase of the melting depth. The melting depth with account only of the temperature dependence of indium thermophysical values differs from the melting depth under constant values just by 1–2 %.

With account of the porosity the lifetime of the melt increases. An increase in the melting depth of a porous film as compared to the solid one, takes place, probably, because of the re-distribution of energy in the film and the decrease in the heat extraction to the substrate due to the drop of the heat conductivity.

In order to find such energy density that the film would be melted at full depth, the melting depth for different energy densities have been calculated (Fig. 3.8). In Fig. 3.8 the melting depth are given in dimensionless units z_m/h. The solid line corresponds to the initial energy density; the dashed line corresponds to the energy 3 times as much as the initial one; the dot-dash line corresponds to the case when the energy is 5 times greater than the initial energy.

The dot-dash line plateau comes from the fact that at the given energy density the film of 2.7 μm thickness is completely melted and the substrate material heating is observed. The substrate is not melted, since the melting temperature of the film and the substrate material essentially differ.

The obtained results allow one to make a conclusion that the material porosity is an important factor, which essentially influences the process of material treatment, and should be considered in theoretical and experimental study.

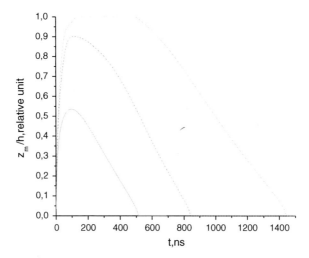

Fig. 3.8 The indium film melting depth at different energy density

The obtained theoretical results are in good agreement with the experimental results [15, 81], and can be used to improve the process of laser treatment of porous materials.

3.3.3 Nanocrystalization Kinetics

The most general approach to describing solidification kinetics during cooling after laser exposure is to examine the distribution of the number of crystalline particles, Z, with respect to the number of atoms per particle, n, at time t Z(n, t) [83]. Knowing the distribution function Z(n, t), one can find the mean size of crystallites formed during cooling and the volume fraction of the new phase.

The rate equation for Z(n, t) was treated in a number of studies [58–60, 85–87]. In particular, the authors of [84] found a solution to a general classic equation for Z(n, t) using an average diffusion coefficient and linearising the rate equation. The solution was used in [85] to approximately determine the mean number of particles per crystalline grain and the fraction of melt solidified for a silicon nitride surface exposed to nanosecond F_2-laser. The available solutions of rate equation used either the mean diffusion coefficient or were numerically simulated. A method for analytically solving the rate equation for the distribution function Z(n, t) was proposed in [58–60].

According to a classical theory the phase transitions of the first order are due to thermofluctuation formation of new atomic structure. New phase nuclei may develop only under the presence of thermonuclear stimulation of phase transformation defined by energy changes associated with the occurrence of the nuclei. The rate equation for the distribution function $Z(n, t)$ takes into account both the variation of the function due to its fluctuations and the increasing or decreasing of the nuclei size under the influence of thermodynamic parameters:

$$\frac{\partial Z(n, t)}{\partial t} = vn^{2/3} \exp\left(-\frac{U}{kT(t)}\right)\left\{\frac{\partial^2 Z}{\partial n^2} + \frac{1}{kT(t)}\frac{\partial}{\partial n}\left[\frac{\partial \Delta \Phi(n)}{\partial n}Z(n, t)\right]\right\}$$

$$\Delta \Phi^\upsilon = \Delta fn, \quad \Delta f = \Delta h\frac{T_0 - T(t)}{T(t)}, \quad \Delta \Phi^S = \chi \sigma n^{2/3} \qquad (3.2)$$

where $\Delta \Phi$ is the change of free specific energy at the occurrence of a new phase nucleus; the indices v, s, stand for the volume and superficial terms, respectively; Δh, the latent heat of phase transition; σ, the surface energy at phase boundary; χ, the nuclei shape factor; U, the activation energy for the transfer of an atom across the interface; v, the Debye characteristic frequency; k, the Boltzmann constant; T_0, the solidification temperature.

The first term to the right of the equation describes the diffusion, i.e. the blur of initially given distribution of the particles, and the second one—the drift or the shift of distribution in the particle number space. Take as the initial condition the condition of system completeness that means the conservation of the number of particles:

$$Z(n, 0) = 2N\delta(n - 1), \quad \int_1^\infty \delta(n - 1)dn = \frac{1}{2} \qquad (3.3)$$

The boundary conditions are given as follows:

$$Z(n, t)|_{n=1} = N, \quad Z(n \to \infty, t) = 0 \qquad (3.4)$$

The second condition in (3.4) excludes consideration of very large nuclei of a new phase (Becker-Dering boundary condition). This is true, of course, for short cooling times realized in pulsed laser processing of metals, when large forming structures fail to grow and are not experimentally observed.

Ultrarapid cooling rates provide the conditions under which a sufficient number of supercritical-sized nuclei are present, i.e. all supercritical-sized nuclei automatically enter the system. Analogous situation takes place in considering the kinetics of polyamorphous transformation in iron under rapid laser heating. Actually, in the analysis of the solidification process the attention is paid to the largest terms under the change of free specific energy at the occurrence of a new phase, namely the volume and the surface ones. They define the radius (or the number of atoms n_c) of the nuclei which corresponds to the maximum of free energy variation. When a spherical nuclei are produced one has

$$r_c(t) = \frac{2\sigma}{\Delta h}\frac{T_0}{T_0 - T(t)} \qquad (3.5)$$

As seen from the (3.5), due to ultrarapid cooling rates the nuclei of under-critical size become supercritical in small fraction of the cooling time, and the radius of critical-sized nuclei may comparable with the lattice parameter. So, the kinetics of ultrarapid cooling depends not on the rate of nucleation but is determined by the growth rate of crystalline nuclei.

Metallographic analysis [59, 88] has revealed a needlelike dendritic structure of crystalline grains forming at ultrarapid cooling. Therefore, we consider the existence of crystalline nuclei of a plate form as primary. In this case the pre-exponential factor in (3.2) will be proportional to n, and not to the value $n^{2/3}$. The latter circumstance allows one to analytically find a solution of the rate equation in an explicit form.

In the above assumptions the (3.2) can be written in the form:

$$\frac{\partial Z}{\partial t} = nv \exp\left(-\frac{U}{kT(t)}\right)\left[\frac{\partial^2 Z}{\partial n^2} - \frac{\Delta h}{kT(t)}\frac{T_0 - T(t)}{T_0}\frac{\partial Z}{\partial n}\right] \tag{3.6}$$

Equation (3.6) is solved using the operator method applied for the solution of the Schrödinger equation [59].

One should find the system evolution operator $\hat{S}$:

$$Z_f = \hat{S}Z_{in} \qquad Z_{in} = Z\left(n, \vartheta = 0\right) \qquad Z_f = Z\left(n, \vartheta \to \infty\right) \tag{3.7}$$

where Z_{in} and Z_f are the initial and finite distribution functions. An explicit form of the evolution operator allows one to analyze its effect on the initial distribution function.

In the initial notations the size distribution function of the nuclei may be presented in an integral form:

$$Z_f(\alpha, \beta, n) = \frac{\sqrt{n}}{\alpha}\int\limits_0^\infty d\xi Z_{in}\left(\frac{1}{4}\xi^2 e^\beta\right)\exp\left(-\frac{4n+\xi^2}{4\alpha}\right)I_1\left(\xi\frac{\sqrt{n}}{\alpha}\right) \tag{3.8}$$

$$\alpha(t) = v\exp\left[-\beta(t)\right]\int\limits_0^t \exp\left[\beta(t) - \frac{U}{kT(\tau)}\right]d\tau$$

$$\beta(t) = -\frac{\Delta hv}{kT_0}\int\limits_0^t \frac{T_0 - T(\tau)}{T(\tau)}\exp\left(-\frac{U}{kT(\tau)}\right)d\tau$$

I_1 is a modified Bessel function of the first order. One can find the final form of $\alpha(t)$ and $\beta(t)$ determined by the variation of temperature during cooling: linear, power-law, and exponential:

$$T(t) = \begin{cases} T_0 - \varepsilon t, \ 0 \le t \le T_0/\varepsilon \\ T_0\left(\frac{\tau}{t}\right)^m, \ \tau < t < \infty \\ T_0\exp(-pt), \ 0 \le t < \infty \end{cases} \tag{3.9}$$

Fig. 3.9 Distribution function Z(n, t) at different instants of time: numerical (-) and analitical (·) solution, $U/kT_0 = 15$

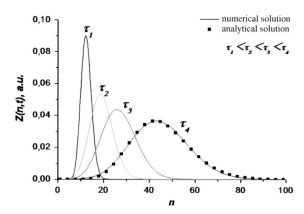

where ε, m and p are the laser radiation parameters. It was shown [58] that, at very high cooling rates, α, β and, accordingly, the distribution function reach a plateau in a small fraction of the total cooling time. The cooling rate determines the explicit form of the distribution function. For example, with

$$Z_{in} = Z_0 \delta(n - n_0) \tag{3.10}$$

We obtain for linear and power-law cooling

$$Z_f = \left(\frac{n}{\alpha^2} e^{-\beta}\right)^{3/4} \exp\left[-\frac{\left(\sqrt{n} - \sqrt{n_0} e^{-\beta/2}\right)}{\alpha}\right] \tag{3.11}$$

Figure 3.9 illustrates the time variation of Z(n, t). It is seen that, in the initial stage of cooling, the function shifts significantly, whereas at the end of cooling it remains essentially unchanged.

One can show that (3.11) meets the conservation condition (3.4). Actually,

$$\int_0^\infty Z_f(\alpha, \beta, n) dn \tag{3.12}$$

$$= \int_0^\infty \frac{\sqrt{n}}{\alpha} dn \int_0^\infty d\xi Z_{in}\left(\frac{1}{4}\xi^2 e^\beta\right) \exp\left(-\frac{4n + \xi^2}{4\alpha}\right) I_1\left(\xi\frac{\sqrt{n}}{\alpha}\right) = N e^\beta$$

In fact, this is the number of particles in the new phase at $t \to \infty$.

3.3.4 Volume of the Phase Crystalline and Average Number of Nuclei Particles

We can find in an explicit form the relative change in the volume of the forming phase:

$$\frac{\Delta V}{V} = \frac{V_{in} - V_f}{V_{in}} = \frac{N\Delta V - Ne^{\beta}\Delta V}{N\Delta V} = 1 - e^{\beta} \tag{3.13}$$

Formula (3.13) corresponds to the Kolmogorov—Avraami equation [83]. The Kolmogov-Avraami equation was derived to describe the metal solidification kinetics, and describes the phase transitions kinetics in an infinite volume assuming a large number of nuclei of the new phase randomly distributed in space and time. The average size of the new phase areas was assumed to be small as compared to the sample size.

Another important parameter is the average size of the crystalline nuclei in the forming phase, or the average number of atoms per nucleus. This determines the size of the forming crystallites. By definition,

$$\langle n \rangle = \frac{1}{N} \int_0^{\bar{n}} Z(n, t) n \, dn \tag{3.14}$$

where $\langle n \rangle$ is a sufficiently large, but limited number of particles. Taking into account that, at high cooling rates, there is possible a transition from $\bar{n}$ to ∞ within the integration limits, and the average number of particles in a crystalline can be found in explicit form:

$$\langle n \rangle = \frac{3\alpha}{2} \left[\alpha - 1 - \frac{1}{\alpha} \ln\left(1 - \frac{\alpha}{3}\right) \right] \tag{3.15}$$

Note that $\langle n \rangle$ is completely determined by α, i.e. by the broadening of the particle initial distribution, and that the $\Delta V/V$ ratio depends only on β, i.e. on the drift of the distribution function.

The results from simulation of $\Delta V/V$ and $\langle n \rangle$ for exponential cooling were presented in [3, 58] for three metals. It was shown that, at relatively slow cooling rates $(\exp(\beta) \to 0)$ the melt crystallizes almost completely $(\Delta V/V \to 1)$. The corresponding cooling rates are: $\upsilon \sim 1.8 \times 10^6$ K/s for Fe, $\upsilon \sim 2.8 \times 10^5$ K/s for Al, $\upsilon \sim 1.73 \times 10^5$ K/s for Ni. With increasing cooling rate, only part of the melt crystallizes: in the $t \to \infty$ limit we have $\exp(\beta) \to 1$ and $\Delta V/V \to 0$. We can assume that the rest part of the melt tends to form amorphous phase (a solidified liquid). In particular, we find in the case of Fe that, even at cooling rate approaching 1.8×10^9 K/s, approximately 0.996 of the melt may have the form of a solidified liquid. Amorphous phases of Al and Ni may form at $\upsilon \sim 9.3 \times 10^8$ K/s and $\upsilon \sim 7.3 \times 10^7$ K/s, respectively. One can see that the calculated critical cooling rate for Al amorphisation is significantly less than that realized in the experiment [59] $(\sim 10^{10}$ K/s) discussed in

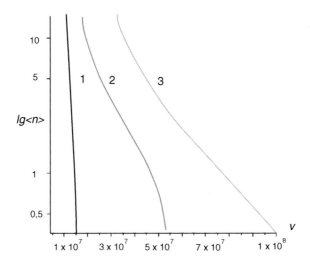

Fig. 3.10 Average number of atoms per crystallite <n> against cooling rate (*1 Fe, 2 Al, 3 Ni*)

Sect. 3.2.2. This has made it possible to produce a thin aluminum amorphous layer. For Fe, as noted in II, the melts with glass formers have been used, and this allowed one to decrease the real critical cooling rate down to $\sim 10^6$ K/s. This is essentially smaller than the calculated critical rate of amorphisation for pure Fe. In [60], in Fe melts with glass formers (B, Si, P and C) the amorphous layers of 5–10 µm thickness have been produced by CO_2 laser pulses (intensity, 10^5–10^6 W/cm^2, pulse duration, 1 ns).

Figure 3.10 shows log-log plots of the average number of atoms per crystallite $\langle n \rangle$ against cooling rate. It is seen that even a slight decrease in cooling rate leads to a sharp drop (by several (two) orders of magnitude) in $\langle n \rangle$. So, the size of the forming superdispersedultrafine crystalline phase depends on the cooling rate. The cooling rates, realized under metal irradiation by solid-state diode laser pulses described above [18, 19] (laser energy density, 0.1 J/cm^2; pulse duration, 6.5 ns), have been estimated as the value $\sim 10^{10}$ K/s. Using (3.11), we can find the average size of the forming grains. The grain average size is 100–400 nm.

Such crystalline structures have been actually observed in the described experiment. On the other hand, the cooling rates are sufficient to produce an amorphous phase in indium. Apparently, a combined structure (amorphous indium with crystalline grains) has been observed in the discussed experiment. In a more detailed analysis one should take into account the indium porosity (indium porosity makes up a value of about 50 %).

When femtosecond pulses are used, the cooling rate may reach 10^{12} E/s, the crystallite size is 20–40 nm (estimates made by (3.17)), and this is in good agreement with experimental data given in Sect. 3.2.2 (see e.g., [50]).

Note that detailed calculations and a quantitative comparison with experimental data are hindered by the lack of accurate thermodynamic data for the system.

Moreover, when femtosecond laser pulses are used the structure of the solidified melt and the presence and size of nanostructures may be governed by the dynamics of the processes that occur in the melt during cooling.

3.3.5 Criterion for Laser Amorphisation

The kinetic criterion for laser amorphisation has been examined in [3, 58]. As shown above, at a certain cooling rate the melt may remain in an amorphous state. In other words, there is a critical cooling rate when a crystalline phase is not formed during cooling, and the cooled material presents an amorphous structure (laser glassing). Note that the condition

$$\upsilon \geq \upsilon_{cr} \tag{3.16}$$

is *a kinetic criterion for amorphisation.* We have established an amorphisation criterion from the condition that an amorphous phase is formed when its volume fraction approaches a value, P, close to unity. The experimentally observed crystalline content is $\sim 10^{-6}$ [88]: $exp[\beta(t \rightarrow \infty)] = P$.

On the other hand, the rate of heat transfer from the melt to the material during solidification is determined by its thermophysical properties and sets the upper limit of the cooling rate. In the case of pulsed laser exposure, the cooling rate has a maximum near the solidification temperature, and is as follows [56]:

$$\upsilon_{\max} = \frac{Aq}{\lambda\sqrt{k/\pi\tau}} \tag{3.17}$$

where A, is an effective absorptance; $\ae$ is the thermal diffusivity; λ, is the thermal conductivity of the material; q and τ, the laser intensity and pulse duration. Note that (3.18) for the cooling rate needs to be refined when thin layers of a material are processed and the cooling rate depends significantly on the thickness of the melt pool. Relation (3.14) was used to determine the critical cooling rate at which a nanocrystalline (amorphous) phase can be obtained for $U/kT_0 > 5$ [58]:

$$\upsilon_{cr} = \frac{\nu\Delta h(1-c_0)}{U\ln P}\exp\left(-\frac{U}{kT_0}\right) \tag{3.18}$$

It can be seen that whether an amorphous phase can be obtained depends on both the laser pulse parameters and the thermodynamic properties of the material.

3.4 Conclusions

Research into the process underlying surface micro- and nanostructuring of metals and other materials by short and ultrashort laser pulses has been reviewed. Particular attention is paid to direct laser irradiation involving melting of the material (with or without ablation), followed by ultrarapid surface solidification after the laser pulse.

The results of theoretical studies of the solidification kinetics of molten metals have been analyzed for ultra-high cooling rates in materials processing with ultrashort laser pulses. A method for analytically solving the rate equation for the distribution function $Z(n, t)$ was considered. The method is based on the operator method that is commonly used to solve the Schrodinger equation. This method builds on the use of specific physical features of metal solidification at ultra-high cooling rates (a large number of supercritical nuclei, absence of very large nuclei and a nearly planar shape of forming crystallites). This allows one to explicitly describe the physical behavior of kinetic parameters such as the crystallite size and the fraction of crystallised phase as functions of laser exposure parameters.

Determination of the volume fraction of crystallised phase allows one to find the critical cooling rate at which crystallization is impossible and the structure amorphises, i.e. to establish a kinetic amophisation criterion.

The model of nanoporous material melting by means of nanosecond laser radiation pulses was considered. The model takes into account the material porosity and the temperature dependence of its thermophysical properties. The porosity has been taken into account in the approximation of an "instantaneous collapse" of the pores under the action of the surface tension force. The melting depth of the material has been defined. The results obtained allow one to make a conclusion that the material porosity is an important factor, which essentially influences the process of material treatment, and should be considered in theoretical and experimental study.

The presented data can be used to optimize direct laser micro- and nanostructuring conditions and to ensure process control and reproducibility.

References

1. P.N. Prasad, *Introduction to Biophotonics* (Wiley, Boston, 2003)
2. E.G. Gamaly, A.V. Rode, in Encyclopaedia of Nanoscience and Nanotechnology, ed. by Stevenson Range (American Scientific Publishers, 2004), v. 7, p. 783.
3. I.N. Zavestovskaya, Laser Part. Beams **28**, 437 (2010)
4. I.N. Zavestovskaya, Quantum Electron. **40**(11), 942 (2010)
5. A.V. Kabashin, Ph Delaporte, A. Pereira, D. Grojo, R. Torres, Th Sarnet, M. Sentis, Nanoscale Res. Lett. **5**, 454 (2010)
6. N.P. Lyakishev, M.I. Alymov, Ross Nanotechnol. **1**, 71 (2006)
7. A.Y. Vorobyev, C. Guo, Phys. Rev. B **72**, 195422(1-5) (2005)
8. A.Y. Vorobyev, C. Guo, Appl. Phys. Lett. **92**, 041914-3 (2008)
9. A.I. Denisyuk, in *Problemy kogerentnoi I nelineinoi optiki* (Topics in Coherent and Nonlinear Optics), (SpbGU ITMO, St. Petersburg, 2008), p. 52. (in Russian)

10. P.G. Eliseev, H.-B. Sun, S. Juodkazis, T. Sugahara, S. Sakai, H. Misawa, Jpn. J. Appl. Phys. **38**, 839 (1999)
11. J. Koch, F. Korte, T. Bauer, C. Fallnich, A. Ostendorf, B.N. Chichkov, Appl. Phys. A **81**, 325 (2005)
12. S. Barcikowski, A. Hahn, A.V. Kabashin, B.N. Chichkov, Appl. Phys. A **87**, 47 (2007)
13. A. Kanavin, N. Kozlovskaya, O. Krokhin, I. Zavestovskaya, AIP Conf. Proc. **1278**, 111 (2010)
14. V.V. Bezotosny, VYu. Bondarev, V.I. Kovalenko, O.N. Krokhin, V.F. Pevtsov, YuM Popov, YuN Tokarev, E.A. Cheshev, Quantum Electron. **37**, 1055 (2007)
15. V.Bezotosny, I. Zavestovskaya, A. Kanavin, N. Kozlovskaya, O. Krokhin, V. Oleshenko, Yu. Popov, E.Cheshev. in *Trudy II Simpoziuma po kogerentnomu opticheskomu izlucheniyu poluprovodnikovyh soedinenii I struktur*. Proceedings II Symposium on Coherent Optical Radiation from Semiconductor Materials and Structures, (RIIS FIAN, Moscow, 2010), p. 165 (in Russian)
16. S.I. Anisimov, B.S. Luk'yanchuk, Usp. Fiz. Nauk **172**, 301 (2002)
17. E.G. Gamaly, A.V. Rode, B. Luther-Davies, V.T. Tikhonchuk, Phys. Plasmas, **9**(3), 949 (2002)
18. I.B. Anan'in, Yu.V. Afanasiev, Yu.A. Bykovskii, I.N. Erokhin, in *Lazernaya plazma*. Fizika i primeneniya (Laser Plasma: Physics and Applications), (MEPHI, Moscow, 2003) (in Russian)
19. N.G. Basov, I.N. Erokhin, Zh Eksp, Teor. Fiz. **46**, 171 (1964)
20. YuV Afanasiev, I.N. Erokhin, Zh Eksp, Teor. Fiz. **52**, 966 (1967)
21. Yu.V. Afanasiev, I. N. Erokhin, Fiz. Inst. Im. P.N. Lebedeva, Acad. Nauk SSSR **52**, 118 (1970)
22. YuV Afanasiev, I.N. Erokhin, in *Fizika vysokih plotnostei energii* (Physics of High Energy Densities) (Mir, Moscow, 1974)
23. YuV Afanasiev, V.A. Isakov, I.N. Zavestovskaya, B.N. Chichkov, F. Von Alvensleben, H. Welling, Appl. Phys. **A64**, 561 (1997)
24. A.P. Kanavin, I.V. Smetanin, V.A. Isakov, Yu.V. Afanasiev, B.N. Chichkov, Phys. Rev. B **57**, 14698 (1998)
25. YuV Afanasiev, N.N. Demchenko, I.N. Zavestovskaya, V.A. Isakov, A.P. Kanavin, S.A. Uryupin, B.N. Chichkov, Izv. Akad. Nauk **63**, 667 (1999)
26. YuV Afanasiev, B.N. Chichkov, N.N. Demchenko, V.A. Isakov, I.N. Zavestovskaya, J. Russ. Laser Res. **20**, 89 (1999)
27. YuV Afanasiev, B.N. Chichkov, V.A. Isakov, A.P. Kanavin, S.A. Uryupin, J. Russ. Laser Res. **20**, 189 (1999)
28. YuV Afanasiev, B.N. Chichkov, N.N. Demchenko, V.A. Isakov, I.N. Zavestovskaya, J. Russ. Laser Res. **20**, 489 (1999)
29. YuV Afanasiev, V.A. Isakov, I.N. Zavestovskaya, B.N. Chichkov, F. Von Alvensleben, H. Welling, Laser Part. Beams **17**(4), 585 (1999)
30. L.V. Keldysh, Zh Eksp, Teor. Fiz. **47**, 1945 (1964)
31. YuA Il'inskii, L.V. Keldysh, *Electromagnetic Response of Material Media* (Plenum Press, New York, 1994)
32. I.N. Zavestovskaya, P.G. Eliseev, O.N. Krokhin, Appl. Surf. Sci. **248**, 313 (2005)
33. I.N. Zavestovskaya, P.G. Eliseev, O.N. Krokhin, N.A. Men'kova, Appl. Phys. A **92**, 903 (2008)
34. P.G. Eliseev, N.A. Kozlovskaya, J.N. Krokhin, I.N. Zavestovskaya, AIP Conf. Proc. **1278**, 143 (2010)
35. YuV Afanasiev, B.N. Chichkov, V.A. Isakov, A.P. Kanavin, S.A. Uryupin, J. Russ. Laser Res. **21**, 505 (2000)
36. A.P. Kanavin, S.A. Uryupin, Phys. Lett. A **372**, 2069 (2008)
37. A.P. Kanavin, S.A. Uryupin, Quantum Electron. **38**, 159 (2008)
38. N.G. Basov, B.I. Bertyaev, I.N. Zavestovskaya, V.I. Igoshin, V.A. Eatulin, in *Primenenie lazerov v narodnom khozyaistve*. Trudy vses. Konf. Proceedings All-Union Conference Application of Lasers in the National Economy (Nauka, Moscow 1986), p. 100
39. I.N. Zavestovskaya, V.I. Igoshin, V.A. Eatulin, S.V. Eayukov, A.L. Petrov, Quantum Electron. **14**, 2343 (1987)
40. I.N. Zavestovskaya, V.I. Igoshin, I.V. Shishkovskii, Quantum Electron. **16**, 1636 (1989)

41. A.F. Fedechev V.I. Igoshin, I.N. Zavestovskaya, I.V. Shishkovskii, J. Sov. Laser Res. **12**, 365 (1991)
42. B.I. Bertyaev, V.I. Igoshin, V.A. Eatulin, I.N. Zavestovskaya, I.V. Shishkovskii, J. Russ. Laser Res. **17**, 164 (1996)
43. V.P. Veiko, Q.E. Eieu, Quantum Electron. **37**, 92 (2007)
44. I.N. Zavestovskaya, O.A. Glazov, N.A. Men'kova, Laser-Mater Interaction. Proc. SPIE **6735**, 673512 (2007)
45. M.S. Komlenok, V.V. Kononenko, I.I. Vlasov, V.G. Ralchenko, N.R. Arutyunyan, E.D. Obraztsova, V.I. Konov, J. Nanoelectronics Optoelectron. **4**, 286 (2009)
46. I.N. Zavestovskaya, Pac. Sci. Rev. **12**, 56 (2010)
47. P. Simon, J. Ihleman, Appl. Surf. Sci. **109/110**, 25 (1997)
48. T.V. Kononenko, S.V. Garnov, S.M. Pimenov, V.I. Konov, V. Romano, B. Borsos, H.P. Weber, Appl. Phys. A **71**, 627 (2000)
49. A.Y. Vorobyev, C. Guo, Appl. Surf. Sci. **253**, 7272 (2007)
50. A.Y. Vorobyev, C. Guo, Opt. Express **14**, 2164 (2006)
51. A.V. Bulgakov, I. Ozerov, W. Marine, Thin Solid Films **453**, 557 (2004)
52. I.N. Zavestovskaya, S.D. Makhlysheva, A.P. Kanavin, in *Trudy III Simpoziuma po kogerentnomu opticheskomu izlucheniyu poluprovodnikovyh soedinenii I struktur*. Proceedings III Symposium on Coherent Optical Radiation from Semiconductor Materials and Structures. (RIIS FIAN, Moscow, 2011), p. 225. (in Russian)
53. M. Sanz, R. De Nalda, J.F. Marco, J.G. Izquierdo, L. Banares, M. Castillejo, J. Phys. Chem. C **114**, 4864 (2012)
54. A.P. Kanavin, I.N. Zavestovskaya, S.B. Borin, Pac. Sci. Rev. **13**, 78 (2011)
55. A.I. Il'in, V.S. Eraposhin, Poverkhnost **6**, 5 (1983)
56. N. Rykalin, A. Uglov, I. Zuev, A. Kokora, *Lazernaya i elektronno-luchevaya obrabotka materialov (Elactron-Beam and Laser Processing of Materials)* (Mir, Moscow, 1988). (in Russian)
57. I.V. Hertel, R. Stoian, D. Ashkinazi, A. Rozenfeld, E.B. Campbell, RIKEN Rev. **32**, 23 (2001)
58. I.N. Zavestovskaya, A.P. Kanavin, N.A. Men'kova, Opt. Zh. **75**(6), 13 (2008)
59. I.N. Zavestovskaya, V.I. Igoshin, A.P. Kanavin, V.A. Eatulin, I.V. Shishkovskii, Or. Fiz. Inst. im P. N. Lebedeva, Ross. Akad. Nauk, **217**, 3 (1993). (in Russian)
60. I.N. Zavestovskaya, Pac. Sci. Rev. **10**, 218 (2008)
61. V.G. Gryaznov, A.M. Eaprelov, A.E. RoManov, Pis'ma Zh. Tekh. Fiz. **15**, 39 (1989)
62. R.W. Siegel, G.E. Fougere, Nanostruct. Mat. **6**, 11 (1995)
63. E. Ivid'ko, E. Gutkin, *Fizicheskaya mekhanika deformiruemykh nanostruktur (Physical Mechanics of Deformable Nanostructures)*, vol. 1 (Yanus, S. Petersburg, 2003). (in Russian)
64. V.N. Eashkin, G.S. Zhdanov, L.I. Mirkin, Docl. Acad. Nauk SSSR **249**, 1118 (1979). (in Russian)
65. P. Mazzoldi, G. Della Mea, G. Battaglin, A. Miotello, M. Servidori, D. Bacci, E. Jannitti, Phys. Rev. Lett. **44**, 88 (1980)
66. YuA Skakov, N.V. Enderal, KhA Mazorra, V.S. Eraposhin, Tr. Mosk. Inst. Stali Splavov **147**, 8 (1983). (in Russian)
67. M. Trtica, B. Garovic, D. Batani, T. Desai, P. Panjan, B. Radak, Appl. Surf. Sci. **253**, 2551 (2006)
68. E.E. Lapshin, A.Z. Ibidin, V.N. Ookarev, VYu. Khomich, V.A. Shmakov, V.A. Yamshchikov, Ross. Nanotekhnol. **2**, 50 (2007)
69. Yu.V. Afanasiev, I.N. Zavestobskaya, A.P. Kanavin, S.V. Kayukov, *Laser Interaction and Related Phenomena*. AIP Conference Proceedings (1995), p. 1274
70. S.V. Kayukov, A.A. Gusev, E.G. Zaychikov, A.L. Petrov, YuV Afanasiev, I.N. Zavestobskaya, A.P. Kanavin, Izv. Akad. Nauk **61**, 1546 (1997). (in Russian)
71. V.N. Tokarev, A.F. Kaplan, J. Appl. Phys. **86**, 28362846 (1999)
72. D.S. Ivanov, L.V. Zhigilei, Phys. Rev. B **68**, 064114 (2003)
73. C. Cheng, X. Xu, Phys. Rev. B **72**, 165415 (2006)
74. M. Kandyla, T. Shih, E. Mazur, Phys. Rev. B **75**, 214107 (2007)
75. T. Frank, Sci. Tech. Adv. Mater. **6**, 221 (2005)

76. J.-T. Chen, M. Zhang, T.P. Russell, Nano Lett. **7**, 183 (2007)
77. R. Xia, J.L. Wang, R. Wang, X. Li, X. Zhang, X.-Q. Feng, Y. Ding, Nanotechnology **21**, 085703 (2010)
78. G.Q. Lu, X.S. Zhao, *Nanoporous Materials: Science and Engineering* (Imperial College Press, London, 2004)
79. B.V. Kaludjerovic, M.S. Trtica et al., J. Mater. Sci. Technol. **27**, 979 (2011)
80. Yu. Pu, P. Ma, J. Appl. Phys. **112**, 023111 (2012)
81. M.S. Zolotykh, I.N. Zavestovskaya, A.P. Kanavin, in *Trudy III Simpoziuma po kogerentnomu optiches komu izlucheniyu poluprovodnikovyh soedinenii I struktur*. Proceedings III Symposium on Coherent Optical Radiation from Semiconductor Materials and Structures. (RIIS FIAN, Moscow, 2011), p. 221. (in Russian)
82. S.Y. Ando, S. Shimamura, J. Porous Mater. **13**, 439 (2006)
83. J.W. Christian, *The Theory of Transformation in Metals and Alloys, Part I* (Pergamon press, Oxford, 1975)
84. G.M. Eudinov, V.A. Shmakov, Dokl. Akad. Nauk SSSR **264**, 610 (1982). (in Russian)
85. V.A. Shklovskii, A.A. Motornaya, K.V. Maslov, Poverkhnost **6**, 91 (1986)
86. A.L. Glytenko, V.A. Shmakov, Dokl. Akad. Nauk SSSR **276**, 1392 (1984)
87. V.N. Ookarev, V.Yu. Khomich, V.A. Shmakov, V.A. Yamshchikov, Dokl. Akad. Nauk **419**, 1 (2008). (in Russian)
88. I.S. Miroshnichenko, *Zakalka iz zhidkogo sostoyaniya (Quenching from the Liquid State)* (Metallurgiya, Moscow, 1982)

Chapter 4
Optical Breakdown in Ambient Gas and Its Role in Material Processing by Short-Pulsed Lasers

Sergey M. Klimentov and Vitaly I. Konov

Abstract Formation of gas breakdown plasma, staying apart from the surface of materials exposed to pulsed laser radiation, is shown to have a strong impact to productivity and accuracy of micromachining. Origin and effect of two types of such plasmas are described: one is induced by low threshold breakdown of ambient gas contaminated by charged ablated nanoparticles; which is characteristic of nano- and subnanosecond lased ablation; the second is caused by ionization of gas at the leading edge of focused pico- and femtosecond pulses. The ways to eliminate undesirable plasma effects are discussed and demonstrated.

4.1 Introduction: Origin of Plasma Near Exposed Surface

In the majority of laser material processing operations, irradiation of the work-piece takes place in an ambient atmosphere, focusing of radiation in which may result in breakdown of the gas and formation of plasma at the leading part of laser pulses. Such plasma is able to absorb and scatter essential part of the incident energy. This way, the intensity on the target is reduced, while the beam profile, pulsewidth and spectrum of the incident radiation are significantly modified. Effect of such microplasma imposes limitations to productivity and precision of laser technologies. More generally, the account of it should be taken in any task calling for precision targeting of intense laser pulses in gases.

Origin and location of such plasma differ and depend on irradiation conditions. One type of plasma is ignited at nanoparticles produced by the preceding laser pulses

S. M. Klimentov (✉) · V. I. Konov
A. M. Prokhorov General Physics Institute of Russian Academy of Sciences,
38 Vavilova St., Moscow, Russia 119991
e-mail: kliment@kapella.gpi.ru

V. I. Konov
e-mail: vik@nsc.gpi.ru

V. P. Veiko and V. I. Konov (eds.), *Fundamentals of Laser-Assisted*
Micro- and Nanotechnologies, Springer Series in Materials Science 195,
DOI: 10.1007/978-3-319-05987-7_4, © Springer International Publishing Switzerland 2014

and residing in the ambient inside and around the ablated crater. If the crater is deep enough, their residence lasts for long time achieving several minutes. The threshold of plasma ignition in this case is lower compared to the breakdown threshold of pure gases. Initially, the particles are electrically charged; their discharge in air takes milliseconds. This fact, neglected previously by researchers, can be used to monitor productivity of ablation in-situ and, to some extent, to control their movement. Presence of nanoparticles in the ambient gas has a strong impact to micromachining in the pulsewidth range from tens of picoseconds to nanoseconds, mostly through the screening of incident radiation.

Another type of plasma is of importance in the domain of ultrashort intense pulses (from femtoseconds to tens of picoseconds). Being focused, these pulses are able to modify refraction index of pure gas while propagating through the ambient, increasing it first due to Kerr effect and then abruptly reducing via strong ionization. Such a complex laser self-action results in significant non-linear scattering of the beam induced by the complex transformation of the refraction index both in time and in space within the beam path. Under conventional conditions of processing by ultrashort pulses the beam may be dramatically distorted on its way to the target.

4.2 Effect of Charged Ablated Nanoparticles Long-Residing in the Ambient Gas

4.2.1 Observation of Low-Threshold Air Breakdown

We should note that only phenomena of plasma formation attributed to the particles are discussed in this section. Other mechanisms of plasma setup near the irradiated surfaces, for example those well described for interaction of pulsed CO_2 lasers with metals [1], are beyond the scope of the chapter. The low threshold breakdown induced by the particles can be observed and characterized by measuring optical losses in through channels drilled by short laser pulses in air [2–4]. Diameter of the channels is typically much larger compared to the beamwaist and should not impose limitations to propagation of radiation, while in fact, their transmittance appears to be drastically dependent on energy and time interval between the pulses, which is illustrated by Fig. 4.1. If this time lag is long enough (minutes), radiation propagates through the channel without losses following the dashed line but, if it is short, the transmittance of the same 0.5 mm long channel drops significantly down. The dotted line corresponds to 5 Hz repetition rate of the pulses. The induced losses cannot be attributed to plasma screening caused by ablation of side walls, as such screening would not depend on time interval between the pulses. More over, the surface plasma, at the applied laser intensities, does not expand fast enough to block the channel aperture during the short ablating pulse [5].

The described transmission method allows finding both: the energy threshold of screening, caused by the gas breakdown, and the time lag after which the initial

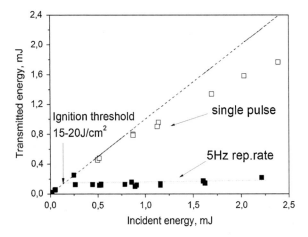

Fig. 4.1 Two branches of output energy illustrating screening due to gas breakdown in 0.5 mm long channel drilled through by 300 ps pulses

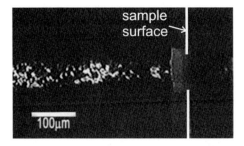

Fig. 4.2 Ignition of microplasma on ablated particles residing in air near the channel drilled in steel

"optical strength" of it is retrieved. The threshold depends on pulsewidth and on the target material. For drilling of steel by 300 ps pulses in air, the screening starts from $15-30 \, \text{J/cm}^{-2}$. For comparison, the breakdown of pure air in this pulsewidth range happens at ten times higher energy densities. The plasma screening, induced by presence of ablated nanoparticles, is especially pronounced inside long and weakly ventilated channels. Recovery of optical transmittance occurs via deposition of particles on the walls or by their removal with convective air flows. The recovery time is mostly dependent on the incident energy density, which defines initial concentrations of the particles, and on geometry of the channel, which delimits possibility of convective removal or determines the mean path for the particle to reach the wall. Employing of vacuum below 200 mbar completely eliminated the screening.

Residence of the ablated particles in air can be monitored by shadowgraphy or laser illumination technique. Particles in Fig. 4.2 are revealed due to the beam of the second harmonics propagating along the path of the ablating infrared pulses. Strobo-scopic methods allow to follow trajectories and velocities of every particle, while the

Fig. 4.3 Polymer replica from the channel drilled in steel by 300 ps pulses (100 J/cm^2)

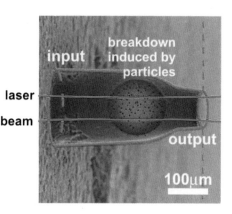

technique of fast interferometry enabled observation of early stages of plasma ignition and the following expansion dynamics [4, 6, 7]. Mechanisms of the gas plasma ignition in the vicinity of a particle are not completely understood. The ignition may start from evaporating or from effective emission of high energy electrons expelled from the surface of the particles by the local electric field significantly amplified at the conductive nano-objects, or via generation of intense evanescent waves [8–10]. Such free electrons, interacting with the incident electromagnetic wave, are able to launch the following gas ionization.

4.2.2 Locations of Plasma and the Resulting Crater Morphology

Deposition of laser energy into plasma of the low threshold air breakdown affects morphology of drilled channels and rates of their pulse-to-pulse formation, which makes location of the gas ignition to be of particular interest. Position of the plasma cloud can be found studying precision longitudinal micro-sections of the channels, or more easily, by replicas of their inside using fine polymer compounds. The typical bottleneck shape in Fig. 4.3 is formed by sub-nanosecond pulses. Location of the gas explosion area, acquiring up to 90 % of incident energy, is indicated by the red sphere. Radial expansion of the plasma results here in widening of the channel and does not contribute to its deepening. The rest of laser energy, braking through this domain, propagates towards the channel bottom and ablates creating conventional surface plasma. As seen, the gap between these two plasmas exceeds 100 μm. Linear rates of ablation in this case depend weakly on the incident energy the excessive level of which contributes mostly to widening of the channel.

A positive aspect of the breakdown consists in creating of plasma flow along the side walls towards the input of the channel. It is often mentioned as "plasma brush" in literature [11, 12]. The flow wipes out the melt minimizing the recast layer and

Fig. 4.4 Positive particles produced by ablation of steel by 300 ps ($80\,J/cm^2$)

making the walls more cylindrical. At a definite depth, the flow results in a moderate erosion of the input hole creating shallow longitudinal ripples. When the drilled depth exceeds $800\,\mu m$ in steel, the melt expulsion becomes a problem, which spoils geometry of the channels and their surface structure. The melt accumulated near the input hole induces further retardation of drilling.

The most pronounced manifestations of the low threshold gas breakdown are observed in the sub-nanosecond range of laser pulses, meanwhile similar features evolve at deep drilling of steel in the longer nanosecond pulsewidth range. The difference makes the higher thresholds ($\sim 80\,J/cm^{-2}$) and merging of the mentioned above two plasma domains due to their broader expansion during the laser pulse.

4.2.3 Dimensions, Lifetime and Electrical Properties of Nanoparticles

Parameters of the particles resulting form ablation of carbon steel were investigated in details. The most striking feature was an electric charge they carried floating in air long after the laser pulse. In some cases it was possible to derive dependence between the charge sign and the mechanism of particle formation. Morphology of positive and negative particles produced by 300 ps pulses was significantly different. The positive particles, extracted and observed by optical microscopy or SEM, were ideal spheres (Fig. 4.4a). Statistics of their diameters nearly followed the Poisson law with the maximum around 400 nm (Fig. 4.5). It is interesting to note that simple Einstein's relationships for Brownian motion result in the same average size of the particles at the typical channel diameter ($\sim 100\,\mu m$) for the measured recovery time of optical strength of air ($\sim 60\,s$). The negative fragments were much larger and irregularly shaped. We suppose those to be produced by explosion of the bead surrounding the crater and formed by expulsion of melt along the walls above the surface.

Fig. 4.5 Statistics of positive particles extracted near the steel surface (5 ps, 100 J/cm^2, 10 Hz)

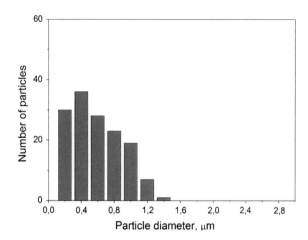

Particles of each sign allowed selective extraction to an external electrode where their volume per pulse could be estimated [13]. The volume of positive ones prevailed being nearly equal to the total ablated volume. So, nearly all components of the ablated plume turned finally into the positively charged particles. Similar features and dependences were obtained in experiments for 300 ps and 2 ns pulses. Conductivity measurements were aimed at characterization of electric properties of the particles [12]. The external voltage in this case was pulsed and came after the incident laser exposures; otherwise, the signal was masked by much higher conductivity of the hot plasma plume. The applied field was limited by $\sim$1 kV/cm to avoid Maxwell screening and multiplication of charges. Alternating of polarity from pulse to pulse, prevented from space-charge accumulation on the electrode and on the sample. Explicit descriptions of the experimental procedure and processing of conductivity signals are given in [13]. Conductivity dynamics is illustrated by Fig. 4.6. To obtain this plot, the peak of current was measured at variable delay between the laser and the electrical pulses. Decay of conductivity revealed two stages. The faster one lasts until $\sim$300 μs; it corresponds mostly to cooling of weakly ionized plasma within so-called "fire ball" which is described below in the text [12, 14]. At the end of this stage, temperature of the gas in the ablation area drops down to $\sim$1500–2000 °C; condensation of the ablated vapor and solidification of the particles start. The following slower discharge is presumably attributed to interaction of the positively charged particles with negative ions in atmosphere. Similar relaxation dependences were obtained for the total collected electric charge.

It is interesting to note that conductivity, measured during the first stage, appears to be nearly proportional to the ablated volume in the broad range of incident laser energies (Fig. 4.7). Such an invariant of the specific ablated charge can be used as a tool for in-situ monitoring of productivity of laser micromachining. The obtained conductivity plots and size distributions in Fig. 4.5 allow estimation of the average electric charge per a particle. This charge typically amounts to $\sim$1.5 $\times$ 10^{-14} C, when measured 200 μs after 2 ns laser pulses at the energy density of 80 J/cm^2.

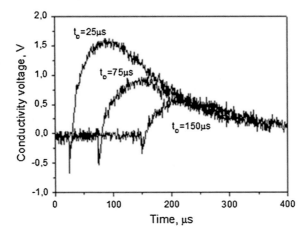

Fig. 4.6 Conductivity traces measured at variable time delay t_D in a cloud of charged particles (ablation of steel, 2 ns pulses, 90 J/cm², U = 200 V)

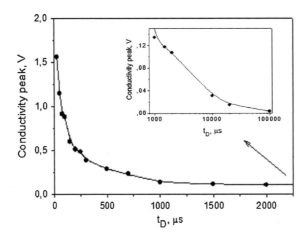

Fig. 4.7 Decay of conductivity due to discharge of ablated particles in air (U = 200 V, 90 J/cm²)

4.2.4 Removal of Charged Particles by Application of Electric Field

Application of an external electric field, within the ablated area, brings the particles into the drift movement, velocity of which can be estimated from simple hydrodynamic relationships for a spherical object driven by Coulomb forces in a gas with dynamic viscosity depending on temperature. For the applied voltage of 200 V and the typical gap between the steel target and the electrode of several millimeters, the average velocity achieves 25 m/s. According to these estimations, the particles should reach the extracting electrode within ~100 μs, which indeed corresponds to the rise

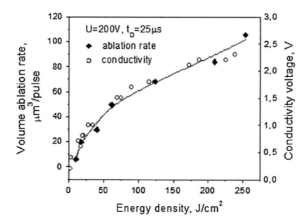

Fig. 4.8 Comparison of volume rates of ablation and conductivity induced by charged particles

time of conductivity peaks in Fig. 4.6a. The rear part of the signal is formed by the discharge of the cable and depends on its capacitance and the electrical resistance of an oscilloscope (see Fig. 4.6).

Such velocity would be sufficient for removal of the particles from shallow craters, and this way, for elimination of plasma screening caused by repetitive laser pulses. Meanwhile, enhancement of drilling rates in deeper channels is a more difficult issue calling for higher electric fields near the surface i.e. for higher voltages and shorter distances between the electrode and the drilled sample. To avoid electrical discharge in air between the electrodes, the gap in this case was filled with dielectric material except for a hole of 3 mm in diameter, through which the focused radiation reached the surface of the material (Fig. 4.8). Electrical strength of the device was additionally improved by a Teflon tube installed tightly inside the hole and extended above the electrode, to ensure from shortcuts caused by deposition of the abated material. That allowed running the drilling experiments at the positive DC voltage up to 3 kV. The applied field could not affect the removal of ablation products by the plasma plume but was able to reduce concentration of the particles residing in atmosphere between laser pulses. The plot in Fig. 4.9 demonstrates enhancement of drilling rates by nearly 50 times, which approaches to the rates achieved in vacuum. Morphology and geometry of channels drilled with the electric field was also similar to that typically obtained in vacuum conditions. Drilling of shallow craters usually was not in need of the field assistance due to sufficient ventilation via air convection. The most noticeable effect of the electric field was observed for steel plates 0.2–0.7 mm thick, ablated by 300 ps pulses in the energy density range 50–150 J/cm^2 [12, 13]. One may note that penetration of the field inside channels of conductive material should be limited by their several diameters (100 ÷ 150 μm in our case) due to equal potential on the metal surface. That would not allow noticeable enhancement of drilling rates in deeper ablated craters. We assume the field in this case to act indirectly. Application of the voltage slightly distorted neutrality of the plasma plume so that a part of

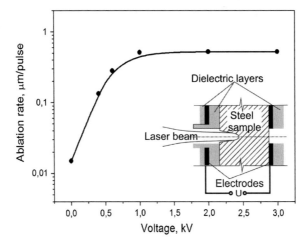

Fig. 4.9 Enhancement of ablation rates due to extraction of particles by electric field: external voltage dependence (0.5 mm thick steel, 300 ps pulses); and the electrode assembly used in the experiment

positive electric charge, carried by vapor and particles, was accumulated on the walls near the input hole and behaved like a "transducer" of the applied electric field deeper inside the channel. The induced surface charge remained here between pulses, presumably, due to a dielectric layer of oxides covering the particles and the walls. Such a surface charge configuration could reproduce itself with deepening of the crater. That explains why the field effect is so critically dependent on parameters of ablative radiation (pulsewidth, energy density, repetition rates).

4.2.5 Relaxation of Plasma in Atmospheric Air and Its Role in High Repetition Rate Micromachining

It seems trivial that rarefying of the ambient atmosphere eliminates the low threshold gas breakdown and the screening during micromachining, but conventional methods of vacuum pumping are inconvenient in technological applications due to difficulties with large processed objects and increased time of preparatory operations, except one rather complex solution based on setting up an ultrasonic gas flow near the surface [15]. Meanwhile, knowledge of dynamics of thermal relaxation in plasma provides us an approach to formation of so called "quasi-vacuum" conditions allowing to alude the mentioned above complications and to boost productivity only by engineering the parameters of laser exposures. The approach is based on the phenomenon of high-temperature conservation in a small hemispherical gas domain at the surface between the ablating laser pulses [4, 12, 16]. Formation of it in the case of breakdown in gases is predicted by theory of point explosion [17], where it is mentioned as "fire

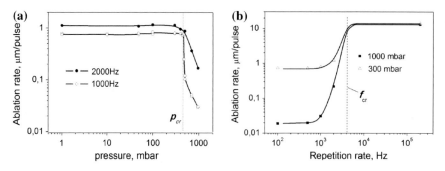

Fig. 4.10 Pressure (**a**) and repetition rate dependences (**b**) for drilling of steel (20 ns pulses, 100 J/cm^2)

ball" [18]. According to this theory, applied to the case of micro-ablation, the gas heated by the pulsed laser energy absorbed within a small spot starts expanding driven by the excessive inner pressure. Expansion stops when the pressure inside the hemisphere is equal to the ambient pressure, but the temperature inside it stays still much higher than in the ambient. This inner temperature then drops slowly down, within hundreds of microseconds. If the following laser pulse happens to reach the target while the gas is enough hot, the density of it may not be sufficient to cause the optical breakdown resulting in noticeable plasma screening. This way, the drilling rates can be boosted up two orders of magnitude while the depth of the channels reaches several centimeters at the aspect ratio exceeding 200 [14, 19, 20]. Basically, the domain of low density atmosphere, around the craters or inside the channels, can be achieved either through the increased repetition rates [14] or by generation of couples of laser pulses [21] and their more complex successions [19].

To reach the quasi-vacuum conditions, two clue parameters have to be defined: the rarefaction level sufficient in terms of plasma screening and enhancement of drilling rates, and the time range within which the following laser pulse finds such rarefaction conditions at his arrival. Both parameters were found experimentally.

Pressure dependences of drilling rates (Fig. 4.10a) showed a distinct plateau below the sharp step at 200–400 mbar, position of which slightly varied depending on the laser pulsewidth. The step-like shape is less pronounced at higher repetition rates of 2 kHz for which the elevated residual temperature of air near the crater comes already into play. The observed retardation of drilling above 400 mbar corresponds to drastic transformation of the channel dimensions and morphology. Diameters of the channels formed at high pressure and low repetition rates are much larger (150 μm compared to 30–40 μm in vacuum) for the same laser energy and the focused spot of ∼20 μm, which clearly indicates their shaping by the ablative effect of laterally expanding plasma. The inside surfaces, formed this way, are smooth due to the "brushing" effect of such plasma.

The second parameter, namely the lifetime of the quasi-vacuum conditions, can be obtained from the repetition rate dependences of drilling (Fig. 4.10b). Two Nd:YAG lasers were used in the measurements: the conventional 20 ns laser with variable repetition rates and the laser generating trains of pulses nearly equal in energy (up to 20 pulses) at the equivalent rate of 200 kHz [19]. The incident energy density per a pulse was equal in both cases. The plots in Fig. 4.10b were obtained at the normal (1000 mbar) and the reduced pressure (300 mbar) corresponding to the vacuum conditions of ablation, according to the pressure dependences in Fig. 4.10a. If quasi-vacuum is achieved at some critical repetition rate of the pulses, we expect these two curves to merge indicating equivalent rarefaction in both cases. As seen, it happens at 4 kHz for the energy density of 100 J/cm^2. Exactly the same critical rate was obtained later in ablation experiments with couples of nanosecond pulses [21]. One may note that ablation rates measured at 300 mbar start also growing above ∼1 kHz which indicates noticeable accumulation of heat not only in the surrounding gas but also in the processed material.

Realistic estimations of the quasi-vacuum lifetime can be made taking into consideration several stages of the process, including the plasma plume expansion during the laser pulse, further growth of the "fire hemisphere" driven by high inner temperature and pressure, and the following cooling of the hot gas inside a distinctly outlined long-living "bubble" [14, 18, 22]. The last longest stage is most interesting for analysis. It begins ∼2 μs after the laser pulse and is featured by formation of a stable "fire hemisphere" with the radius of ∼0.6 mm and the inner temperature of 5000–8000 K [18]. According to Grashof criterion, convective cooling via formation of vortexes in air is impossible for such a small area at the given temperature difference, and the conductive heat transfer into the walls of the drilled channel becomes dominating [23].

The result of calculations of the gas temperature and density, for typical conditions of drilling in steel, is shown in Fig. 4.11. The temperature drops down with time while the density grows correspondingly up. The two density curves in the plot embrace an uncertainty range of the initial temperature T_F inside the hemisphere. After the rarefaction threshold is crossed by these curves, the temperature becomes too low (1500–2300 K) and enhancement of drilling rates can no longer be expected. The estimated lifetime of quasi-vacuum conditions is in good agreement with the experimental values in Fig. 4.10a and the data of [21].

The described experiments and calculations allow definition of the time window within which the following laser pulse should come to allow enhancement of drilling rates by, at least, one order of magnitude. For the conventional terms of micromachining, this window is roughly between 50–100 ns and 200–300 μs. When coming earlier than 50 ns, the second pulse meets the expanding plasma plume and can be absorbed at the front of the shock wave. When it is coming later than 300 μs, the gas near the surface is already too dense which launches the mechanism of plasma screening. It worth reminding that we are speaking of sub-nanosecond and nanosecond pulses at the energy densities ∼100 J/cm^2.

Conditions for the sufficient rarefaction can also be met generating tandem laser pulses or longer and more complex successions of those. Technically, it is achieved by extension of pumping time in diode or flash lamp pumped lasers and multiple Q-switching [19, 24]. Drilling rates achieved by such tandems of 20 ns pulses (Fig. 4.12) exceed even those obtained in conventional vacuum conditions at low

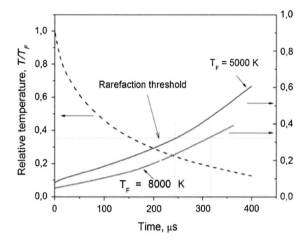

Fig. 4.11 Lifetime of quasi-vacuum conditions: dynamics of gas cooling and gas density in drilled channels

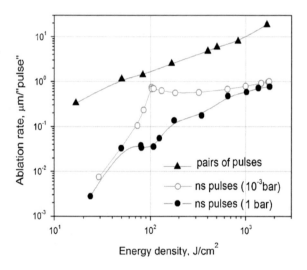

Fig. 4.12 Comparison of ablation rates in air, vacuum and quasi-vacuum conditions obtained using tandem pulses (steel, 20 ns pulses)

repetition rates. The additional gain in productivity is ascribed to the already mentioned residual heat effect in the material [25]. Longer trains of high energy pulses were successfully used for fast drilling of extremely deep channels in metals and ceramics [20, 26]. Their length achieved several centimeters at the diameters of ∼100 μm demonstrating insignificant deviations from the cylindrical shape. One should keep in mind that laser radiation undergoes multiple reflections on its way through such long channels which appears to be similar to the waveguide propagation,

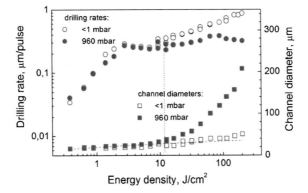

Fig. 4.13 Drilling rates and hole diameters with respect to incident energy and the ambient gas pressure (steel, 150 fs pulses)

so that insignificant plasma "brushing" of the walls may result in a pronounced polarization effect on the drilled geometry, especially in ceramics and other dielectric materials [19].

4.3 Self-Scattering of Focused Ultrashort Pulses

Micromachining by femtosecond laser pulses demonstrates significant advantages due to effective energy deposition into the electron system of solids and the minor thermal effect on the surrounding material, which results in predictable and high precision ablation [27–29]. At the same time, intensities in femtosecond beams focused on a target in conventional micromachining conditions achieve 10^{13} W/cm^2, i.e. the level when Kerr effect and ionization in the ambient gas may not be disregarded. Both phenomena are known to induce fast modification of the refractive index of the gas at the leading edge of the ultrashort laser pulses. The rest of the pulse then propagates through the domain with a complex spatial profile of Δn, undergoing strong scattering and dramatic time-spectral modification via self-phase modulation. Accuracy and productivity of micromachining suffer in this case from defocusing and distortions of the ablating beams, which is illustrated by difference of drilling rates and diameters of the resulting craters in air and in vacuum (Fig. 4.13) [30]. This scattering and deformations affects not only deep laser drilling but also micromachining of thin films, especially when the focal plate is located in front of the ablated surface. Morphology of the crater produced in such high scattering regime is shown in Fig. 4.14. The white dashed circle delimits the theoretical, or undisturbed, beam waist boundaries on the surface demonstrating how far can reach the beam metamorphosis. These transformations, spectral and spatial, can be easily observed by bare eyes. A screen placed behind the beamwaist reveals in this case a complex structure of colored rings [30].

Fig. 4.14 Steel surface
ablated by the beam
distorted due to scattering
(120 fs pulses, 260 J/cm²);
the *dashed circle* ounlines the
beam waist without scattering

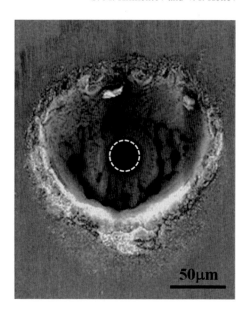

4.3.1 Threshold Conditions

The non-linear scattering of ultrashort pulses in the focused area can be integrally characterized by a portion of laser energy no more propagating within the initial spatial profile of the beam, Gaussian or hat-top. A number of techniques were used to measure this scattered energy in several gases at different incident wavelengths in the pulsewidth range from 50 fs to 15 ps [31, 32]. The plots in Fig. 4.15 characterize scattering and absorption in femtosecond beams with respect to the incident energy density which would be reached at focusing in vacuum. We see that the absorbed energy is typically less than 10 %, while the scattered one happens to exceed 30 %, which indicates divergence and distortions in the beam inacceptable for the tasks of precision micromachining or exact targeting of laser radiation.

All the gases in Fig. 4.15 except He, standing apart due to higher ionization potential, show similar behavior at a particular incident wavelength, 800 or 400 nm, while the shorter wavelength of these two (Fig. 4.15b) is always featured by the smaller scattered energy and higher thresholds of scattering, marked in the graphs by vertical dashed lines. Obviously, operation below the threshold in air (5 J/cm² for 800 nm and 20 J/cm² for 400 nm) ensures from significant beam distortions without special precautions. To achieve the broader energy range without scattering (up to 100 J/cm²), one should combine helium atmosphere and the doubled frequency of Ti:Sa laser radiation. The detailed explanation of the observed dependences is given below.

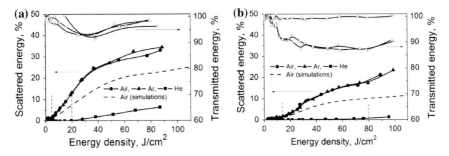

Fig. 4.15 Scattering and absorption of femtosecond pulses in gases: **a** 800 nm, 120 fs; **b** 400 nm, ~90 fs

4.3.2 Contributing Mechanisms

To reduce the unwanted effects of scattering on productivity of micromachining, and more generally on accuracy of targeting of ultashort pulses, one has to define main factors contributing to this complex phenomenon in the case of focusing geometry typical for material processing (~0.5 cm beam diameters, ~5 cm focal length). Propagation of picoseconds and femtosecond pulses in air within relatively low intensity collimated beams ($<10^{13}$ W/cm^2) was earlier extensively investigated aiming at effects of filamentation, long distance self-channeling and formation of conductive plasma strings in air in terms of a fine balance between self-focusing caused by the optical Kerr effect and defocusing via diffraction and feeble ionization [33–36]. Theoretically, the account of both can be taken by introducing the corresponding self-action terms in the nonlinear Schrödinger equation for propagation of ultrashort optical pulses in transparent media.

A contrast case of tight focusing geometry, characterized by much shorter interaction lengths and higher intensities of radiation, is regarded in the chapter. Numerical calculations become challenging in such conditions calling for effective codes, powerful computers and precise knowledge of physical constants which, in many cases, are known only approximately (Kerr and ionization related constants depending on the irradiation wavelength and pulsewidth). The data presented here were obtained mostly experimentally at variation of a large set of parameters such as the incident laser wavelength, pulsewidth, the ambient gas, the beam profile and geometry of focusing. Numerical estimations were used for analysis of the obtained dependences.

The impact of self-focusing on the scattering was estimated making a comparison of the scattered energies to Kerr constants in different gases (air, helium, argon and CO_2). The constants, in turn, were either taken from literature or directly collated measuring efficiency of the third harmonics generation for 120 fs pulses in the same focusing conditions. It is known that Kerr effect and the third harmonics generation imply the same kind of nonlinearity. Such a direct experimental comparison allowed to avoid discrepancies attributed to difference of evaluation methods, conditions of measurements, temperature dependences etc. Surfing through literature makes one to expect the lowest Kerr constant in helium, being the atomic gas with the compact spherical electron orbital, and the highest in argon [37, 38]. Indeed, the measured efficiency curves for the third harmonics showed qualitatively the same order, which did not correlate with the scattering data in Fig. 4.15. The argon and air gases appear to be very contrast in terms of Kerr effect while their scattering plots nearly coincide. This observation allows us to derive that in the case of focused geometry the optical Kerr effect plays minor role in scattering. The brief positive modification of Δn at the leading edge of the focused laser pulse is quickly overcompensated by a negative Δn induced by the following gas ionization, the defocusing effect of which lasts during the longer part of the pulse. This way, the fragile balance of self-focusing and defocusing, attained in collimated beams at a particular level of intensity, through their reciprocal influence, cannot be achieved in the sharply converging focused beams due to fast variation of intensity along the beam axis. That is why the distinct signs of filament formation were not observed in the described experiments.

Namely the ultrafast ionization in the focused beams was shown to provide the largest contribution to the scattering. Using of the interferometry technique in the femtosecond time domain with spatial resolution of several micrometers allows to reveal peculiarities of photo-ionization dynamics for the fundamental and the second harmonics of Ti:Sa laser [32]. Comparison of two incident wavelengths in an equivalent focusing geometry is presented in Fig. 4.16, where time profiles of the pulses are shown for the reference. We see that the ionization rate, given by the slope of the phase modification curve, is approximately twice higher for IR radiation. The same are electron densities achieved during the 800 nm pulses. It takes place in spite of longer pulsewidth and lower intensity compared to the visible radiation. The following growth of the phase, seen in both plots when the pulses are already over, does not contribute to the scattering and is caused by multiplication of high-energy electrons, initially generated and heated by intense femtosecond radiation. That happens via impact ionization of their hosts or neighboring atoms [39]. A hint to positive Δn induced by Kerr effect is barely seen at the beginning of the phase plot only for the visible pulses.

The same ionization trends can be qualitatively traced by spectra of scattered radiation significantly modified through the effect of self-phase modulation [40]. This spectral modification is caused by self-induced variation of the phase inside the wave packet, corresponding to positive or negative changes of Δn during the pulse. As seen in Fig. 4.17, the abrupt ionization during the IR pulse result in broader deformations of the spectrum towards the shorter wavelengths. Reaching the visible range, the scattered light is seen as "a rainbow" of colored ring around the beam. Smaller ionization rates at 400 nm correspond to the narrower spectra of self-phase modulation (Fig. 4.17b).

The observed difference of ionization rates in Fig. 4.16 explains the wavelength dependences of scattering (Fig. 4.15) only partially. Another important reason for this is attributed to dispersion of light in the breakdown plasma. Such plasma, at a given electrons density, provides higher refractive ability at longer incident wavelengths i.e. in the case of IR pulses. According to classical relationships, the negative Δn is directly proportional to electron concentrations N and to the square of the laser

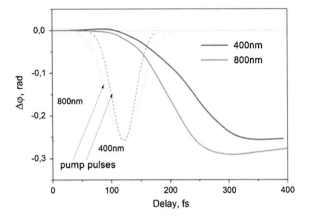

Fig. 4.16 Dynamics of ionization in the beam waist area induced by 800 and 400 nm femtosecond pulses (shown by *dashed curves*)

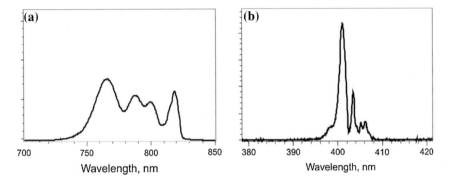

Fig. 4.17 Spectra of scattered radiation for 800 nm 150 fs pulses (**a**) and 400 nm 120 fs pulses (**b**) broadened due to self-phase modulation in similar conditions (100 J/cm^2)

wavelength λ, which makes the level of scattering for IR and UV pulses so different, even for equal electron concentrations:

$$\frac{\Delta n_{IR}}{\Delta n_{UV}} = \frac{N_{IR}\lambda_{IR}^2}{N_{UV}\lambda_{UV}^2} \, .$$

The result of numerical estimations, shown by dashed lines in Fig. 4.15, illustrates the predicted spectral trends. In these simplified calculations the variables in charge of propagation and self-induced modification of the refractive index were divided. Details of this approach are described in [32]. Taking account of the accuracy of ionization calculations using ADK model [41] and the assumptions, the obtained values demonstrate fairly good agreement with the experiment.

4.3.3 Optimization of Exposure Conditions to Eliminate the Scattering

In spite of fundamental nature of the scattering, its unwanted effect to accuracy of focusing can be significantly reduced or even eliminated in a broad range of incident energy, as it was already demonstrated for combination of 400 nm incident radiation and the helium ambient gas. It is shown below how the level of scattering can be reduced taking account of the pulsewidth dependences, focusing conditions and the spatial profile of the beam.

4.3.3.1 Pulsewidth

Fast ionization of the ambient gas was shown to be the main factor contribution to the scattering. Assuming the effect of oscillating electric field as a main ionization mechanism [41, 42], the corresponding rates are exponentially dependent on the field value and should slow down at increased duration of laser pulses. This approach cannot be applied to ablation of dielectrics calling for sufficient intensity of radiation due to the highly non-linear initial stage of free electron formation via multiphoton transitions across the band gap [43]. These electrons should be created fast enough for the following portion of the laser pulse had chances to heat them effectively through the induced intraband absorption [44, 45], which results finally in perfectly localized small scale ablation [28]. Ablation of metals is an easier issue; free electrons are already present in the conduction band. At the same time, relaxation of energy acquired by electrons from laser radiation is known to take several picoseconds due to weak electron-phonon coupling [46–48].

Indeed, plots of scattered energy in Fig. 4.18 demonstrate much higher thresholds in all the gases compared to Fig. 4.15. The same way, using of visible pulses enhances the situation even more by extending the range ensured from scattering beyond 150 J/cm^2. Benefits of the shorter wavelength for productivity of micro-drilling in steel by the short picoseconds pulses are illustrated by Fig. 4.19. An additional advantage of visible light in this case is related to smaller reflection from metal surfaces. Application of short picoseconds pulses (5–12 ps) to micromachining of metals was extensively investigated aiming at high accuracy and improved productivity compared to femtosecond pulses [30]. As a result, using of less expensive, simpler and generally more reliable picoseconds lasers was broadly recommended for processing of metals.

4.3.3.2 Beam Profile and Focusing

Deviation of the scattered beam from axial direction is obviously driven by the radial gradient of the induced phase, which is given by product of Δn and the length of the ionized domain. This way, characteristics of scattering should be sensitive to the initial beam profile and focusing conditions.

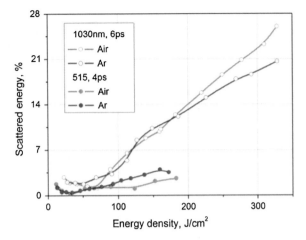

Fig. 4.18 Scattering of short picosecond pulses in gases: effect of the incident wavelength

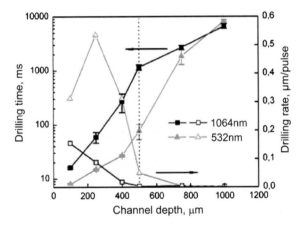

Fig. 4.19 Effect of wavelength on productivity of drilling by 10 ps pulses in steel

Gaussian and hat-top beams are typically used for micromachining. Comparison of scattering induced by these profiles is shown in Fig. 4.20 for equal beam waist diameters of 25 μm. The plots nearly coincide in the energy range below 30 J/cm² then diverge demonstrating higher scattered energies for the Gaussian profile, which is explained by a combination of factors. One is attributed to shifting of the scattering threshold domain in the direction to the focusing lens, which happens at growth of the incident energy. This shift takes place obviously for both beams: for the Gaussian and for the hat-top. The difference makes transformation of the focused profile along the axis. In the vicinity of the beam waist, both focused beams keep similar bell-like shape which results in comparable scattering characteristics. When, at the growing pulsed energy, scattering threshold conditions are met at a distance exceeding ∼3 mm

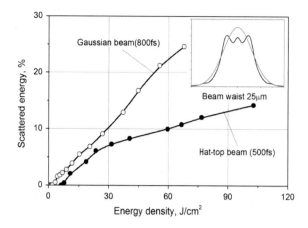

Fig. 4.20 Scattered energy for the Gaussian and the hat-top beams focused into the same focal spot. Profiles of the beams (Gaussian and flat-top) are shown 3 mm in front of the beam waist

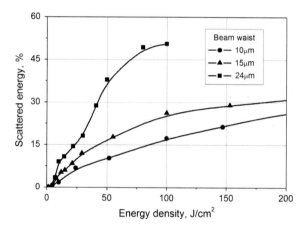

Fig. 4.21 Scattered energy with respect to the beam wais diameter (hat-top beams, 60 fs, 800 nm)

from the focal plane, which happens here at $\sim$30 J/cm^2, the profiles start significantly differ, which is also shown in Fig. 4.20. The non-Gaussian beam becomes flattened. The energy inside the profile is redistributed sidewise reducing intensity of radiation near the axis. So, the scattering threshold conditions are achieved here at higher incident energy compared to the Gaussian beam, known to maintain its shape at focusing.

The similar effect is observed at variation of the initial diameter of the hat-top beam. The scattered energies in Fig. 4.21 are noticeably less for sharper focusing or for the larger initial beams. The bell-like shape, in the last case, disappears closer to the focal plane i.e. at the lower incident energy. Shifting of the ionization and scattering domain was observed experimentally using the mentioned above interferometry technique [32]. We see, that using of the hat-top beams of larger diameters, typically shaped by installing apertures in the beam path, provide a simple approach to reducing the scattering. Going this way, one should take special care of aberrations of the focusing optics. Objectives and lenses implying significant level of aberrations were shown to increase additionally the scattered energy by more than 10–15 %.

4.4 Discussion and Conclusions

Let us briefly summarize similar and different features of the two types of plasma described in the chapter. In both cases, it is formed by the front part of the laser pulse and located noticeably apart from the target surface. So, the following part of the pulse propagates through the plasma, which modifies its energy, the spatial profile, pulse width and the spectrum. In the case of femtosecond and short picosecond pulses, the plasma density profile does not have time to expand. It stays within the beam path, and the induced gradient of Δn results in scattering or rather defocusing of the beam. In the case of longer nanosecond and sub-nanosecond pulses, it happens in two steps and only in the vicinity of the target. The preceding pulses modify the ambient, seeding the gas by plasma ignition centers in the form of long-living nanoparticles. Multiple "grains" of microplasma, ignited by the following pulses, have chance to expand and merge during the longer laser pulses [4], energy of which is mostly reflected and absorbed in the long and broad plasma cloud. In the intermediate pulse-width domain, between 10 and 300 ps, the plasma "grains" stay apart from each other having the typical size of 1.5–3.0 μm which should result in scattering, diagram of which varies chaotically from pulse to pulse.

Apparent analogy to the described scattering effect of free electron plasma exists in transparent solids, dielectrics and semiconductors, where comparable concentrations of carriers ($> 10^{19}$ cm^{-3}) can be easily reached at focusing of ultrashort pulses in the bulk of the materials [49, 50]. One should expect more pronounced manifestations of Kerr effect in this case due to higher nonlinearities of solids. Dominating mechanisms of fast ionization here are multiphoton transitions through the band gap followed by heating of free electrons, ending up, in some beneficial cases, in their impact multiplication, taking less time due to higher frequency of electron collisions with atoms [51].

Tailoring parameters of laser radiation can reduce distortions in the beams and the energy losses in plasma. Shorter incident wavelengths are beneficial by two reasons. On the one hand, they are shown to generate less dense plasmas, which is attributed to lower rates of high order ions formation, due to smaller ponderomotive energy of free electrons in the oscillating electric field. On the other hand, the plasma of given density induces smaller modifications to the refractive index of gas and, this way, smaller reflectivity and refractivity of the ionized area. Space and time shaping of laser beams and pulses is also of help. Using of hat-top profiles was demonstrated to reduce the length of the ionization domain near the beams waist. Following the same approach, generation of ultrashort pulses rectangular in time would allow to reduce the peak intensity and to increase the energy threshold of scattering. At the same time, constant intensity level near the pulse maximum could result in more simple and regular profiles of scattered radiation. Optimization of the ambient was shown to reduce scattering and screening drastically. The ambient gas can be cleaned out using electric properties of the residing particles or significantly rarefied by means of laser heating near the exposed surface. One can imagine a similar rarefaction effect in the beam path of femtosecond pulses generated at the typical repetition rates of $\sim 100\,\mathrm{MHz}$. Choice of the ambient gas is also of importance. Helium is the best due to the beneficial combination of two parameters: the highest ionization potential and the smallest number of electrons per atom (hydrogen does not count because of the explosion hazard). Using of other noble gases does not demonstrate advantages due to effective formation of high order ions.

References

1. A.I. Barchukov, F.V. Bunkin, V.I. Konov, A.A. Lyubin, JETP **39**, 469 (1974)
2. S.M. Klimentov, T.V. Kononenko, P.A. Pivovarov, S.V. Garnov, V.I. Konov, A.M. Prokhorov, D. Braitling, F. Dausinger, Quantum Electron. **31**, 378 (2001)
3. S.M. Klimentov, T.V. Kononenko, S.V. Garnov, V.I. Konov, P.A. Pivovarov, F. Dausinger, Bull. Russ. Acad. Sci: Phys. **65**, (2001)
4. S.M. Klimentov, S.V. Garnov, V.I. Konov, T.V. Kononenko, P.A. Pivovarov, O.G. Tsarkova, D. Breitling, F. Dausinger, Phys. Wave Phenom. **15**, 1 (2007)
5. S.M. Klimentov, T.V. Kononenko, P.A. Pivovarov, V.I. Konov, A.M. Prokhorov, D. Breitling, F. Dausinger, Quantum Electron. **32**, 433 (2002)
6. S.V. Garnov, V.I. Konov, A.A. Malyutin, O.G. Tsarkova, I.S. Yatskovsky, F. Dausinger, Laser Phys. **13**, 386 (2003)
7. R.E. Russo, X.L. Mao, H.C. Liu et al., Appl. Phys. A **69**, S887 (1999)
8. M. Schenk, M. Krüger, P. Hommelhoff, Phys. Rev. Lett. **105**, 257601 (2010)
9. S.E. Irvine, A. Dechant, A.Y. Elezzabi, Phys. Rev. Lett. **93**, 184801 (2004)
10. H.A. Sumeruk, S. Kneip et al., Phys. Plasmas **14**, 062704 (2007)
11. D. Breitling, A. Ruf, F. Dausinger, Proc. SPIE **5339**, 49 (2004)
12. S.M. Klimentov, V.I. Konov, P.A. Pivovarov, S.V. Garnov, T.V. Kononenko, F. Dausinger, Proc. SPIE. **6606**, 0H1 (2007)
13. S.M. Klimentov, P.A. Pivovarov, V.I. Konov, D.S. Klimentov, F. Dausinger, Laser Phys. **18**, 774 (2008)
14. S.M. Klimentov, P.A. Pivovarov, V.I. Konov, D. Breitling, F. Dausinger, Quantum Electron. **34**, 537 (2004)
15. P. Berger, D. Breitling, F. Dausinger, Ch. Föhl, H. Hügel, S. Klimentov, T. Kononenko, V. Konov, German Patent DE 102 03 452 B4 (2007)
16. F. Colao, V. Lazic, R. Fantoni, S. Pershin, Spectrochim. Acta B **57**, 1167 (2002)
17. KhS Kestenboim, G.S. Roslyakov, L.A. Chudov, *Tochechniy vzryv (Point Explos.)* (Nauka, Moscow, 1974). [in Russian]
18. A.M. Prokhorov, V.I. Konov, I. Ursu, I.N. Mikheilesku, *Interaction of Laser Radiation with Metals* (Nauka, Moscow, 1988). [in Russian]
19. S.M. Klimentov, S.V. Garnov, T.V. Kononenko, V.I. Konov, P.A. Pivovarov, F. Dausinger, Appl. Phys. A **69**, S633 (1999)
20. T.T. Basiev, S.V. Garnov, C.M. Klimentov, P.A. Pivovarov, A.V. Gavrilov, S.N. Smetanin, S.A. Slolokhin, A.V. Ftdin, Quantum Electron. **37**, 956 (2007)

21. X.D. Wang, X. Yuan, S.L. Wang, J.S. Liu, A. Michalowski, F. Dausinger, in *Advanced Design and Manufacture to Gain a Competitive Edge*, 2008, p. 759
22. I.A. Bufetov, S.B. Kravtsov, V.B. Fyodorov, Quantum Electron. **26**, 520 (1996)
23. V.P. Ageev, A.I. Barchukov, F.V. Bunkin, V.I. Konov, S.B. Puzhaev, A.S. Silenok, N.I. Chapliev, Sov. J. Quantum Electron. **9**, 43 (1979)
24. S.M. Klimentov, V.I. Konov, P.A. Pivovarov, S.V. Garnov, T.V. Kononenko, F. Dausinger, Proc. SPIE. **6606**, 66060H (2007)
25. N.M. Bulgakova, V.P. Zhukov et al., Appl. Phys. A **92**, 883 (2008)
26. T. Riesbeck, Laser Phys. Lett. **5**, 240 (2008)
27. R.S. Taylor, C. Hnatovsky, E. Simova, D.M. Rayner, V.R. Bhardwaj, P.B. Corkum, Opt. Lett. **28**, 1043 (2003)
28. A.P. Joglekar, H. Liu, G.J. Spooner, E. Meyhöfer, G. Mourou, A.J. Hunt, Appl. Phys. B **77**, 25 (2003)
29. J. Krüger, W. Kautek, Laser Phys. **9**, 30 (1999)
30. F. Dausinger, F. Lichtner, H. Lubatschowski (eds.), *Femtosecond Technology for Technical and Medical Applications* (Springer, Heidelberg, 2004)
31. S. Klimentov, P. Pivovarov, V. Konov, D. Walter, M. Kraus, F. Dausinger, Laser Phys. **19**, 1282 (2009)
32. S.M. Klimentov, P.A. Pivovarov, N. Fedorov, S. Guizard, F. Dausinger, V.I. Konov, Appl. Phys. B **105**, 495 (2011)
33. A. Braun, G. Korn, X. Liu, D. Du, J. Squier, G. Mourou, Opt. Lett. **20**, 73 (1995)
34. V.P. Kandidov, O.G. Kosareva, A.A. Koltun, Quant. Electron. **33**, 69 (2003)
35. S.C. Rae, Opt. Commun. **104**, 330 (1994)
36. A. Couairon, G. Mechain, S. Tzorzakis et al., Opt. Commun. **222**, 177 (2003)
37. H. Koch, Ch. Hatting, H. Larsen et al., J. Chem. Phys. **111**, 10108 (1999)
38. E. Inbar, A. Arie, Appl. Phys. **B70**, 849 (2000)
39. V.V. Bukin, S.V. Garnov, A.A. Malyutin, V.V. Strelkov, Quant. Electron. **37**, 961 (2007)
40. Y.R. Shen, *Principles of Nonlinear Optics* (Wiley-Interscience, New York, 1984)
41. N.V. Ammosov, N.B. Delone, V.P. Krainov, J. Exp. Theor. Phys. **94**, 2008 (1986)
42. V. Popov, Phys.-Usp. **47**, 855 (2004)
43. F. Quéré, S. Guizard, Ph Martin, Europhys. Lett. **56**, 138 (2001)
44. B. Rethfeld, Phys. Rev. Lett. **92**, 187401 (2004)
45. H. Bachau, A.N. Belsky, P. Martin, A.N. Vasil'ev, B.N. Yatsenko, Phys. Rev. B **74**, 235215 (2006)
46. B. Rethfeld, K. Sokolowski-Tinten, D. von der Linde, S.I. Anisimov, Appl. Phys. A **79**, 767 (2004)
47. M. Bonn, D. Denzler, S. Funk, M. Wolf et al., Phys. Rev. B **61**, 1101 (2000)
48. M. Ligges, I. Rajkovic, P. Zhou, O. Posth, C. Hassel, G. Dumpich, D. von der Linde, Appl. Phys. Lett. **94**, 101910 (2009)
49. V.V. Kononenko, E.V. Zavedeev, M.I. Latushko, V.I. Konov, Laser Phys. Lett. **10**, 036003 (2013)
50. V.V. Kononenko, V.V. Konov, E.M. Dianov, Opt. Lett. **37**, 3369 (2012)
51. A. Mouskeftaras, S. Guizard, N. Fedorov, S. Klimentov, Appl. Phys. A **110**, 709 (2013)

Part II
Nanoparticles Related Technologies and Problems

Chapter 5
Laser Generation and Printing of Nanoparticles

A. Barchanski, A. B. Evlyukhin, A. Koroleva, C. Reinhardt, C. L. Sajti, U. Zywietz and Boris N. Chichkov

Abstract Different laser-based methods for the fabrication of nanoparticles and ordered nanoparticle structures, including possibilities for their functionalization and replication in polymeric materials, are discussed. Nanoparticles made from noble metals, supporting collective electron oscillations, and low absorbing dielectric nanoparticles, having large permittivity values, can both be resonantly excited by external electromagnetic fields which make them attractive for biophotonic and sensing applications. For applications in biomedicine especially polymeric nanoparticles, as drug delivery systems, are very important. Fabrication of all these types of nanoparticles can be realized with laser technologies, which are briefly reviewed in this chapter.

5.1 Introduction

In this chapter, we discuss different laser-based methods for the fabrication of nanoparticles and ordered nanoparticle structures, including possibilities for their functionalization and replication in polymeric materials. Research fields involving nanoparticles and their applications in photonics, biomedicine, and sensorics are rapidly growing. At present, "nanoparticles" search in Google provides approximately 8 million references.

Nanoparticles made from noble metals support collective electron oscillations, which can be resonantly excited by external electromagnetic fields and are known as localized surface plasmon resonances. The frequency of these resonances strongly depends on the size, shape, and environment of the nanoparticles, making them very

A. Barchanski · A. B. Evlyukhin · A. Koroleva · C. Reinhardt · C. L. Sajti · U. Zywietz · B. N. Chichkov (✉)
Nanotechnology Department, Laser Zentrum Hannover e.V, Hollerithallee 8, 30419 Hannover, Germany
e-mail: b.chichkov@lzh.de

V. P. Veiko and V. I. Konov (eds.), *Fundamentals of Laser-Assisted Micro- and Nanotechnologies*, Springer Series in Materials Science 195, DOI: 10.1007/978-3-319-05987-7_5, © Springer International Publishing Switzerland 2014

attractive for different practical applications, for example real-time sensor technologies. Alternatively, low absorbing dielectric nanoparticles having large permittivity values can also resonantly interact with light and support strong Mie resonances. Such particles represent dielectric resonators or antennas, which can trap external electromagnetic fields. It is well-known from Mie theory that the first and second lowest frequency resonances of dielectric spheres correspond to the magnetic and electric dipole contributions. Scattering diagrams of the resonant dielectric nanoparticles are determined by interference of electromagnetic waves generated by the electric and magnetic dipoles, which can result in directional back or forward light scattering.

For applications in biomedicine especially polymeric nanoparticles, as drug delivery systems, are very important. Fabrication of all these types of nanoparticles can be realized with laser technologies, which are briefly reviewed below.

5.2 Laser Printing of Nanoparticles and Nanoparticle Arrays

5.2.1 Laser Printing of Nanoparticles

Femtosecond lasers have opened new possibilities for the generation of spherical nanoparticles with predefined sizes and positions, based on laser-induced melting, fluid dynamics, and molten material transfer. In this section, a printing process for the controllable generation of noble metal and semiconductor nanoparticles by laser-induced transfer of liquid material droplets is described. The liquid droplets are captured on a receiver substrate, e.g. glass, where they obtain a near-spherical shape after solidification, due to the surface tension of molten material. Resonant optical properties of the nanoparticles generated by this method can be applied for the development of novel sensor concepts.

During the laser printing of metallic nanoparticles, a thin material layer deposited on a glass substrate, acting as a donor substrate, is irradiated by single femtosecond laser pulses. Due to the ultrashort pulse duration, the pulse energy is absorbed by electrons in the metallic film and is slowly transferred to the lattice. Processes which are important for structural changes of the laser-excited solid, such as carrier–phonon scattering and thermal diffusion, occur after the laser pulse absorption [1]. Therefore, only the material within the laser beam spot size on the target is affected and a highly controlled and reproducible material transformation is possible. Due to the solid-liquid phase transition, the irradiated material changes its density and its volume. This effect induces strong temperature and pressure gradients, leading to a complex fluid dynamics [2]. Noble metals, such as gold, silver, and copper expand during melting. The liquid metal then forms a protrusion with a back-jet structure in its center [3, 4]. One example of the irradiation of a thin gold film by a single tightly focused femtosecond laser pulse is shown in the SEM image in Fig. 5.1 [4, 5]. When the amount of melted material increases, the surface tension induces the material to form spherical droplets and a spherical nanodroplet appears on top of the

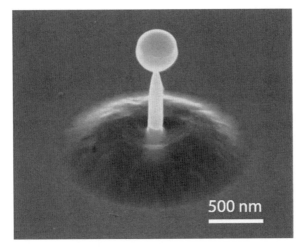

Fig. 5.1 Side-view SEM image of a typical back-jet structure generated on a 60-nm gold film after irradiation with a single 30 fs laser pulse

back jet. With higher laser pulse energy the nanodroplet starts to separate from the back-jet and is ejected upwards [6]. By placing the receiver substrate on top of the thin gold layer the ejected nanoparticles can be collected [5]. A schematic illustration of this laser printing process is shown in Fig. 5.2. Each laser pulse is producing one nanoparticle, which is deposited on the receiver substrate. By focusing laser pulses on different positions on the donor substrate it is possible to generate large nanoparticle arrays with precisely arranged nanoparticles [7]. Furthermore, the printing process allows generating nanoparticles with reproducible diameters by using the same laser pulse energy for every nanodroplet ejection process [8]. The right part of Fig. 5.2 shows SEM images of gold nanoparticles which are deposited on a receiver substrate by the described process. The diameter of these nanoparticles can controllably be influenced by the experimental parameters such as the layer thickness of the material film, laser pulse energy, and focusing conditions of the laser pulse. To demonstrate the precise deposition of nanoparticles on the receiver substrate and its capabilities, an example is given by the SEM and dark field microscopic images in Fig. 5.3. Here, the nanoparticles with a diameter of 400 nm are deposited at a distance of 5 μm, forming the word "NANO". The observed strong light scattering is due to multipole plasmon resonances in the visible spectral range [8].

This process, however, cannot be used with materials (e.g. silicon) reducing their volume during melting. In this case, no back-jet is generated and the ejection of molten material is different, as will be discussed below.

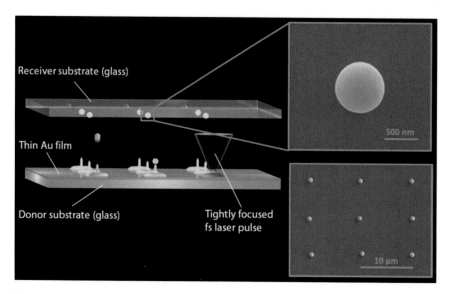

Fig. 5.2 Schematic illustration of femtosecond laser printing of nanoparticles. A thin gold layer was used as a target material to transfer spherical nanoparticles towards the transparent glass receiver substrate

Fig. 5.3 SEM and dark field microscopic images of precisely deposited gold nanoparticles

5.2.2 Laser Fabrication of Large-Scale Nanoparticle Arrays

For sensing applications of nanoparticle structures and arrays, the main technological drawback is the high cost and low-throughput of existing fabrication methods. Most of the experimentally studied structures are fabricated by electron or ion beam lithography, which are not suitable for large-scale and low-cost production.

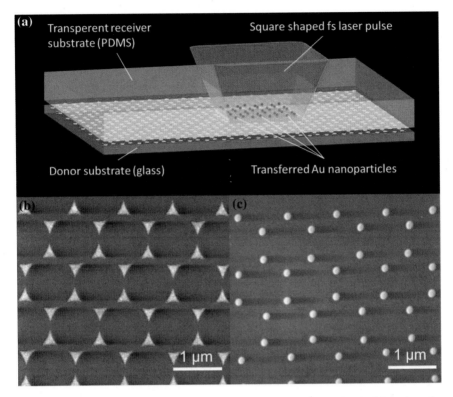

Fig. 5.4 **a** Schematic representation of laser induced transfer procedure for the fabrication of an Au nanoparticle array. **b** SEM image of hexagonal arrays of triangular prism structures fabricated by the nanosphere lithography. **c** SEM image of a nanoparticle array fabricated by a single laser pulse

This stimulates the development of novel high-throughput methods for the fabrication of nanoparticle structures with particular optical properties, for example, biosensors or metamaterials [9, 10]. In this section, a method for high-speed low-cost fabrication of large-scale nanoparticle arrays is introduced [11]. This method is based on a combination of nanosphere lithography and femtosecond laser-induced transfer. The metallic film, e.g. gold, on the donor substrate is pre-structured by nanosphere lithography. Both, interparticle distance and particle size can be independently controlled, which allows engineering of optical properties of such arrays. A scheme of high-speed fabrication of large-scale nanoparticle arrays based on a combination of the nanosphere lithography and laser-induced transfer is shown in Fig. 5.4. The final structure consists of hexagonal arrays of spherical gold nanoparticles partially embedded into a polymeric substrate. First, submicron sized spheres of silica are deposited onto a glass plate forming a large area closely packed monolayer. Subsequently, gold is evaporated onto the substrate and the silica spheres are removed, leaving triangular metallic islands on the donor substrate. After laser irradiation, the triangles are melted and forming spherical droplets. This process is accompanied by a raise in the center-of-mass, resulting in an upward acceleration. The particles are finally captured in a receiver substrate consisting of a thin layer of PDMS (poly-dimethoxysiloxane) [11].

In this metallic nanoparticle array, a collective plasmonic mode with diffractive coupling between the nanoparticles can be excited. The excitation of this mode leads to the appearance of a narrow (fwhm = 14 nm) Fano-type resonance dip in the optical transmission spectra. The spectral position of this dip is sensitive to the refractive index changes of the local environment, allowing the realization of novel sensing concepts [11].

In principle, this process can also be applied with other materials, for example, silicon. The magnetic Mie resonances of Si nanoparticles would allow the realization of novel type sensors for magnetic fields at optical frequencies. The optical properties of silicon nanoparticles are briefly discussed in the next section.

5.3 Resonant Electric and Magnetic Response of Silicon Nanoparticles

Metal nanoparticles are characterized by a strong resonant response to the electric field of light [12, 13], which appears due to excitation of collective electron oscillations known as localized surface plasmons. The main drawback of using plasmonic particles in the visible spectral range is their intrinsic Ohmic loss, which strongly affects their overall performance and limits practical applications. One of the possible ways to avoid such limitation and still to have similar resonant properties is to use high-refractive index dielectric nanoparticles [14, 15]. Nanoparticles of crystalline silicon (Si), the basic material of silicon photonics, provide a promising choice [16, 17]. Scattering properties of Si nanostructures with Mie resonances are attracting growing interest due to their potential applications for solar cells [18, 19], for tuning optical responses of nanostructured systems [20, 21], and for field-enhanced surface spectroscopy [22]. Spherical silicon nanoparticles with sizes of few hundred nanometers exhibit unique optical properties due to their strong electric and magnetic dipole responses in the visible spectral range [23, 24]. The spectral position of these resonances can be tuned throughout the whole visible spectral range from violet to red by changing the nanoparticle size in the range of 100–200 nm. Experimentally measured scattering intensities of spherical Si nanoparticles with diameters between 100 and 150 nm are shown in Fig. 5.5. For every nanoparticle in this range, two resonance peaks in the visible spectral region are present. These two resonance peaks have their origin in the excitation of magnetic and electric dipole resonances. As it can be seen in Fig. 5.5, resonances are red-shifted for increased nanoparticle radius.

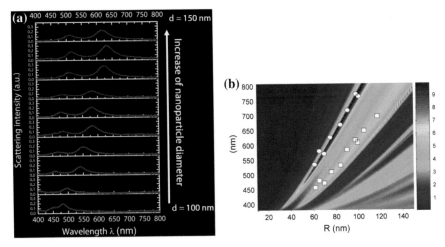

Fig. 5.5 **a** Experimentally measured scattering intensities of Si nanoparticles with different diameters. **b** Extinction efficiencies of Si nanoparticles with the radius R calculated by Mie theory (along vertical axis measured wavelengths of scattered light). *Circles* and *squares* correspond to magnetic and electric dipole resonances, respectively, of fabricated Si nanoparticles measured experimentally

5.4 Generation of Silicon Nanoparticles from Bulk Silicon

Silicon nanoparticles, as they have been used in the measurements discussed in the previous section, can be generated by focussing single femtosecond laser pulses onto the surface of a silicon wafer. However, as already pointed out, silicon material reduces its volume during melting, leading to a depression on the material surface. Melted silicon forms a smooth toroidal ring around this depression. When the amount of material increases the formerly smooth ring gets instable and the surface tension contracts liquid material into a number of small spherical droplets. These droplets are ejected from the molten zone and can be captured on a receiver glass substrate. Since the contraction of the molten material into droplets is a statistical process, the sizes and the positions of the ejected particles are centered around some average value, but cannot be controlled precisely.

A schematic illustration of this process is given in Fig. 5.6. In this case, every single femtosecond laser pulse generates a small group of silicon nanoparticles with sizes of 100–300 nm in diameter. An example of Si nanoparticles, which are generated by this method, is shown in the darkfield microscopic image in Fig. 5.7. The silicon nanoparticles deposited on the glass receiver substrate scatter white light in different colors, depending on the sizes of these particles. As already shown in the previous section, the reason for these colors is in the strong electric and magnetic Mie resonances depending on the particles size and located in the visible spectral range.

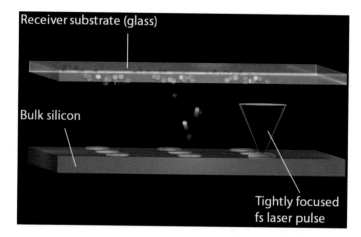

Fig. 5.6 Schematic illustration of femtosecond laser printing of silicon nanoparticles. A silicon wafer is used as a target material to transfer spherical nanoparticles towards the transparent glass receiver substrate

Fig. 5.7 Dark field microscopic image of silicon
nanoparticles deposited on
a glass receiver substrate

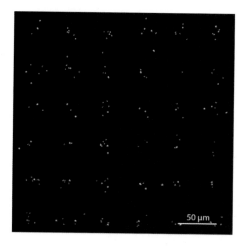

As it can be seen in Fig. 5.7, the nanoparticles are grouped around a position above the ejection point. The number of particles per group varies from 1 to 7. The contraction of the molten material into spherical nanodroplets is the driving force transferring particles onto the receiver substrate. This process might in future enable formation of Si nanodroplets from pre-structured silicon films, which will allow achieving better control over their sizes and positions.

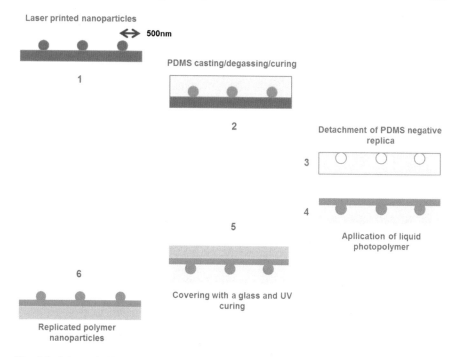

Fig. 5.8 Schematic illustration of soft-lithography replication of laser generated nanoparticles: *1* laser transferred gold nanoparticles on the glass substrate are *2* casted with liquid PDMS; after removal of air gaps in the vacuum chamber and curing *3* PDMS mold is detached from the master substrate; *4* Liquid photomonomer is coated onto the PDMS mold; *5* it is covered with a glass coverslip and cured with UV light

5.5 Microreplication of Laser-Transferred Gold Nanoparticles/Nanomolding

Laser printed nanoparticles can be reproduced in a polymer by applying the so-called soft lithography technique (see Fig. 5.8). This procedure represents a non-photolithographic method and is based on replica molding of micro- and nanostructures. It provides a fast, effective and low-cost strategy for the manufacturing of complex nanostructures and 3-D features. In soft lithography, an elastomeric template (stamp) with patterned features on its surface is applied to fabricate replica patterns and structures on the target surface with the sizes ranging from tens of nanometers to hundreds of micrometers. Such elastomeric stamps with patterned relief structures are the basis of soft lithography [25].

The replication procedure involves the fabrication of negative poly-dimetylsiloxane (PDMS) molds of the original (master) structure, which are then used as templates to form structures in polymer (see Fig. 5.8). The elastomeric mold is manufactured by casting liquid PDMS onto a master structure consisting from an array of metal nanoparticles. After curing PDMS at room temperature or by heating, the PDMS mold can be peeled-off of the master structure.

PDMS forms a conformal contact with all surfaces and therefore, superiorly encloses convex structures and penetrates into cavities. Its elastomeric properties allow PDMS to be easily released from a master without damaging either the master or the mold itself. PDMS's thermal and chemical stability enables most polymers to be patterned. The material is optically transparent down to 300 nm what permits molding of photocurable polymers. The PDMS is not hygroscopic; it does not swell with humidity [26]. For replication of laser-generated gold nanoparticles we used photochemically curable resist NIL 6000.2 (Microresist Technology). The unique features of this polymer, like low viscosity, non-harmful solvents in its composition, and room temperature conditioning enable the fabrication of smallest feature sizes down to 50 nm with a very low residual layer thickness <10 nm. Moreover, NIL 6000.2 has an excellent film quality after curing. The replication results are shown in Fig. 5.9.

Even though in general PDMS is chemically inert, it readily swells in a number of nonpolar organic solvents such as toluene and hexane [26], which can limit the application of such molds for polymers containing nonpolar organic solvents. Rolland et al. applied photocurable perfluoropolyether (PFPE) for mold fabrication. PFPE-based molds are both non-wetting and non-swelling in contact with both inorganic and organic materials [27]. Such physicochemical properties of the mold material are essential for repetitive fabrication of nanoobjects using the same mold, which is essential in industrial fabrication. The same group has developed a general technique of Particle Replication In Nonwetting Templates (PRINT) for the fabrication of monodispersed particles with simultaneous control over their structure and function. Using highly fluorinated PFPE surfaces that are non-wetting to organic materials, they have fabricated isolated objects with superior shape and composition control without harsh processing steps. Although the PRINT technique is similar to the soft lithography methods, it is unique because it can produce isolated free standing particles instead of particles anchored to the film [28].

Micro- and nanospheres play important role in many applications, such as drug delivery systems [29], optical materials [30], cosmetics [31], chemical and biological diagnostics [32] and other biomedical fields [33]. Applying soft lithography technique, nanoparticles can be fabricated also from synthetic hydrogels. These hydrogel nanoparticles, usually called "nanogels", are very promising as drug-delivery carriers [34]. The nano-replication soft lithography and PRINT methods enable strict control over the particle size, shape, composition and permit the loading of delicate units, including pharmaceutical drugs and biomacromolecules. For example, monodisperse 200 nm PEG-based swellable particles were fabricated with the PRINT method by UV-induced copolymerization of several vinyl monomers such as PEG triacrylate, PEG monomethyl ether monomethacrylate, and p-hydroxystyrene [35].

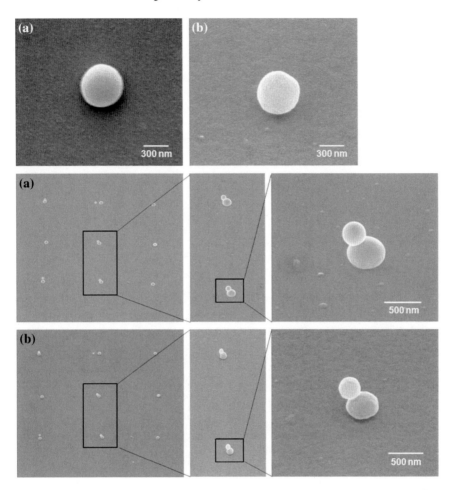

Fig. 5.9 Nanomolding of laser printed gold nanoparticles. **a** Original master structure; **b** nanoreplicated NIL 6000.2 nanoparticles

5.6 Laser-Based Synthesis of Nanoparticles and Surface Modified Nanoconjugates

5.6.1 Ultrapure Nanoparticles by Pulsed Laser Ablation in Liquids

Besides the direct deposition of single nanoparticles by laser-printing, ultrapure nanoparticles can also be produced using laser irradiation of solids in liquid media [36]. This green technology allows generation of high amounts of spherical nanoparticles by complex ablation phenomena depending on pulse duration, pulse energy,

Fig. 5.10 Laser-generated
ultrapure nano-colloids
obtained by picosecond laser
ablation in acetone

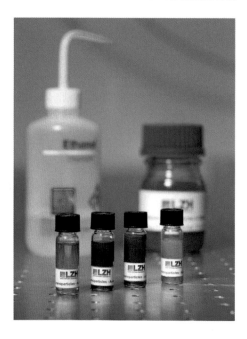

irradiation wavelength, and liquid-material combination. Compared to conventional manufacturing approaches of nanoparticles such as wet chemical synthesis, mechanical milling and grinding, laser ablation does not require usage of any additional material such as chemical precursors or reducing agents [37] and lacks any other form of contaminations arising from application of mechanical or abrasive components. Therefore, laser ablation in liquid environments attracts more and more attention, enabling the generation of nanoparticle colloids of a great variety of materials with outstanding purity, as illustrated in Fig. 5.10. Taking benefits of the simplicity and flexibility of this technique, a vast diversity of functional materials have been already prepared including metal nanoparticles [38, 39], semiconductors [40, 41], ceramics, alloys [42] and highly photosensitive bioconjugates [43]. Working in dense environments, such as liquid media, results in less efficient and more complex ablation process than laser ablation in air or under vacuum conditions. In the femtosecond and picosecond time regimes [37], nanoparticle generation is mainly limited by relatively low pulse energy of currently available laser sources. Employing nanosecond lasers reveals that the ablation efficiency can be further increased under rigorous control of laser and process parameters. Note that generation of nanoparticles based on continuous wave laser ablation has also been reported [44].

Common for all liquid-based nanoparticle generation methods applying laser ablation is that high reproducibility was only achieved using axial separation of the incoming laser beam and the generated air bubbles. Usually, this is obtained by horizontal beam guidance, by transmitting the laser beam through an entrance media that allows accurate control of the liquid layer. This way also allows elimination of the liquid meniscus which considerably reduces inaccuracy in laser focusing. Besides, ablation efficiency strongly depends on the deposited laser energy into the target material involving linear and non-linear absorption regimes [45, 46]. The thickness of applied liquid layer is another major parameter determining and limiting nanoparticle generation due to absorption and scattering of laser radiation by previously ablated nanoparticles. Flow rate of the applied liquid media has only recently been considered to impact ablation efficiency due to removal of ablated particles and generated air bubbles from the ablation zone. Without liquid circulation, the ablated nanomaterials disperse into the entire liquid volume by slow diffusion and Brownian motion, hence after each ablation sequence a dense particle cloud is ejected, having a relatively long residual living time in front of the target leading to significant absorption of the subsequent laser beam. Therefore, considerable increase in ablation rate can be observed by increasing liquid flow rate from stationary to several hundreds of mL/min using the same laser parameters. Optimal flow rate however strongly depends on the laser fluence and pulse duration. Besides these evident parameters, recent studies demonstrated that the interpulse distance has a dramatic effect on the material removal rate, when laser ablation takes place in liquid environment [47]. Figure 5.11 (top) shows ablation rate of alumina as a function of interpulse distance for 4 mm liquid layer, 4.6 mJ pulse energy at 4 kHz repetition rate, and a focal spot size of 50 μm. The term interpulse distance defines the distance between two pulses, from one pulse center to the center of the neighboring pulse. As expected, material removal rate and nanoparticle productivity strongly depends on overlapping of spatially separated laser pulses. By adjusting position of separated laser pulses using a galvanometric laser scanner, optimal interpulse distance of 125 μm for 4 kHz repetition rate has been identified. Two competitive effects influence the nanoparticle generation rate. First, for strongly overlapping laser pulses, the pulse interaction with previously ablated nanoparticles and previously generated cavitation bubbles are the most important mechanisms, preventing higher material ablation. Referring to the literature, laser ablation in liquids using Nd:YAG nanosecond laser irradiation at 36 J/cm^2 laser fluence generates cavitation gas bubbles which last around 300 μs. This cavitation bubble contains primary nanoparticles of extreme high local concentration which can scatter, reflect or absorb subsequent laser pulses. This absorption and scattering effects become less important with increasing interpulse distance. On the other hand, increasing interpulse distance also results in significant temperature variations and temperature gradients in the solid target. Above the optimal pulse overlap, the regime of thermally isolated ablation areas is reached, that negatively influences the achievable material removal. This hypothesis of cavitation bubble and heat accumulation affected laser ablation has been clearly confirmed by comparison of nanoparticle generation in gas and liquid media [48]. By changing the repetition rate with equal interpulse distances at constant pulse energy of 3.3 mJ and pulse duration of 40 ns FWHM, experimental results shown in Fig. 5.11 (bottom) are obtained. Investigations revealed a strong decrease in the ablation rate when the repetition rate was increased from 0.5 to 20 kHz. An enhancement of almost two orders of magnitude in the ablation rate was identified by decreasing the laser frequency from 20

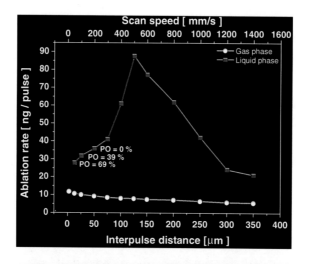

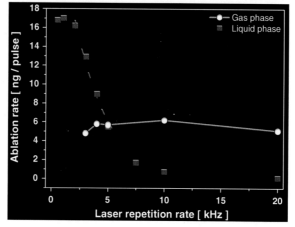

Fig. 5.11 Comparison of ablation rate in gas and liquid phase varying interpulse distance for 40 ns laser pulses with 4 kHz repetition rate and constant pulse energy (*top*) and as a function of the pulse repetition rate using 125 μm fixed interpulse distance (*bottom*)

to 0.5 kHz resulting in 0.26 ng/pulse and 17 ng/pulse, respectively. The maximum ablation rate of 17 ng/pulse was achieved at 2 kHz repetition rate. Nanosecond laser ablation initiates a complex sequence of events occurring both during and after the laser pulse. Nanoparticle formation starts at the nanosecond to sub-microsecond time scales during plasma plume expansion. Due to rapid cooling, which can occur faster than condensation, exceptionally high saturation ratios can be achieved. Therefore, even at high laser fluences relatively small nanoparticles can be generated by laser ablation in liquids. For nanoparticle generation, the subsequent events are even more critical, since on the time scale of several microseconds a cavitation bubble is growing on the target surface due to the local heating of solvent in the vicinity of ablated spot. In the highly confined region of the cavitation bubble, large amount of primary nanoparticles are formed and trapped. They are ejected when the bubble collapses on the time scale of 200–300 µs (or even later depending on the applied pulse energy). During the lifetime of the cavitation bubble, the target cannot be ablated efficiently due to light scattering. As shown in Fig. 5.11 (bottom), maximum nanoparticle generation rate is obtained at the laser repetition rate of 2 kHz, which corresponds to 500 µs time delay between laser pulses [47]. This particular time delay is in the order of the cavitation bubble lifetime. It is likely, that when the time delay between two laser pulses is high enough, the cavitation bubble collapses between subsequent laser pulses, resulting in a saturation plateau in the ablation rate. A similar behavior has been identified for picosecond laser ablation, however with shorter cavitation bubble lifetimes.

5.6.2 Surface-Functionalized Nano(Bio)Conjugates

Small gold and other plasmonic nanoparticles, characterized by enhanced resonant absorption and scattering properties, are particularly useful in numerous biomedical applications, including cell-targeted drug delivery [49], high resolution bioimaging [50], biomedical diagnostics and therapeutics, when conjugated with functional molecules such as DNA, RNA, oligonucleotides, peptides, drugs, etc. Pulsed laser ablation in liquids offers an alternative single step surface functionalization, allowing size-controlled generation of stable nanoparticle colloids with outstanding purity and novel surface chemistry, not possible by conventional manufacturing methods. Due to the presence of partial oxidation into high oxidation states, such as Au^+ and Au^{3+}, laser-generated gold nanoparticles act as electron acceptors [51]; hence, they are easily coordinated by molecules bearing electron donor moieties such as thiol, amine or carboxyl groups. A (bio)molecule, having a particular functional group, added to ablation media prior (in-situ conjugation) or after (ex-situ conjugation) the laser process will be chemically or physically bound to the surface. However, in-situ conjugation leads to higher conjugation efficiencies than ex-situ functionalization. Figure 5.12 shows absorption spectra of ligand-free gold nanoparticles and gold nanoconjugates generated by 100 µJ, 7 ps laser pulses at 2 kHz repetition rate in aqueous media and in 5 µM Cy5-tagged model peptide (TAT) derived from the human immunodeficiency virus type-1. High resolution scanning electron microscopy, in SEM and STEM modus, can reveal the presence of cohesive organic molecules around the nanoparticle core as shown in Fig. 5.13, whereas the ligand-free nanoparticles lack such organic shell.

Although ultrashort pulsed laser ablation presents a promising tool for in-situ bioconjugation, due to minimal thermal impact to ablated material and surrounding media when working at laser fluences close to the ablation threshold, it induces similar residual thermal effects as nanosecond lasers in the high fluence regime, including melting and photothermal reshaping of ablated particles. During laser ablation

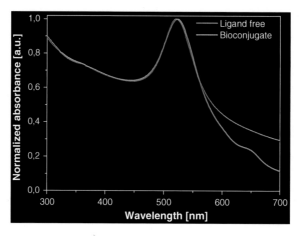

Fig. 5.12 Absorption spectra of ligand-free and bioconjugated gold nanoparticles generated by in-situ picosecond laser ablation and laser-induced conjugation using a fluorescence-tagged model peptide

(especially during ultrashort-pulsed ablation), the ejected material is in a chemically activated state of growth which allows to significantly influence the nanoparticle size by the amount of conjugative agents in solution. Reciprocal dependency of the primary nanoparticle size as a function of the concentration of active molecules was reported by several groups using the term growth-quenching. As described above, ablation efficiency and nanoparticle yield increase with the laser pulse energy, although at the same time, heat impact and the risk of molecule degradation also increase. In order to investigate whether the ablation process itself, or presence of nanoparticles trigger biomolecule degradation during in-situ conjugation of biomolecules to nanoparticles by femtosecond pulsed laser ablation, we ablated gold in the presence of a photo-sensitive model biomolecule (fluorescence-tagged single stranded oligonucleotide with $1.5 \, \mu M$ concentration) using various pulse energies in stationary solution and in biomolecule flow using the experimental setup schematically illustrated in Fig. 5.14 (top). It was clearly observed that femtosecond laser ablation in stationary biological media induces drastic degeneration of the sensitive molecular compartments by denaturing the fluorophore tag, while ablation in a biomolecule flow with the same laser parameters minimizes degradation [52]. The degree of degradation reduces considerably with higher flow rates. Since only the total residence time of ablated species in the ablation zone is varied by the liquid flow rate, it was assumed that biomolecule disintegration or laser/heat-induced DNA depurination by femtosecond laser irradiation is mainly induced by absorption and scattering of subsequent laser pulses on previously ablated, suspended nanoparticles/nanoparticle conjugates in the colloidal solution and is not triggered directly by the ablation itself.

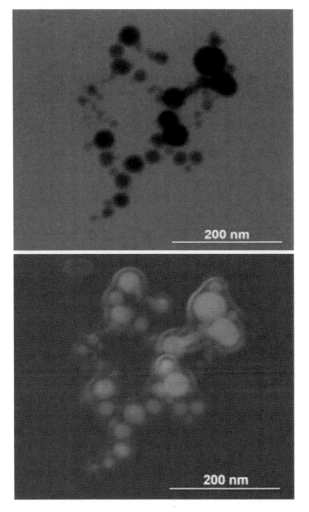

Fig. 5.13 STEM picture of the core gold particles of gold nanobioconjugates obtained by laser ablation (*top*). SEM picture of the identical nanoparticle revealing the presence of organic molecules around individual nanoparticles (*bottom*)

5.7 Novel Laser-Based Conjugation Concepts

Both ex-situ and in-situ laser-based bioconjugation techniques have major disadvantages The ex-situ method does not allow precise nanoparticle size control as colloidal species might have already been aggregated in the moment of ligand addition. In contrast, the in-situ method generates stable and size-controlled bioconjugates due to rapid size-quenching, but has limited productivity and induces photo-degradation

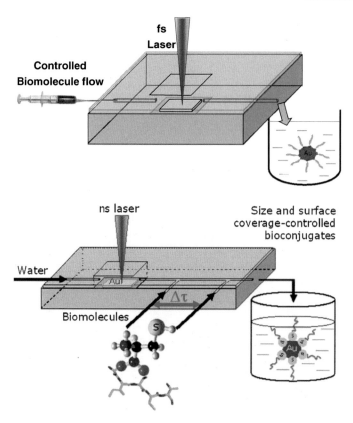

Fig. 5.14 Schematic illustration of the in-situ bioconjugation in liquid flow for femtosecond irradiation to minimize biodegradation (*top*) and the fast ex-situ bioconjugation with cascade conjugation
steps allowing multiple conjugation even by nanosecond laser ablation in liquid flow (*bottom*)

to biomolecules at high laser fluences. In order to overcome limitations of the
existing methodologies, a combination of both techniques (referred below as fast
ex-situ synthesis) represents a novel single-step method [53] for design of highly-
controlled biofunctionalized nanoparticle surfaces. In this method the sensitive molecules are not irradiated but promptly added to the formed nanoparticles, presented
schematically in Fig. 5.14 (bottom). Our investigations demonstrate that longer time
delays for biomolecule introduction (in the range of 200 ms to 120 s) induce exponential increase in bioconjugate sizes, varied from about 20 to 50 nm, under conditions
when gold nanoparticles are conjugated with fluorophore-labeled model peptides
using 6 mJ, 40 ns laser pulses at 3 kHz and 1,064 nm wavelength. The central position of the plasmon resonance band follows the size of nanoparticle bioconjugates. Conjugation efficiency and bioconjugate stability also correlates with the
observed size-variation. By identifying the amount of non-conjugated biomolecules,
it was found that the conjugation efficiency depends exponentially on the delay

time. At 200 ms delay time the smallest nanoparticle size is 20 nm, and the highest conjugation efficiency of 93 % is reached. The importance of delay time for the quality of synthesized colloids is thus obvious. Our results confirm that primary nanoparticles, synthesized by laser ablation in solution, continue to grow on the multisecond time scale, until achieving their final sizes.

5.8 Conclusion

We have focused here on laser printing, replication, generation, and conjugation of nanoparticles. Nanoparticles and ordered nanoparticle arrays can be fabricated from metal, semiconductor, dielectric and polymer materials. The described technologies are very simple and allow high throughput fabrication of nanomaterials for different applications.

References

1. S. Sundaram, E. Mazur, Inducing and probing non-thermal transitions in semiconductors using femtosecond laser pulses. Nat. Mater. **1**, 217–224 (2002)
2. F. Korte, J. Koch, B.N. Chichkov, Formation of microbumps and nanojets on gold targets by femtosecond laser pulses. Appl. Phys. A **79**, 879–881 (2004)
3. J. Koch, F. Korte, T. Bauer, C. Fallnich, A. Ostendorf, B.N. Chichkov, Nanotexturing of gold films by femtosecond laser-induced melt dynamics. Appl. Phys. A **81**, 325–328 (2005)
4. A.I. Kuznetsov, J. Koch, B.N. Chichkov, Nanostructuring of thin gold films by femtosecond lasers. Appl. Phys. A **94**, 221–230 (2009)
5. A.I. Kuznetsov, J. Koch, B.N. Chichkov, Laser-induced backward transfer of gold nanodroplets. Opt. Express **17**(2), 18820–18825 (2009)
6. A.I. Kuznetsov, C. Unger, J. Koch, B.N. Chichkov, Laser-induced jet formation and droplet ejection from thin metal films. Appl. Phys. A **106**, 479–487 (2012)
7. A.I. Kuznetsov, A.B. Evlyukhin, C. Reinhardt, A. Seidel, R. Kiyan, W. Cheng, A. Ovsianikov, B.N. Chichkov, Laser-induced transfer of metallic nanodroplets for plasmonics and metamaterial applications. J. Opt. Soc. Am. B **26**, B130–137 (2009)
8. A.B. Evlyukhin, A.I. Kuznetsov, S.M. Novikov, J. Beermann, C. Reinhardt, R. Kiyan, S.I. Bozhevolnyi, B.N. Chichkov, Optical properties of spherical gold mesoparticles. Appl. Phys. B **106**, 841–848 (2012)
9. S. Aksu, A.A. Yanik, R. Adato, A. Artar, M. Huang, H. Altug, High-throughput nanofabrication of infrared plasmonic nanoantenna arrays for vibrational nanospectroscopy. Nano Lett. **10**(7), 2511–2518 (2010)
10. J. Henzie, M.H. Lee, T.W. Odom, Multiscale patterning of plasmonic metamaterials. Nat. Nanotechnol. **2**(9), 549–554 (2007)
11. A.I. Kuznetsov, A.B. Evlyukhin, M.R. Gonc-alves, C. Reinhardt, A. Koroleva, M.L. Arnedillo, R. Kiyan, O. Marti, B.N. Chichkov, Laser fabrication of large-scale nanoparticle arrays for sensing applications. ACS Nano **5**(6), 4843–4849 (2011)
12. L. Novotny, N. van Hulst, Antennas for light. Nat. Photonics **5**(2), 83–90 (2011)
13. A. Alù, N. Engheta, Theory, modeling and features of optical nanoantennas. IEEE Trans. Antennas Propag. **6**, 1508–1517 (2013)

14. N. Liu, H. Guo, L. Fu, S. Kaiser, H. Schweizer, H. Giessen, Three-dimensional photonic metamaterials at optical frequencies. Nat. Materi. **7**(1), 31–37 (2007)
15. N. Liu, H. Liu, S. Zhu, H. Giessen, Stereometamaterials. Nat. Photonics **3**(3), 157–162 (2009)
16. A.B. Evlyukhin, C. Reinhardt, A. Seidel, B.S. Luk'yanchuk, B.N. Chichkov, Optical response features of Si-nanoparticle arrays. Phys. Rev. B **82**, 045404 (2010)
17. A.B. Evlyukhin, C. Reinhardt, B.N. Chichkov, Multipole light scattering by nonspherical nanoparticles in the discrete dipole approximation. Phys. Rev. B **84**, 235429 (2011)
18. L. Cao, P. Fan, A.P. Vasudev, J.S. White, Z. Yu, W. Cai, J.A. Schuller, S. Fan, M.L. Brongersma, Semiconductor nanowire optical antenna solar absorbers. Nano Lett. **10**(2), 439–445 (2010)
19. P. Spinelli, M. Verschuuren, A. Polman, Broadband omnidirectional antireflection coating based on subwavelength surface mie resonators. Nat. Commun. **3**, 692 (2012)
20. L. Cao, P. Fan, E.S. Barnard, A.M. Brown, M.L. Brongersma, Tuning the color of silicon nanostructures. Nano Lett. **10**(7), 2649–2654 (2010)
21. K. Seo, M. Wober, P. Steinvurzel, E. Schonbrun, Y. Dan, T. Ellenbogen, K.B. Crozier, Multi-colored vertical silicon nanowires. Nano Lett. **11**(4), 1851–1856 (2011)
22. S.M. Wells, I.A. Merkulov, I.I. Kravchenko, N.V. Lavrik, M.J. Sepaniak, Silicon nanopillars for field-enhanced surface spectroscopy. ACS Nano **6**(4), 2948–2959 (2012)
23. A.B. Evlyukhin, S.M. Novikov, U. Zywietz, R.L. Eriksen, C. Reinhardt, S.I. Bozhevolnyi, B.N. Chichkov, Demonstration of magnetic dipole resonances of dielectric nanospheres in the visible region. Nano Lett. **12**(1), 3749–3755 (2012)
24. A.I. Kuznetsov, A.E. Miroshnichenko, Y.H. Fu, J. Zhang, B. Luk'yanchuk, Magnetic light. Scientific reports, vol. 2 (2012)
25. Y. Xia, G.M. Whitesides, Soft lithography. Annu. Rev. Mater. Sci. **28**, 153–84 (1998)
26. S.J. Clarson, J.A. Semlyen (eds.), *Siloxane Polymers* (Prentice Hal, Englewood Cliffs, 1993)
27. J.P. Rolland, R.M. Van Dam, D.A. Schorzman, S.R. Quake, J.M. DeSimone, J. Am. Chem. Soc. **126**, 8349–8349 (2004)
28. J.P. Rolland, B.W. Maynor, L.E. Euliss, A.E. Exner, G.M. Denison, J.M. DeSimone, Direct fabrication and harvesting of monodisperse, shape-specific nanobiomaterials. J. Am. Chem. Soc. **127**, 10096–10100 (2005)
29. K.J. Pekarek, J.S. Jacob, E. Mathiowitz, Double-walled polymer microspheres for controlled drug-release. Nature **367**, 258–260 (1994)
30. M.C.W. van Boxtel, R.H.C. Janssen, D.J. Broer, H.T.A. Wilderbeek, C.W.M. Bastiaansen, Polymer-filled nematics: a new class of light-scattering materials for electro-optical switches. Adv. Mater. **12**, 753–757 (2000)
31. M.N.V.R. Kumar, A review of chitin and chitosan applications. React. Funct. Polym. **46**, 1–27 (2000)
32. Y.J. Zhao, X.W. Zhao, J. Hu, J. Li, W.Y. Xu, Z.Z. Gu, Z.Z. Multiplex, Label-free detection of biomolecules with an imprinted suspension array. Angew. Chem. Int. Ed. **48**, 7350–7352 (2009)
33. F. Danhier, E. Ansorena, J.M. Silva, R. Coco, A. Le Breton, V. Préat, PLGA-based nanoparticles: an overview of biomedical applications. J. Controlled Release **161**, 505–522 (2012)
34. A.V. Kabanov, S.V. Vinogradov, Nanogels as pharmaceutical carriers: finite networks of infinite capabilities. Angew. Chem. Int. Ed. **48**, 5418–5429 (2009)
35. S.E. Gratton, P.D. Pohlhaus, J. Lee, J. Guo, M.J. Cho, J.M. DeSimone, Nanofabricated particles for engineered drug therapies: a preliminary biodistribution study of PRINT nanoparticles. J. Controlled Release **121**, 10–18 (2007)
36. P.T. Anastas, J.C. Warner, *Green Chemistry: Theory and Practice* (Oxford University Press, New York, 1998), p. 160
37. S. Besner, A.V. Kabashin, M. Meunier, Two-step femtosecond laser ablation-based method for the synthesis of stable and ultra-pure gold nanoparticles in water. Appl. Phys. A. **88**, 269 (2007)
38. F. Mafuné, J. Kohno, T. Takeda, T. Kondow, H. Sawabe, Formation of gold nanoparticles by laser ablation in aqueous solution of surfactant. J. Phys. Chem. B **105**, 5114 (2001)

39. A.V. Kabashin, M. Meunier, Synthesis of colloidal nanoparticles during femtosecond laser ablation of gold in water. J. Appl. Phys. **94**, 7941 (2003)
40. L. Sajti, S. Giorgio, V. Khodorkovsky, W. Marine, Femtosecond laser synthesized nanohybrid materials for bioapplications. Appl. Surf. Sci. **253**, 8111 (2007)
41. H. Usui, Y. Shimizu, T. Sasaki, N. Koshizaki, Photoluminescence of ZnO nanoparticles prepared by laser ablation in different surfactant solutions. J. Phys. Chem. B. **109**, 120 (2005)
42. A. Hahn, S. Barcikowski, Production of bioactive nanomaterial using laser generated nanoparticles. J. Laser Micro/Nanoeng. **4**, 51 (2009)
43. A. Barchanski, N. Hashimoto, S. Petersen, C.L. Sajti, S. Barcikowski, Impact of spacer and strand length on oligonucleotide conjugation to the surface of ligand-free laser-generated gold nanoparticles. Bioconj. Chem. **23**(5), 908–915 (2012)
44. A. Abdolvand, S.Z. Khan, Y. Yuan, P.L. Crouse, M.J.J. Schmidt, M. Sharp, Z. Liu, L. Li, Generation of titanium-oxide nanoparticles in liquid using a high-power, high-brightness continuous-wave fiber laser. Appl. Phys. A. **91**, 365 (2008)
45. R. Kelly, A. Miotello, Comments on explosive mechanisms of laser sputtering. Appl. Surf. Sci. **96–98**, 205 (1996)
46. A. Miotello, R. Kelly, Laser-induced phase explosion: new physical problems when a condensed phase approaches the thermodynamic critical temperature. Appl. Phys. A. **69**, 67 (1999)
47. L. Sajti, R. Sattari, B. Chichkov, S. Barcikowski, Gram scale synthesis of pure ceramic nanoparticles by laser ablation in liquid. J. Phys. Chem. C. **114**, 2421 (2010)
48. C.L. Sajti, R. Sattari, B. Chichkov, S. Barcikowski, Ablation efficiency of alpha-Al_2O_3 in liquid phase and ambient air by nanosecond laser irradiation. Appl. Phys. A. **100**, 203–206 (2010)
49. M. Bruchez, M. Moronne, P. Gin, S. Weiss, A.P. Alivisatos, Semiconductor nanocrystals as fluorescent biological labels. Science **281**, 2013 (1998)
50. K. Sokolov, J. Aaron, B. Hsu, D. Nida, A. Gillenwater, M. Follen, Optical systems for in vivo molecular imaging of cancer. Technol. Cancer Res. Treat. **2**, 491 (2003)
51. J.P. Sylvestre, S. Poulin, A.V. Kabashin, E. Sacher, M. Meunier, J.H.T. Luong, Surface chemistry of gold nanoparticles produced by laser ablation in aqueous media. J. Phys. Chem. B. **108**, 16864 (2004)
52. L. Sajti, S. Petersen, A. Menéndez-Manjón, S. Barcikowski, In situ bioconjugation in stationary media and in liquid flow by femtosecond laser ablation in liquid. Appl. Phys. A. **101**, 259–264 (2010)
53. L. Sajti, A. Barchanski, P. Wagener, S. Klein, S. Barcikowski, Delay time and concentration effects during bioconjugation of nanosecond laser-generated nanoparticle in liquid flow. J. Phys. Chem. C. **115**(12), 5094–5101 (2011)

Chapter 6
Light Scattering by Small Particles and Their Light Heating: New Aspects of the Old Problems

Michael I. Tribelsky and Boris S. Luk'yanchuk

Abstract A survey of recent results in light scattering by nanoparticles is presented. Special attention is paid to the case of particles from weakly dissipating materials, when the radiative damping prevails over the dissipative losses. It makes the scattering process completely different from the Rayleigh one. Peculiarities of the energy circulation in the near field zone are inspected in detail. The problem of optimization of the energy release in the particle is discussed. The chapter is concluded with consideration of laser heating of a metal particle in liquid important for biological and medical applications.

6.1 Introduction

Since the first quantitative study by Lord Rayleigh [1], the problem of light scattering by small particles has remained one of the most important and appealing issues of electrodynamics. There are thousands of articles and numerous monographs devoted to this subject, see, e.g., [2–5] and references therein. Plasmon (polariton) resonances and their role in the light scattering, as well as a related issue of interplay between radiative and dissipative damping are not new too [6], but they still remain topics of intense study [7–9].

M. I. Tribelsky
Faculty of Physics, Lomonosov Moscow State University, Moscow 119991, Russia

M. I. Tribelsky (✉)
Moscow State Institute of Radioengineering, Electronics and Automation (Technical University MIREA), Moscow 119454, Russia
e-mail: tribelsky@mirea.ru

B. S. Luk'yanchuk
Data Storage Institute, Agency for Science, Technology and Research,
Singapore 117608, Singapore
e-mail: Boris_L@dsi.a-star.edu.sg

V. P. Veiko and V. I. Konov (eds.), *Fundamentals of Laser-Assisted Micro- and Nanotechnologies*, Springer Series in Materials Science 195, DOI: 10.1007/978-3-319-05987-7_6, © Springer International Publishing Switzerland 2014

Regarding absorption of light by small plasmonic nanoparticles, it is a key effect for numerous applications of nanostructures in data storage technology, nanotechnology, chemistry, medicine, biophysics and bioengineering. The absorption characteristics depend on the material of the particle and its shape. It is sufficed mentioning absorption enhancement in amorphous silicon nanocone arrays [10]. Properties of such black silicon are useful for a wide range of commercial devices. Recently the problem of laser heating of plasmonic nanoparticles has attracted a lot of attention too (see, e.g., [11–20] and references therein). A similar problem arises in astrophysics with thermal noise in interstellar dust, where temperature fluctuations for small particles of interstellar dust may be about 1000 K [21]).

Various aspects of laser and laser-enhanced production of such particles, their properties and applications are discussed in Chaps. 4, 5 and 8 of the present monograph. In this chapter a survey of recent results in light scattering by small (relative to the wavelength of the incident light λ) spatially uniform nonmagnetic spherical nanoparticles and their laser heating is presented. We reveal a number of paradoxical, counterintuitive features of the problem, which shed new light on these important phenomena.

First, the problem of the so-called *anomalous light scattering* is discussed. The anomalous scattering may be realized close to the plasmon resonance frequencies, provided the dissipative losses are small enough. It is shown that despite the smallness of the particle the phenomenon has very little in common with the Rayleigh scattering. The most attention is paid to the discussion of the Poynting vector near-field structure, which occurs rather complicated. It includes a number of singular points, while the energy flow is divided into various branches of different shapes and orientations. In this case fine variations of the incident light frequency may result in global changes of the near-field structure.

Then, the problem of optimization of light absorption by a nanopartile is discussed in detail. It is shown that counter-intuitively the maximal absorption is achieved for a particle with a *small* value of the dissipative constant. A simple universal formula describing the absorption line shape as a function of complex dielectric permittivity of the particle is presented. Close to plasmon resonances the particle acts as a funnel, collecting the incident light from rather a broad area and delivering it to the near field zone. As a result the local field inside the particle (and hence the dissipation rate) may increase dramatically relative to the ones for the same material in bulk. We introduce a new quantity, the *effective volume absorption coefficient* α_{eff} of the particle, which allows to compare the dissipation rates in the particle and the corresponding bulk material quantitatively. Such a comparison is performed for a number of metals. It allows to find the optimal size of the particle and its optimal material to achieve the maximal energy release.

Next, we consider the general problem of laser pulse heating of a spherical metal particle embedded in a host medium, taking into account heat transfer from the particle to the environment. We employ the exact Mie solution of the diffraction problem and solve heat-transfer equations to determine the maximum temperature rise at the particle surface as a function of the optical and thermometric parameters of the problem. The main attention is paid to the case when the thermal diffusivity of the particle is much larger than that of the environment (metal particles in liquids, e.g., in water and alike). We show that in this case at any given duration of the laser pulse the maximal temperature rise as a function of the particle size reaches an absolute maximum at a certain finite size of the particle. Simple approximate analytical expressions for this dependence, which cover the entire range of the problem parameters and agree well with the direct numerical simulation are presented.

In conclusion we summarize the main points of the issues discussed.

6.2 Anomalous Light Scattering

6.2.1 General Principles

The conventional reasoning employed to describe light scattering by a small uniform nonmagnetic spherical particle, which may be found in any textbook, is as follows. If the particle is small relative to the wavelength of the incident light, the electric field of the latter, which "feels" the particle, is practically spatially homogeneous. The homogeneous field produces just a dipole polarization, oscillating in time with the frequency of the incident light ω. Any oscillating dipole emits electromagnetic waves. In our case these waves are precisely what the scattered light is. Then, recollecting the well known expression for the polarizability of a sphere with a given permittivity by a uniform electric field [22], we immediately arrive at the famous Rayleigh formula for the scattering cross section σ_{sca}:

$$\sigma_{\text{sca}} = \frac{8}{3}\pi R^2 q^4 \left|\frac{\varepsilon - 1}{\varepsilon + 2}\right|^2 ; \quad q = \frac{2\pi R}{\lambda} \ll 1, \tag{6.1}$$

where R is the particle radius and $\varepsilon = \varepsilon_p/\varepsilon_m$ stands for the relative permittivity of the particle. Here ε_p and ε_m mean the absolute permittivities of the particle and host medium, respectively. Note, that while the permittivity of the particle, generally speaking, is complex, the one for the host medium is supposed to be a purely real positive quantity.

Divergence of the denominator in (6.1) at $\varepsilon = -2$ is a well known fact. It corresponds to a resonant excitation of localized plasmons, whose eigenfrequency ω_1 (the meaning of subscript 1 will be clear later on) is defined through the dispersion law $\text{Re } \varepsilon(\omega_1) = -2$, while $\text{Im } \varepsilon \neq 0$ provides a finite cutoff for the divergence.

Note now, that though the mentioned cutoff always prevents the divergence of σ_{sca}, it is not always meaningful. If $\text{Im } \varepsilon$ is very small $\sigma_{\text{sca}}(\omega_1)$ may become extremely large, so that a nanoparticle may have the scattering cross section equal to, say, several square kilometers, which, of course, cannot be the case. The point is that apart dissipative losses the problem in question has an additional cutoff mechanism, namely the inverse transformation of localized plasmons into traveling electromagnetic waves or, in other words, the radiative damping. Its grounds are related to the fact that the scattered light is the radiation emitted owing to oscillations of the eigenmodes excited in the particle by the incident light. Since the emitted waves take off energy, the eigenmodes are damped even if the dissipative losses do not exist at all. This non-dissipative damping is a very weak effect, which usually may be neglected. The neglect corresponds to the Rayleigh approximation and yields (6.1). However, if the dissipation is weak itself, the radiative damping may become the major mechanism of the cutoff.

It should be stressed that the radiative damping and its role in the cutoff is well known for a very long time. We could find a reference to it in paper [23] published in 1951 (!), though we are not sure that this is the very first mentioning of the effect. However, what has been overlooked is the fact that when the radiative damping begins to prevail over the dissipative one, the entire scattering process undergoes drastic changes. For this reason we have singled out this type of light scattering into a separate class, naming it the *anomalous scattering*. The goal of the present section is to elucidate various features of the anomalous scattering and their consequences.

To this end, we have to go beyond the Rayleigh approximation. Fortunately, there is the exact solution to the problem of light diffraction by a sphere, known as the Mie solution, see e.g. [2, 3]. According to the solution the scattered electromagnetic wave is presented as a superposition of waves emitted by an infinite number of multipoles excited in the sphere by the incident wave. The net extinction, scattering and absorption efficiencies (the corresponding cross sections normalized over the geometrical cross section of the sphere) in this case equal to

$$Q_{\text{ext, sca, abs}} = \sum_{l=1}^{\infty} Q_{\text{ext, sca, abs}}^{(\ell)}, \qquad (6.2)$$

where the *partial efficiencies* $Q_{\text{ext, sca, abs}}^{(\ell)}$ are expressed in terms of the scattering coefficients a_ℓ, b_ℓ:

$$Q_{\text{ext}}^{(\ell)} = \frac{2(2\ell + 1)}{q^2} \text{Re}(a_\ell + b_\ell), \quad Q_{\text{sca}}^{(\ell)} = \frac{2(2\ell + 1)}{q^2}(|a_\ell|^2 + |b_\ell|^2). \qquad (6.3)$$

Each partial efficiency corresponds to the radiation of the ℓth order multipole, and terms proportional to a_ℓ and b_ℓ in (6.3) describe the radiation related the to electric and magnetic polarizabilities, respectively.

Regarding the absorption efficiency, in accord with the energy conservation law $Q_{\text{abs}} = Q_{\text{ext}} - Q_{\text{sca}}$. Since radiation of each multipole is independent, the same relation is valid for the partial efficiencies too, i.e.,

$$Q_{\text{abs}}^{(\ell)} = Q_{\text{ext}}^{(\ell)} - Q_{\text{sca}}^{(\ell)} \qquad (6.4)$$

Thus, for the problem in question the key quantities are a_ℓ, b_ℓ. They may be presented in the form:

$$a_\ell = \frac{F_\ell^{(a)}(q, \varepsilon)}{F_\ell^{(a)}(q, \varepsilon) + iG_\ell^{(a)}(q, \varepsilon)}, \quad b_\ell = \frac{F_\ell^{(b)}(q, \varepsilon)}{F_\ell^{(b)}(q, \varepsilon) + iG_\ell^{(b)}(q, \varepsilon)}, \quad (6.5)$$

where $F_\ell^{(a, b)}$, $G_\ell^{(a, b)}$ are expressed in terms of the Bessel $[J_{l+1/2}(\zeta)]$ and Neumann $[N_{l+1/2}(\zeta)]$ functions. The corresponding general expressions are rather cumbersome and are not be presented here. They may be found, e.g., in [2, 3].

In what follows we will focus on light scattering by a small particle, when $q \ll 1$, see (6.1). In this case the general expressions for $F_\ell^{(a, b)}$, $G_\ell^{(a, b)}$ may be expended in powers of small q. The expansion yields

$$F_\ell^{(a)}(q, \varepsilon) \simeq q^{2\ell+1} \frac{\ell + 1}{[(2\ell + 1)!!]^2} (\varepsilon - 1) + \cdots \quad (6.6)$$

$$G_\ell^{(a)}(q, \varepsilon) \simeq \frac{\ell}{2\ell+1} \left\{ \varepsilon + \frac{\ell+1}{\ell} - q^2 \frac{\varepsilon-1}{2} \left[\frac{\varepsilon}{2\ell+3} + \frac{\ell+1}{\ell(2\ell-1)} \right] + \ldots \right\}, \quad (6.7)$$

where ellipses denote omitted higher order in q terms. We do not need the corresponding expressions for $F_\ell^{(b)}$, $G_\ell^{(b)}$ because estimates show that for the case in question $|b_\ell|$ is always small relative to $|a_\ell|$, so that the radiation of magnetic multipoles may be neglected [2, 3].

Note, that $|F_\ell^{(a)}| \ll 1$, while $|G_\ell^{(a)}|$, generally speaking, is of the order of unity. Then it seems, that $F_\ell^{(a)}$ in the denominator of (6.5) may be neglected relative to $iG_\ell^{(a)}$. The neglect yields the estimates $Q_{\text{ext}}^{(1)} \gg Q_{\text{ext}}^{(2)} \gg Q_{\text{ext}}^{(3)} \gg \cdots$ and the same for $Q_{\text{sca,abs}}^{(\ell)}$, see (6.3)–(6.5). Thus, we have obtained that the scattering efficiencies are overwhelmingly determined by the electric dipole mode with $\ell = 1$, which in the leading approximation eventually brings about (6.1).

However, this reasoning becomes invalid at the vicinity of the points $\varepsilon = \varepsilon_\ell$, where ε_ℓ are roots of the equations $G_\ell^{(a)}(\varepsilon) = 0$. At any ℓ there is at least one root of this equations,[1] namely

$$\varepsilon_\ell = -\frac{\ell + 1}{\ell} + O(q^2), \quad (6.8)$$

see (6.7). As well as it has been discussed above for the dipole resonance at $\ell = 1$, the quantities ε_ℓ at any ℓ through the dispersion law $\varepsilon(\omega)$ define frequencies of the ℓth order plasmon resonance ω_ℓ. The neglect of $F_\ell^{(a)}$ in the denominator of (6.5) in the vicinity of the plasmon resonance results in divergence of a_ℓ at $\varepsilon = \varepsilon_\ell$, cf. (6.1). In contrast to that the exact expression (6.6) yields $a_\ell(\varepsilon_\ell) = 1$, which gives rise to

---

[1] Actually, at any ℓ the equation $G_\ell^{(a)}(\varepsilon) = 0$ has an infinite number of roots. The roots different from (6.8) correspond to large values of ε and lie beyond the validity range of (6.7). Interference of modes with the same ℓ related to different such roots may result in new interesting phenomena, including cloaking of the particle (the complete suppression of the scattering, which makes the particle invisible), see [24]. However, discussion of these matters lies beyond the scope of the present chapter.

the following finite value for the partial cross sections $\sigma^{(\ell)} = \pi R^2 Q^{(\ell)}$:

$$\sigma_{\text{ext}}^{(\ell)} = \sigma_{\text{sca}}^{(\ell)} = (2\ell + 1)\frac{2\pi}{k^2}, \tag{6.9}$$

see (6.3). As it has been already mentioned above, the physical meaning of the cutoff of the divergency is related to the inverse transformation of resonant localized plasmons, excited in the particle by the incident electromagnetic wave, into scattered light.

Regarding the values of all other off resonant partial extinction (scattering) cross sections, these quantities are defined by the usual Rayleigh approximation, which corresponds to the mentioned neglect of $F^{(a)}$ in the denominator of (6.5). Then, it is seen straightforwardly that at $\varepsilon = \varepsilon_\ell$ the contribution of expression (6.9) to the net cross section is overwhelming.

Equation (6.9) exhibits what we call *the inverted hierarchy of resonances* [25, 26]. In the conventional cases the partial cross sections of high order resonances for a small particle are dying off sharply with an increase in ℓ as has been mentioned above: $\sigma_{\text{ext}}^{(\ell)} \sim q^{2(2\ell+1)}$, see (6.3)–(6.7). In contrast, (6.9) yields an increase of the resonant cross section with an increase in ℓ, so that the resonant dipole cross section occurs smaller than the one for the quadrupole resonance, the latter is smaller than the resonant octupole cross section, etc.[2]

In the vicinity of the resonance we may write $F_\ell^{(a)}(\varepsilon) \simeq F_\ell^{(a)}(\varepsilon_\ell)$, $G_\ell^{(a)}(\varepsilon) \simeq (dG_\ell^{(a)}/d\varepsilon)_{\varepsilon_\ell}(d\varepsilon/d\omega)_{\omega_\ell}\delta\omega$, where $\delta\omega = \omega - \omega_\ell$. It immediately brings about the conventional Lorenzian profile for $\sigma_{\text{ext}}(\omega)$ with the full-width at half-maximum (FWHM) [26]

$$\gamma_\ell = \frac{2q^{2\ell+1}(\ell+1)}{[\ell(2\ell-1)!!]^2(d\varepsilon/d\omega)_{\omega_\ell}}, \tag{6.10}$$

see (6.3). Note a sharp decrease in the linewidth with an increase in ℓ.

The next point to be made is independence of the resonant cross section (6.9) of the particle size R. It gives rise to the paradoxical conclusion that a particle with $R = 0$ still has a finite extinction cross section. We face the paradox because in our consideration we have neglected dissipative processes entirely. Meanwhile dissipation never can vanish completely, and hence the linewidth (6.10) related to the radiative damping cannot be smaller than the natural linewidth determined by the dissipation. When the latter becomes comparable with the former, (6.9) becomes invalid and the conventional Rayleigh scattering is restored. It results in the following applicability conditions for the anomalous scattering to come into being [27][3]:

---

[2] The inverted hierarchy does not affect convergence of the multipole expansion because each resonance takes place at its own resonant value of ε, so that at a given order of the resonance we have just a single partial cross section describing by (6.9).

[3] Usually, taking into account sign of strong inequality in (6.11), $(\ell + 1)^\ell$ in its right-hand-side is replaced by 1. Here we do not do that because it is important for the anomalous absorption, which will be discussed in the next section.

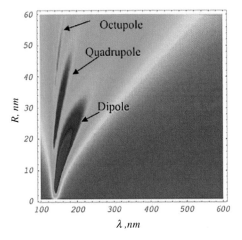

Fig. 6.1 The extinction cross section of a spherical aluminum particle in a vacuum. The exact Mie solution and actual optical properties of aluminum

$$\varepsilon''(\omega_\ell) \ll q^{2\ell+1} \frac{\ell+1}{[\ell(2\ell-1)!!]^2}, \tag{6.11}$$

where ε'' stands for Im ε.

Thus, to observe the anomalous scattering Im ε at the resonant frequency should be small. However, no matter how small Im ε is, at small enough q the condition (6.11) is violated, the Rayleigh scattering is restored, and the extinction cross section tends to zero at $q \to 0$, as it should be.

It is worthwhile stressing that, actually, the condition (6.11) is too strict. Thus, while for a particle with $q = 0.3$ at $\ell = 1$ (6.11) reads $\varepsilon'' \ll 0.03$, in fact the complete restoration of the Rayleigh scattering occurs only at $\varepsilon'' > 0.6$ [28].

As an example, the extinction cross section of a spherical aluminum particle in a vacuum calculated based upon the exact Mie solution is presented in Fig. 6.1. Aluminum is selected owing to small ($\sim$0.1) values of ε'' in the range of the frequencies of its plasmon resonances. To obtain these results the following model has been employed [25]. The empirical dependence ε on ω (known as a table [29]) is approximated by the Drude formula:

$$\varepsilon = 1 - \frac{\omega_p^2}{\omega^2 + \gamma^2} + i \frac{\gamma \omega_p^2}{\omega(\omega^2 + \gamma^2)}. \tag{6.12}$$

To enhance the accuracy, quantities ω_p and γ, regarded as functions of ω, are calculated at every point in the table with a polynomial interpolation between the points. To take into account collisions of free electrons with the particle surface, $\gamma(\omega)$ found for bulk aluminum is replaced by $\gamma_{\text{eff}} = \gamma(\omega) + v_F/R$, where $v_F = 10^8$ cm/s stands for the Fermi velocity of the free electrons.

In accord with what has been said above, while at $R > 30$ nm the extinction cross section at the quadrupole resonance is larger than that at the dipole one (the inverted hierarchy at the anomalous scattering) the conventional hierarchy of the resonances is restored with a decrease in R. Finally, at $R \to 0$ the cross section vanishes in agreement with the dependence describing the Rayleigh scattering.

6.2.2 Near Field Effects

However, the most appealing manifestation of the anomalous scattering takes place in the near field zone. The key point is that the dramatic changes in both the modulus and phase of the complex amplitude a_ℓ in the vicinity of the plasmon resonances at the anomalous scattering relative to that at the conventional Rayleigh approximation brings about the corresponding dramatic changes in the near-field structure.

We recall that at the anomalous scattering the dissipative losses are negligible, the extinction cross section approximately equals the scattering one and both do not depend on R, see (6.9). On the other hand, by definition the scattering cross section is the overall scattered power [W] normalized over the intensity of the incident wave [W/cm^2]. If for a given incident wave we decrease the geometrical size of a scatterer, and the decrease does not affect the overall scattered power, it means that the characteristic value of the electromagnetic field in the particle should increase to provide the same emitted power from a smaller volume. In other words, at the anomalous scattering the electromagnetic field in the particle and its immediate vicinity should be *singular in q*. This is the case indeed. Utilizing the exact Mie solution and bearing in mind that at the point of the resonances the corresponding $a_\ell = 1$, we can readily obtain the following estimates for the components of the electric $\boldsymbol{E}(\boldsymbol{r})$, magnetic $\boldsymbol{H}(\boldsymbol{r})$ fields and the time-averaged Poynting vector $\langle \boldsymbol{S}(\boldsymbol{r}) \rangle$ in the particle and its near field zone [30]:

$$\frac{E_{r,\varphi,\theta}}{E_0} \sim q^{-(\ell+2)}; \quad \frac{H_r}{E_0} \sim q^{\ell+1}; \quad \frac{H_{\varphi,\theta}}{E_0} \sim q^{-(\ell+1)}; \quad \frac{\langle S_{r,\varphi,\theta} \rangle}{\langle S_0 \rangle} \sim q^{-(2\ell+3)}, \quad (6.13)$$

where subscripts r, φ, θ designate the corresponding components and E_0, $\langle S_0 \rangle$ stand for the values of E, $\langle S \rangle$ in the incident wave. The time-averaged Poynting vector $\langle \boldsymbol{S}(\boldsymbol{r}) \rangle$, as usual, is defined as follows [2, 3]:

$$\langle \boldsymbol{S} \rangle = \frac{c}{8\pi} \langle \text{Re}[\boldsymbol{E} \times \boldsymbol{H}^*] \rangle. \quad (6.14)$$

We remind for reference that at the conventional scattering (the Rayleigh approximation) $a_\ell \sim q^{2\ell+1}$ [2, 3]. For the near field zone instead of (6.13) it yields

$$\frac{E_{r,\varphi,\theta}}{E_0} \sim q^{\ell-1}; \quad \frac{H_r}{E_0} \sim q^{\ell+1}; \quad \frac{H_{\varphi,\theta}}{E_0} \sim q^{\ell}; \quad \frac{\langle S_{r,\varphi,\theta} \rangle}{\langle S_0 \rangle} \sim q^{2\ell}. \quad (6.15)$$

Let us stress the dramatic difference between (6.13) and (6.15). If for the latter the electromagnetic field in the particle and its vicinity vanishes sharply when $q \rightarrow 0$, for the former it (but component H_r) *increases sharply* with a decrease in q, as long as the approximation of the anomalous scattering holds.[4] Physical grounds for this unusual behavior will be clear if we elucidate the detailed structure of the electromagnetic field in the near field zone. To this end, we need a certain geometrical description of the field structure.

The characteristic spatial scale of the problem is much smaller than the wavelength of the incident light. Therefore, we cannot use the convenient and visual way to show the propagation of light by optical rays, applicable only in the opposite limit of the geometrical optics. However, instead of that we can show the energy circulation in the near field, plotting field lines of the time-averaged Poynting vector. The spatial dependence of electric $E(r)$ and magnetic $H(r)$ fields, entering in (6.14), are given by the exact Mie solution. Thus, drawing the field lines of $\langle S(r) \rangle$ is a straightforward but quite laborious matter. Details of the corresponding algorithms may be found, e.g., in [2, 28].

Study of the structure of the vector Poynting field in the near field zone [25, 26, 28, 31–33] shows that in the vicinities of the plasmon resonances it is very complicated and includes singular points, whose number and positions are very sensitive to the detuning of the ω from ω_ℓ. The structure depends significantly on the order of the resonance and the dissipation rate in the particle. Though the general picture of the dependence of the structure on the entire set of the problem parameters is not clear yet, the results obtained allow to conclude that the rate of complexity of the structure decreases with an increase in $\varepsilon''(\omega_\ell)$. Thus, the extreme case is the near field structure in the non-dissipative limit.

As an example, the field lines of $\langle S(r) \rangle$ at the vicinity of the dipole resonance at $q = 0.3$ in the non-dissipative limit are presented in Fig. 6.2 [33]. Generally speaking, the field lines are essentially 3D curves. However, owing to the problem symmetry, the plane $y = 0$ is invariant, i.e., if any non-singular point of a field line belongs to this plane, the entire line belongs to the plane too. It reduces the field structure in this plane to a 2D picture.

It is seen from Fig. 6.2 that the particle acts as a funnel, collecting the energy flux from a large "upstream" area bounded by the red separatrixes of saddle points 1,2 and delivering the energy to the immediate vicinity of the particle. The smaller the particle, the greater the ratio of the diameter of the "inlet" of the funnel to the one of the "outlet" and hence the greater the field concentration at the "outlet." This is the physical reason explaining the mentioned singular in q dependence of

---

[4] The conventional dependence $H_r(q)$ at the anomalous scattering is explained by the fact that this type of the scattering corresponds to a resonant excitation of eigenmodes related to electric polarizability of the particle. For these modes $H_r = 0$ [2, 3]. Non-zero values of H_r in the near field correspond to the contribution of the magnetic modes, related to the magnetic polarizability of the particle by the electromagnetic field of the incident wave. For a non-magnetic particle these modes always are non-resonant and therefore have the same amplitude both at the anomalous and Rayleigh scattering.

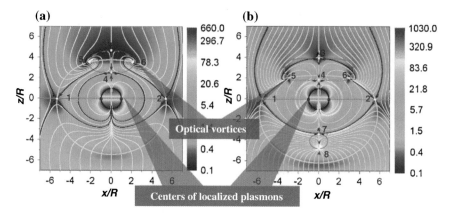

Fig. 6.2 Color density plot of modulus of the time-averaged Poynting vector $\langle S(r)\rangle$ in the vicinity of the dipole resonance in the invariant plane $y = 0$ for a spherical particle with $q = 0.3$. The non-dissipative limit Im $\varepsilon'' = 0$. The wave vector of the incident plane linearly polarized wave is parallel to the z axis. Its E vector oscillates along the x axis. Field lines (*white*) are described by equation $dr/d\theta = r\langle S_r\rangle/\langle S_\theta\rangle$. Null isoclines $\langle S_\theta(r,\theta)\rangle$ and $\langle S_r(r,\theta)\rangle$ are shown in *yellow*, and *pink*, respectively. Numerals indicate different singular points. The exact resonance corresponds to $\varepsilon = -2.22$; **a** $\varepsilon = -2.17$; **b** $\varepsilon = -2.20$. Note different scales of panels (**a**) and (**b**)

the characteristic values of the electromagnetic field achieved in the particle and its vicinity at the anomalous scattering.

The discussed 2D picture of the field lines is very informative, but at the same time it may be misleading in somewhat. The point is that looking at it, one may expect that the density of the field lines (the number of lines crossing a straight unite-length segment aligned normal to them, i.e., the 2D flux density) is proportional to the modulus of the Poynting vector $\langle |S(r)|\rangle$. Actually, the former has nothing to do with the latter. Due to the Gauss theorem and condition div$\langle S(r)\rangle = 0$ in any non-singular point outside the particle (inside the particle it is true only if $\varepsilon'' = 0$) the density of the field lines is connected with $\langle |S(r)|\rangle$ indeed. But it is in the 3D space! In invariant plane $y = 0$ y-component of $\langle S(r)\rangle$ vanishes, but this is not the case for $\langle \partial S_y/\partial y\rangle$. Therefore, two-dimensional divergence $\langle \partial S_x/\partial x\rangle + \langle \partial S_z/\partial z\rangle$ in the invariant plane, generally speaking, does not equal to zero, and the 2D flux in the plane is not conserved along "stream tubes" of the Poynting vector.

To understand the full picture of the energy circulation in the near field zone, even in the vicinity of the invariant plane $y = 0$, we have to get out of this plane, considering actual 3D field lines. Two such lines, are shown in Fig. 6.3. Both of them enter the vicinity of focus 6 in Fig. 6.2a. One of the lines approaches the focus from a direction transversal to plane $y = 0$, then sharply transforms into an unwinding spiral, lying almost parallel to this plane, and after a few rotations around the focus goes downstream to $z \to \infty$. The other is deviated by the focus toward the center of the localized plasmon, approaches it, making a lot of rotations as a winding spiral with a very small pitch, then it is ejected from the center, transforming into an unwinding

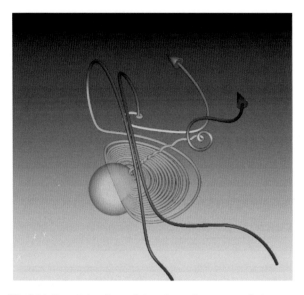

Fig. 6.3 Two 3D field lines lying beyond invariant plane $y = 0$ (*courtesy* of Andrey E. Miroshnichenko). The other details of the problem are identical to those in Fig. 6.2a

spiral with an increasing pitch, and eventually also goes downstream. It means, that the resonant plasmons collect energy not only from the upstream area—a part of it comes from the optical vortexes (points 5 and 6), which lie downstream with respect to the localized plasmons.

Thus, in the vicinity of the resonance the energy circulation corresponds to collection of the energy flux from a large area and delivery of the electromagnetic radiation to the centers of the localized plasmons. The delivery occurs in rather a narrow layer close to invariant plane $y = 0$ and results in huge concentration of the energy in the resonant plasmons. Then, this energy is radiated from the plasmons along directions transversal to the invariant plane, which brings about a powerful emission of electromagnetic field in narrow solid angles, i.e., the plasmons act as very powerful subwavelength floodlights.

All these phenomena take place in the near field zone exclusively and have the characteristic scale of the order of the particle size R. When the distance from the particle increases to the far field zone the scattered field transforms into the conventional field structure typical for the corresponding point dipole. Qualitatively the same picture is valid for other orders of the resonances (quadrupole, octupole, etc.) too.

6.3 Anomalous Absorption

For the time being to inspect the extreme case of the anomalous scattering the main focus of our consideration has been put on the non-dissipative limit. On the other hand, for a number of applications the maximization of the energy release in the particle is important. A naïve answer to the question "How to maximize the absorption cross section of a small particle?" is "To do that maximize the dissipative constant of the particle material, i.e., Im ε." However, the correct answer is just opposite: to achieve the maximal energy release in the particle its dissipative constant should be small. To reveal the physical grounds for that seemingly paradoxical conclusion let us discuss a toy auxiliary problem of the power dissipated at forced oscillations of a pendulum. The equation of motion is as follows:

$$\ddot{x} + 2\nu\dot{x} + \omega_0^2 x = (f/m)\exp(i\omega t),$$

with a steady solution of the form:

$$x = \frac{(f/m)}{\omega_0^2 - \omega^2 + 2i\nu\omega}\exp(i\omega t). \tag{6.16}$$

The mean dissipated power P is the work of the friction force $-2m\nu\dot{x}$ in a unite of time averaged over a period of oscillations. Trivial algebra results in the following expression:

$$P = m\nu\langle\dot{x}\dot{x}^*\rangle = \frac{|f|^2\omega^2\nu}{m[(\omega_0^2 - \omega^2)^2 + 4\nu^2\omega^2]}. \tag{6.17}$$

Suppose that the dissipative constant ν and the driving frequency ω are fixed. Maximization of $P(\omega_0)$ in this case readily yields

$$\max P(\omega_0) = P(\omega) = \frac{|f|^2}{4m\nu} \to \infty \quad \text{at} \quad \nu \to 0. \tag{6.18}$$

The reason for the divergence is obvious. While the dissipated power is proportional to the dissipative constant indeed, it is also proportional to square of the amplitude of oscillations, see (6.16), (6.17). At $\omega_0 = \omega$ the latter depends on ν as $1/\nu^2$. Thus, at the point of resonance the product square of the amplitude times the dissipative constant occurs proportional to $1/\nu$ and tends to infinity at $\nu \to 0$.

Precisely the same happens with the dissipated power at the plasmon resonances. The only difference with the discussed toy problem is that the damping rate at the plasmon resonances can never vanish owing to the radiative losses and hence the dissipated power always remains finite. A detailed study of light absorption in this case is presented in [30]. It is shown that the maximal absorption takes place exactly at the frequencies of the plasmon resonances ω_ℓ, corresponding to the roots of equation $G_\ell^{(a)}(\text{Re}\,\varepsilon(\omega)) = 0$. Regarding Im ε, to provide the maximal absorption it should have such a value that the dissipative damping occurs equal to the radiative one exactly. The corresponding value of Im ε is determined by (6.11), where sign of strong inequality should be replaced by equality.

The physical grounds for this result are very simple. As it has been mentioned, the power dissipated in the particle is proportional to the actual ε'', while the cutoff of a_ℓ at the vicinity of the resonance occurs owing to the effective $\varepsilon''_{\text{eff}}$ which is a sum of the actual ε'' and the non-dissipative term related to the radiative damping, see (6.5). As long as the dissipative losses are smaller than the radiative damping, an increase in the former increases the dissipative constant, but practically does not affect $\varepsilon''_{\text{eff}}$ and hence the amplitude of the resonant field. Then, the overall effect is an increase in the dissipation. In the opposite limit, when the radiative damping is negligible relative to the dissipative losses the case is analogous to the discussed toy model, and an increase in ε'' results in a decrease in the dissipated power, see (6.18). The crossover between the two regimes takes place at the point where the dissipative and radiative dampings are equal each other.

The scattering cross section at this point occurs equal to the absorption one and equal to 1/4 of that given by (6.9). Thus, for the problem in question the quantity

$$\sigma^{(\ell)}_{\text{abs max}} = \frac{\pi}{k^2}\left(\ell + \frac{1}{2}\right) \tag{6.19}$$

corresponds to the absolute theoretical maximum for the partial absorption cross section, which never can be exceeded.[5]

This resonant absorption inherits many features of the anomalous scattering, which allows naming it the *anomalous light absorption*. For example, according to (6.19) $\sigma^{(\ell)}_{\text{abs max}}$ is a certain universal quantity, which does not depend on R and the optical constants of the particle, cf. the corresponding behavior of the extinction cross section at the anomalous scattering, etc.

Estimates show that, as usual, as long as the particle is small ($q \ll 1$) at the ℓth resonance point the contributions of the entire set of the off resonant multipoles is negligible with respect to the resonant partial cross section, and the net resonant absorption cross section is determined by (6.19) with great accuracy.

Thus, in the vicinity of the resonance $\sigma^{(\ell)}_{\text{abs}} = \sigma^{(\ell)}_{\text{abs}}(q, \varepsilon', \varepsilon'')$. However, it is possible to show that introduction of the proper selected dimensionless variables scales out the dependences on ℓ and q, while the dependence on the two remaining variables is reduced to the following simple universal form [30]:

$$\varsigma = \frac{\chi''}{(1 + \chi'')^2 + \chi'^2}; \quad \chi'' \geq 0,$$

---

[5] We stress that (6.9), (6.19) should be regarded as the upper theoretical limit for the corresponding cross sections just for the problem in question (a small spatially uniform non-magnetic particle). In other cases these limits may be exceeded considerably, see, e.g., [34].

where ς, χ' and χ'' stand for the rescaled dimensionless partial absorption cross section, real and imaginary parts of ε, respectively.

6.4 Optimization of Laser Energy Release in Real Plasmonic Nanoparticles

However, application of the results discussed in the previous section to real experiments is not so straightforward. In any real case ε' and ε'' cannot be regarded as two independently varying parameters. The actual tuning parameters are ω and R. To achieve the maximal absorption these parameters must satisfy the set of equations

$$\varepsilon'(\omega, R) = \varepsilon'_\ell(\omega, R), \tag{6.20}$$

$$\varepsilon''(\omega, R) = \varepsilon''_\ell(\omega, R), \tag{6.21}$$

where the left-hand sides follow from the dispersion properties of the particle material[6] while the right-hand sides are given by the conditions for the anomalous absorption: $G^{(a)}_\ell(\varepsilon'(\omega), q) = 0$ and equality of the radiative and dissipative damping, see (6.7), (6.11), respectively. Solutions of (6.20), (6.21) yield a unique set of pairs (ω_ℓ, R_ℓ), where R_ℓ should satisfy the additional constraint following from restriction $q \ll 1$.

It seems that these conditions are very strict, the entire set of them is extremely difficult to fulfill in any real experiment, and hence the limit corresponding to (6.19) is hardly reachable. Fortunately the case is not so dramatic. The point is that usually the functions $\varepsilon'(\omega)$ and $\varepsilon''(\omega)$ are rather smooth, at least as long as metal particles are a concern. On the other hand, while at $q \ll 1$ the dependence $\varepsilon'_\ell(q)$ is weak, see (6.8), the corresponding dependence of the resonant value of ε'' is sharp, see the right-hand-side of (6.11). Then, the following iterative procedure may be built up. First, for a given material we select in its dispersion law $\varepsilon'(\omega)$ the frequencies of the plasmon resonances, which in the zeroth (in q) approximation are determined by the conditions $\varepsilon'(\omega_\ell^{(0)}) = -(\ell + 1)/\ell$. Next, we check if among the obtained $\omega_\ell^{(0)}$ there are any, satisfying the condition $\varepsilon''(\omega_\ell^{(0)}) \ll 1$. If this is the case, then, equalizing the left- and right-hand-side parts of (6.11), we find q, corresponding to this value of $\varepsilon'(\omega_\ell^{(0)})$. This q is employed to calculate the corrected solution of equation $G^{(a)}_\ell(\varepsilon'(\omega_\ell), q) = 0$. The obtained corrected $\omega_\ell^{(1)}$ is used to calculate corrected $\varepsilon''(\omega_\ell^{(1)})$, etc. The procedure quickly converges to q_ℓ and ω_ℓ corresponding to the conditions of the anomalous absorption.

If none of $\varepsilon''(\omega_\ell^{(0)})$ is small, the optimization should be performed by a standard numerical method of maximization of a function of several variables. In this case the question "To what extent is the reasoning of the previous section applied to the

---

[6] R-dependence may appear here owing to γ_{eff}, see (6.12).

results obtained during the maximizarion?" arises. Naturally, the described iterative procedure has just a methodological meaning. In practical calculations it is convenient to apply a single method of maximization, regardless the values of $\varepsilon'(\omega_\ell^{(0)})$. It is also desirable to have quantitative parameters allowing to compare the effectiveness of the resonant laser energy release in a small plasmonic particle relative to the one in the same bulk material.

Let us begin with the second point. The volume density of laser energy release $\mathcal{E}$ [J/cm^3] in a bulk material at the normal incidence of the laser beam is given by the expression:

$$\mathcal{E} = A\alpha\Phi, \tag{6.22}$$

where A is the absorptivity, Φ stands for the fluence of the laser beam [J/cm^2] and $\alpha = 2k\mathrm{Im}\sqrt{\varepsilon} = 2k\kappa$ is the volume absorption coefficient.

On the other hand, the volume density of the laser energy released in the particle averaged over particle volume is given by the expression

$$\mathcal{E} = \frac{3}{4}\frac{Q_{\mathrm{abs}}}{R}\Phi = \alpha_{\mathrm{eff}}\Phi, \tag{6.23}$$

where $\alpha_{\mathrm{eff}} = 3Q_{\mathrm{abs}}/(4R)$. Comparing (6.22) and (6.23), we see that quantitative measures of the effectiveness of the resonant energy release in the particle relative to the bulk material may be the following coefficients [35]:

$$\beta = \frac{\alpha_{\mathrm{eff}}}{\alpha}, \quad \beta_{\mathrm{eff}} = \frac{\alpha_{\mathrm{eff}}}{A\alpha}. \tag{6.24}$$

We name them the *net absorption enhancement factors*. Since $0 \leq A \leq 1$, the quantity β presents the *minimal* value of the absorption enhancement. If condition $\beta_{\mathrm{eff}} \geq \beta \gg 1$ is met, the nanoparticle absorbs light much more efficiently than the corresponding bulk material. To distinguish the role of partial resonances introduction of the partial enhancement factors $\beta_{\mathrm{eff}}^{(\ell)}$ and $\beta^{(\ell)}$ with the replacement in the definition of α_{eff} in (6.23), $Q_{\mathrm{abs}} \to Q_{\mathrm{abs}}^{(\ell)}$ is also meaningful [35].

To understand to what extent the arguments of the previous section may be applied to natural materials we have optimized numerically β_{eff} for six metals: potassium, aluminum, sodium, silver, gold and platinum, whose properties cover the range from weak dissipation at the optical frequencies (potassium) to the strong one (platinum). The model employed in the calculations was identical to the one discussed in the previous section for the aluminum particle, see (6.12). The results of these calculations obtained for each particle in the vicinity of its dipole resonance ($\ell = 1$) are collected in Table 6.1 [35].

These results exhibit good agreement with the preceding theoretical discussion. All the particles do have the optimal wavelength, lying close to the one for the corresponding plasmon resonance, and the optimal size, lying in the several-nanometer-scale. Larger enhancement of the energy release relative to the corresponding bulk

Table 6.1 Values of the parameters maximizing absorption of nanoparticles made of different metals irradiated by a plane linearly polarized electromagnetic wave in a vacuum; λ_0 corresponds to the frequency, satisfying the plasmon resonance condition at $R \to 0$: $\varepsilon'(\omega) = -2$

Metal	λ_0, nm	$\mathrm{Im}\,\varepsilon(\lambda_0)$	v_F, cm/s	Optimal R, nm	Optimal λ, nm	$\alpha_{\mathrm{eff}}^{\max}$, cm^{-1}	α, cm^{-1} (bulk)	β_{eff}
K	542	0.138	0.86×10^8	14.1	548	5.02×10^6	3.29×10^5	242
Na	377	0.178	1.07×10^8	10.3	381	5.75×10^6	4.74×10^5	151
Al	139	0.16	2.02×10^8	4.4	141	1.38×10^7	1.29×10^6	147
Ag	354	0.6	1.39×10^8	12.2	356	2.40×10^6	5.2×10^5	19.2
Au	485	3.97	1.39×10^8	40.8	505	4.22×10^5	4.83×10^5	1.51
Pt	276	5.64	1.45×10^8	17.9	213	6.48×10^5	9.1×10^5	1.24

materials have particles with smaller values of the dissipative constant at the resonance point, cf. β_{eff} for potassium and platinum.

However, if laser heating of particles is concerned, the discussed optimization of the energy release plays the role of an intermediate result. To find the temperature rise of the particle we must consider the heat transfer from the particle to a host environmental medium. This problem is inspected in the next section.

6.5 Laser Heating of Particles in Liquid

In the present section we discuss heating with a single laser pulse with duration t_u of a metal particle embedded in a transparent liquid. The thermometric constants of the liquid are close to those for water. The initial temperature equals the room one. The temperature rise is limited by several tens Kelvin, so that the boiling temperature is not reached. Such a problem formulation is typical for numerous medical and biological applications, see, e.g., [36–40] and references therein.

Owing to the apparent importance of the problem, it has attracted a great deal of attention of numerous researchers, see, e.g., [41–45]. However, in a standard approach to the theoretical description of the heating (see, e.g., [46]) the problem is formulated as study of heat transfer with a source (energy release in the particle) calculated as a solution of the corresponding diffraction problem. Such a heat transfer problem does not have simple exact analytical solutions, and different studies employ either approximate analytical methods (valid for certain cases only) or direct numerical calculations with the specified parameters. On the other hand, for practical applications it is highly desirable to obtain a simple analytical solution describing the temperature rise at the surface of the particle T_s in a broad domain of variations of the problem parameters, and for the particle sizes ranging from nanometers to millimeters. In the present section we discuss an approximate solution of such a kind presented in recent publication [47].

We suppose that heat diffusion is the only process responsible for energy exchange between the particle and the liquid. Convection processes do not play any role in the problem owing to a large characteristic time required for the convection to arise. We also suppose that the electron and phonon subsystems in the metal particle are in local equilibrium and may be described by single joint temperature. It imposes the restriction on the laser pulse duration $t_u \gg 1/\gamma$ for particles whose size is larger than the electron free path length, where γ stands for electron-phonon collision frequency. For smaller particles relaxation of the electron subsystem occurs because of collisions of the free electrons with the particle's surface. It results in the condition $t_u \gg R/v_F$. Joining of the conditions yields $t_u \gg \min(1/\gamma, R/v_F)$. For typical values of the constants $v_F \sim 10^8$ cm/s and $\gamma \sim 10^{13}$ s^{-1} (room temperature) a crossover from condition $t_u \gg 1/\gamma$ to $t_u \gg R/v_F$ occurs at $R \sim 10^{-5}$ cm.

For the sake of simplicity, we restrict our consideration by the Rayleigh scattering, excluding the anomalous absorption as rather an "exotic" effect. It may be readily incorporated into the discussed approach, if required.

Next, we neglect the angular dependence of the heat sources, supposing that the volume density of the sources inside the particle depends just on radial variable r. It allows to replace the actual 3D heat transfer problem by its spherically-symmetric version.

The grounds for this neglect are in the following. The actual angular inhomogeneity in the heat sources for small particles is rather weak due to diffractive distortions of the incident light. It results even in weaker temperature inhomogeneities, owing to the high rate of heat transfer in metals. On the other hand, in what follows we are interested in *estimates* of the maximal temperature rise at the particle surface, rather than in its exact calculations. For this reason we employ the spherically-symmetric problem formulation even for large particles, when the illuminated part of the particle obviously has a temperature higher than that in the shadow.

Under the approximation made the problem is characterized by four spatial scales, namely the particle size R, skin layer thickness δ, characteristic length of heat diffusion in the particle $2\sqrt{a_p t_u}$ and the one in the surrounding medium $2\sqrt{a_m t_u}$. Here a_p and a_m are the thermal diffusivity for the particle and medium, respectively. Interplay of these scales determines the entire variety of different heating regimes. Note also that according to the problem formulation (a metal particle in a water-like liquid) $a_m \ll a_p$.

Let us consider, for example, two cases: (i) $\delta \ll R \ll \sqrt{a_m t_u} \ll \sqrt{a_p t_u}$ and (ii) $\delta \ll \sqrt{a_m t_u} \ll R \ll \sqrt{a_p t_u}$. Condition $\delta \ll R$ means that absorption of light occurs in a narrow surface layer of the particle. In this case the absorption cross section should be proportional to the surface area, i.e., in equality $\sigma_{abs} = \pi R^2 Q_{abs}$ quantity Q_{abs} is practically R-independent.

Regarding condition $R \ll \sqrt{a_p t_u}$, it means that the temperature field inside the particle is quasi-steady and should satisfy the Laplace equation. In spherical coordinates the only non-singular solution of this equation is a constant profile, so that $T(r, t)$ at $r < R$ is reduced to $T(t)$, where time t should be regarded as a parameter.

As for the field outside the particle, it is essentially different in cases (i) and (ii). In case (i) it is also quasi-steady and satisfies the Laplace equation. The corresponding solution is $T = T_s(t)R/r$. Equalizing the dissipated power of the laser beam $q_{laser} = q_0\sigma_{abs}$ to the one transferring to the liquid $q_{transf} = -4\pi R^2 k_m(\partial T/\partial r)_R$, where q_0 stands for the laser beam intensity [W/cm^2] and k_m is the liquid thermal conductivity, we easily find

$$T_s = \frac{\sigma_{abs}q_0(t)}{4\pi Rk_m} \equiv \frac{Q_{abs}Rq_0(t)}{4k_m}. \tag{6.25}$$

In case (ii) the temperature field in the liquid is t- and r-dependent. To determine T_s we may employ the energy conservation law. For simplicity, we consider a rectangular laser pulse. To a certain moment of time $t \le t_u$ the energy W absorbed by the particle is $\sigma_{abs}q_0 t$.

The absorbed energy is consumed to heat the particle (temperature rise T_s) and to heat an adjacent layer of the fluid. The former requires the energy $(4/3)\pi R^3 C_p\rho_p T_s$, the latter $4\pi R^2 2\sqrt{a_m t}C_m\rho_m T_s/2$, (to enhance the accuracy of the estimate we replace the profile of the temperature in the heated layer by its mean value $T_s/2$). Here C_p and C_m are the corresponding specific heats.

Equalizing W to the consumed energy and considering the equality as an equation for unknown T_s, one easily derives

$$T_s(t) = \frac{\sigma_{abs}q_0 t}{\frac{4}{3}\pi R^3 C_p\rho_p + 4\pi R^2 C_m\rho_m\sqrt{a_m t}} \equiv \frac{Q_{abs}q_0 t}{\frac{4}{3}RC_p\rho_p + 4C_m\rho_m\sqrt{a_m t}}. \tag{6.26}$$

The obtained $T_s(t)$ is a monotonic function of t, so the maximal temperature rise is achieved in the end of the laser pulse. Replacement $t \to t_u$ brings about the corresponding expression for the maximal temperature rise $\max_{0<t<t_u} T_s$ during the entire laser pulse.

Note, that while in case (i) T_s increases with an increase in R, in case (ii) an increase in R results in a decrease in T_s. It means that T_s reaches a maximum at $R \approx 2\sqrt{a_m t_u}$. This is the case indeed, see Fig. 6.4a.

The other cases are treated analogously. As a result simple analytical expressions for every possible heating regime are obtained [47]. A summary of these expressions is presented in Table 6.2.

To understand Table 6.2 we should clarify that, if a parameter is not explicitly specified in a cell of the table, its value may be any permitted for a given column. For example, condition $R \ll \sqrt{a_m t_u}$ in column *Small particles*, $R \ll \delta$ means any of the following three cases: $R \ll \delta \ll \sqrt{a_m t_u}$, $R \ll \sqrt{a_m t_u} \ll \delta \ll \sqrt{a_p t_u}$ and $R \ll \sqrt{a_m t_u} \ll \sqrt{a_p t_u} \ll \delta$. Other cells should be treated in a similar manner.

Surprisingly, these simple expressions provide quite a reasonable accuracy, see Fig. 6.4a, or [47] for more detailed comparison of the analytical and numerical results. Note also that existence of a maximum of T_s at any fixed value of t_u and a certain finite value of R is quite a common feature of the problem, see Fig. 6.4b.

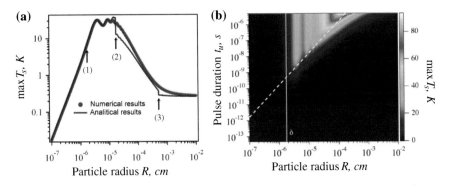

Fig. 6.4 The maximal temperature rise $\max\limits_{0<t<t_u} T_s$ at the surface of a spherical gold particle in water heated by a rectangular laser pulse with wavelength (in vacuum) 532 nm and intensity $q_0 = 5 \times 10^4$ W/cm^2. **a** Comparison of the analytical and numerical results $t_u = 50$ ns ($\delta \ll \sqrt{a_m t_u}$). The points of transitions from one heating regime to another are indicated by *black arrows*: (1) $R = \delta$, (2) $R = 2\sqrt{a_m t_u}$, (3) $R = 2\sqrt{a_p t_u}$; **b** The same quantity as a function of the duration of the pulse t_u and the particle radius R. The *dashed line* corresponds to $R = 2\sqrt{a_m t_u}$. Above this line the temperature rise becomes t_u-independent [47]

Table 6.2 Summary of qualitatively different regimes of the laser heating, for small and large metal particles in liquids [47]

Small particles, $R \ll \delta$		**Large particles, $R \gg \delta$**	
Condition	Maximal surface temperature rise	Condition	Maximal surface temperature rise
		Long pulses	
		$R \ll \sqrt{a_m t_u} \ll \sqrt{a_p t_u}$	$\dfrac{Q_{abs} R q_0}{4k_m}$
$R \ll \sqrt{a_m t_u}$	$\dfrac{Q_{abs} R q_0}{4k_m}$	$\sqrt{a_m t_u} \ll R \ll \sqrt{a_p t_u}$	$\dfrac{Q_{abs} q_0 t_u}{\frac{4}{3} RC_p \rho_p + 4 C_m \rho_m \sqrt{a_m t_u}}$
		$\delta \ll \sqrt{a_p t_u} \ll R$	$\dfrac{Q_{abs} q_0}{2\sqrt{\pi}} \dfrac{\sqrt{a_p a_m t_u}}{k_p \sqrt{a_m} + k_m \sqrt{a_p}}$
		Short pulses	
$\sqrt{a_m t_u} \ll R$	$\dfrac{Q_{abs} q_0 t_u}{\frac{4}{3} RC_p \rho_p + 4 C_m \rho_m \sqrt{a_m t_u}}$	$\sqrt{a_p t_u} \ll \delta \ll R$	$\dfrac{Q_{abs} q_0 t_u}{4(C_p \rho_p \delta + C_m \rho_m \sqrt{a_m t_u})}$

6.6 Conclusion

To summarize the results discussed in the present chapter we may say that recent studies of the old problems of light scattering by small particles and their light heating have revealed many new and sometimes unusual aspects of these problems. Specifically, light scattering by particles with small values of imaginary part of permittivity exhibits in the vicinity of the plasmon resonances the resonant anomalous scattering. Such a scattering is characterized by a giant concentration of the incident light in the particle and its near field zone, so that the characteristic values of the electric and magnetic fields as well as the one of the Poynting vector occur singular in the particle size.

The extinction cross section at the anomalous scattering achieves the maximal possible for this quantity value. It depends just on the frequency of the incident light, the order of the resonance (dipole, quadrupole, octupole, etc.) and does not depend on the particle size and its optical properties.

At the anomalous scattering the energy circulation in the near field zone corresponds to its delivery to the centers of the localized plasmons with further re-emission of the delivered energy to "infinity". However, this general scenario may be realized in a very complicated manner, when the vector Poynting field has a very unusual structure with a number of singular points. Details of this structure are very sensitive to the detuning of the incident light frequency from the exact plasmon resonant value. It provides a unique opportunity of controlled variations of electromagnetic field on a subwavelength scale.

The anomalous scattering exhibits the inverted hierarchy of resonances, when the resonant cross section and the range of complexity of the vector Poynting field in the near field zone increases with an increase in the order of the resonance.

The anomalous scattering may be accompanied by the resonant anomalous absorption. The anomalous absorption is realized at the plasmon resonance frequencies, when imaginary part of the particle permittivity is a small quantity, satisfying a certain resonant condition. The absorption cross section at the anomalous absorption has the maximal value which cannot be exceeded. At the very point of the anomalous absorption the scattering and absorption cross sections equal each other, while the extinction cross section equals half of the maximal extinction cross section at the anomalous scattering. After the proper scale transformation dependence $\sigma_{abs}(\varepsilon', \varepsilon'')$ for the anomalous absorption may be reduced to a certain universal form, which does not depend on the particle size and the order of the resonance.

The optimization of the energy release in particles made from various metals shows that for any of them there is a certain pair of the optimal particle size and wavelength of the incident light, maximizing the energy release, where the optimal size is about 10 nm and the corresponding optimal wavelength for different metals in a vacuum varies from green (potassium) to UV (aluminum, platinum) ranges of the spectrum.

The effectiveness of the energy release in the particles with respect to the one in the same bulk material increases with a decrease in imaginary part of the particle permittivity at the optimal wavelength. Thus, the most effective are the particles made from weakly dissipating metals.

The effect of the optimal size takes place at laser heating of a particle embedded in a transparent liquid too. However, in this case it is related to changes of the heating regimes with an increase in the particle size at a fixed value of the laser pulse duration, depends on the latter, optical and thermometric constants of the problem and may vary in broad limits.

We hope the results discussed in the present chapter will help numerous researchers to select the best for given applications particles and irradiation regimes and stimulate further study of this important and fascinating subject.

Acknowledgments This study was partially supported by RFBR, research project No 12-02-00391_a.

References

1. L. Rayleigh, Phil. Mag. **41**, 107, 274, 447 (1871)
2. C.F. Bohren, D.R. Huffman, *Absorption and Scattering of Light by Small Particles* (Willey, New York, 1998)
3. H.C. van de Hulst, *Light Scattering by Small Particles* (Dover, New York, 2000)
4. M.I. Mishchenko, J.W. Hovenier, L.D. Travis (eds.), *Light Scattering by Nonspherical Particles: Theory, Measurements, and Applications* (Academic Press, San Diego, 2000)
5. M.I. Mischenko, L.D. Travis, A.A. Lacis, *Scattering, Absorption, and Emission of Light by Small Particles* (Cambridge University Press, Cambridge, 2002)
6. R. Fuchs, K.L. Kliewer, J. Opt. Soc. Am. **58**, 319 (1968)
7. J.A. Fan et al., Science **328**, 1135 (2010)
8. J.B. Lassiter et al., Nano Lett. **10**, 3184 (2010)
9. F. Hao et al., Nano Lett. **8**, 3983 (2008)
10. J. Zhu et al., Nano Lett. **9**, 279 (2009)
11. S. Tretyakov, *Analytical Modeling in Applied Electromagnetics* (Norwood, MA, Artech House, 2003)
12. A.O. Govorov, W. Zhang, T. Skeini, H. Richardson, J. Lee, N.A. Kotov, Nanoscale Res. Lett. **1**, 84 (2006)
13. A.O. Govorov, H.H. Richardson, Nano Today **2**, 30 (2007)
14. G. Baffou, R. Quidant, C. Girard, Appl. Phys. Lett. **94**, 153109 (2009)
15. J.B. Khurgin, G. Sun, J. Opt. Soc. Am. B **26**, 83 (2009)
16. H.A. Atwater, A. Polman, Nat. Mater. **9**, 205 (2010)
17. G. Baffou, R. Quidant, F.J. Garcia de Abajo, ACS Nano **4**, 709 (2010)
18. V. Giannini, A.I. Fernandez-Domnguez, S.C. Heck, S.A. Maier, Chem. Rev. **111**, 3888 (2011)
19. G. Baffou, H. Rigneault, Phys. Rev. B **84**, 035415 (2011)
20. X. Li, N.P. Hylton, V. Giannini, K.-H. Lee, N.J. Ekins-Daukes, S.A. Maier, Opt. Express **19**, A888 (2011)
21. S. Barlett, W.W. Duley, Astrophys. J. **464**, 805 (1996)
22. L.D. Landau, E.M. Lifshitz, *Electrodynamics of Contineous Media* (Pergamon Press, Oxford, 1989) §8
23. N. Herlofson, Arkiv foer Fysik **3**, 257 (1951)
24. M.I. Tribelsky, A.E. Miroshnichenko, Y.S. Kivshar, Europhys. Lett. **97**, 44005 (2012)
25. B.S. Luk'yanchuk, M.I. Tribelsky,2 *Anomalous Light Scattering by Small Particles and Inverted Hierarchy of Optical Resonances* in Collection of papers dedicated to memory of Prof. M. N. Libenson (The St.-Petersburg Union of Scientists, Russia, 2005) pp. 101–117 (in Russian)
26. M.I. Tribelsky, B.S. Luk'yanchuk, Phys. Rev. Lett. **97**, 263902 (2006)
27. M.I. Tribel'skiĭ, JETP **59**, 534 (1984)
28. Z.B. Wang et al., Phys. Rev. B. **70**, 035418 (2004)
29. E.D. Palik, *Handbook of Optical Constants of Solids* (AP, New York, 1985–1998)
30. M.I. Tribelsky, Europhys. Lett. **94**, 14004 (2011)
31. M.V. Bashevoy, V.A. Fedotov, N.I. Zheludev, Opt. Express **13**, 8372 (2005)
32. B.S. Luk'yanchuk, M.I. Tribel'skiĭ, V. Ternovskiĭ, J. Opt. Technol. **73**, 7 (2006)
33. B.S. Luk'yanchuk, et al., EEE Photonics Global@Singapore (IPGS), vols. 1&2, pp. 187–190 (2008)
34. Z. Ruan, S. Fan, Phys. Rev. Lett. **105**, 013901 (2010)
35. B.S. Luk'yanchuk et al., New J. Phys. **14**, 093022 (2012)
36. X. Huang, P.K. Jain, I.H. El-Sayed, M.A. El-Sayed, Lasers Med. Sci. **23**, 217228 (2008)

37. I. Brigger, C. Dubernet, P. Couvreur, Adv. Drug Deliv. Rev. **54**, 631651 (2002)
38. R.R. Anderson, J.A. Parrish, Science **220**, 524–527 (1983)
39. G. Han, P. Ghosh, M. De, V.M. Rotello, NanoBioTechnology **3**, 40–45 (2007)
40. A.G. Skirtach, C. Dejugnat, D. Braun, A.S. Susha, A.L. Rogach, W.J. Parak, H. Möhwald, G.B. Sukhorukov, Nano Lett. **5**, 13711377 (2005)
41. A.N. Volkov, C. Sevilla, L.V. Zhigilei, Appl. Surf. Sci. **253**, 6394–6399 (2007)
42. H.H. Richardson, M.T. Carlson, P.J. Tandler, P. Hernandez, A.O. Govorov, Nano Lett. **9**, 1139–1146 (2009)
43. G.W. Hanson, S.K. Patch, J. Appl. Phys. **106**, 054309 (2009)
44. E. Sassaroni, K.C.P. Li, B.E. O'Neill, Phys. Med. Biol. **54**, 5541 (2009)
45. S. Bruzzone, M. Malvaldi, J. Phys. Chem. **113**, 15805 (2009)
46. G. Baffou, H. Rigneault, Phys. Rev. B **84**, 035415-1-13 (2011)
47. M.I. Tribelsky et al., Phys. Rev. X. **1**, 021024 (2011)

Part III
Surface and Thin Films Phenomena and Applications

Chapter 7
Laser-Induced Local Oxidation of Thin Metal Films: Physical Fundamentals and Applications

Vadim P. Veiko and Alexander G. Poleshchuk

Abstract Local laser oxidation of thin metal films allows recording of an optical image on thin films with the highest resolution and high productivity at the same time, and without distortions specific to laser ablation. A technique for writing of diffractive optical elements was developed on the basis of this process. Laser-matter interaction physics and laser technology underlying this method are described in this chapter.

7.1 Introduction

In our opinion, laser oxidation of thin metal films is a successful example of the discovery of a new effect, its explanation and application. Furthermore, although all these developments took place at the beginning of the laser era, the interest in these has not decayed up to now, not to mention a new technique (short-pulse laser writing), at this point, a new physics was also required due to the coming of the "nano" era.

Let us briefly describe these developments in the introduction because the authors were their direct participants. The effect that formed the basis of the research field under discussion was first observed in our experiments aimed to determine the evaporation thresholds q_{ev} of thin chromium films exposed in the optical projection

V. P. Veiko (✉)
Mechanics and Optics Chair of Laser Technologies and Applied Ecology,
St. Petersburg National Research University of Information Technologies, 49 Kronverksky pr.,
St. Petersburg 197101, Russian Federation
e-mail: veiko@lastech.ifmo.ru

A. G. Poleshchuk
Institute of Automation and Electrometry Siberian Branch of Russian Academy of Science,
Academician Koptug ave. 1, Novosibirsk 630090, Russian Federation

V. P. Veiko and V. I. Konov (eds.), *Fundamentals of Laser-Assisted*
Micro- and Nanotechnologies, Springer Series in Materials Science 195,
DOI: 10.1007/978-3-319-05987-7_7, © Springer International Publishing Switzerland 2014

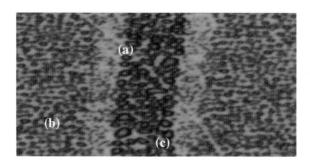

Fig. 7.1 Application of water vapor to the surface of three irradiated chromium strips: **a** film structure at CrO_2-coated areas, **b** unirradiated chromium, and **c** area coated with Cr_2O_3 and covered by a water film

configuration to laser pulses with widths of 10^{-3} to 10^{-7} s [1], (Leningrad, Institute of Fine Mechanics and Optics, Technical University, 1968–1969). Namely, an exposure of the films to radiation with a beam fluence lower than q_{ev} resulted in the formation of a latent image, which was developed by breathing on it and equally rapidly disappeared upon moisture drying.[1] This experiment was understood in half a year, when we attributed the appearance of the selective hydrophilic behavior of the irradiated area to the laser action. In the next half a year, the resistance of the irradiated areas to the conventional solvents for chromium (!) and the nature of the phenomenon we discovered was revealed. In fact, a new lithographic method was discovered, the maskless laser patterning. Originally, recrystallization of the structure, diffusion of the film material into a substrate, and interaction with the atmosphere were regarded as the possible reasons for the selective solubility of laser-heated Cr films.

However, experiments with an exposure preceded by the full annealing, with sublayers of refractory metals, with transparent SiO_2 coatings, etc., led us to conclude that the oxidation of chromium in the atmosphere under laser heating plays the key role here [2].

Similar phenomena, such as the local thermal decomposition of metal-organic compounds, reduction of metal oxides, etc., were successfully demonstrated soon after the discovery of laser oxidation [2]. As a result, there appeared an important publication in "Reports of Academy of Sciences of the USSR" [3] (received by the editorial office in 1971!). Then, the oxidation kinetics was studied by analyzing the variation of the electrical resistance of a film, which allowed us to develop the basic theory of pulsed laser oxidation, published in [4, 5].

The second part of this history proceeded in Novosibirsk where a laboratory of the Institute of Automation and Electrometry, Russian Academy of Sciences was engaged in the development of a technology for manufacture of diffractive optical

---

[1] The reader can now carry out this experiment independently; we repeated it 40 years later (1968–2008) and obtained the same result; however, this time a micrograph was taken with a high-magnification microscope (see Fig. 7.1).

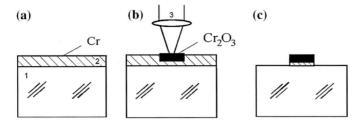

Fig. 7.2 Principle of the lithographic process based on the laser-induced local oxidation of thin metal films: before irradiation (**a**), after irradiation (**b**), after development (etching) (**c**). (*1* substrate, *2* film, *3* laser beam)

elements (DOEs) [6]. The scientists of the laboratory succeeded in finding a very high-precision tool for this purpose, laser generator of ring images, and the above thermochemical idea was granted a second birth here. Unique DOEs for astronomy [7] and other elements were designed and fabricated by this method.

A continuation of this history is being written right before our eyes due to the increasing requirements to DOE resolving power, need for development of UV and x-ray DOEs, and search for new ways to improve the resolution of the thermochemical technology for optical pattern recording.

The principle of the thermochemical writing is illustrated by Fig. 7.2. The oxidation-patterning method takes advantage of a local irradiation of a thin metal film in an oxidative environment (air, oxygen, etc.). The oxide layer to be produced enhances the chemical stability of the irradiated film area. If a sample is placed in a suitable etchant, the unirradiated areas of the metal film quickly dissolve, while the irradiated zones (IZs) remain on the substrate (Fig. 7.2) and form a topological structure on its surface.

In this chapter, the principal characteristics of the chemical processes activated by laser heating will be illustrated by the example of the oxidation reaction that has found wide practical application.

7.2 Physics of Heterogeneous Laser Oxidation: Diffusion, Adsorption and Chemical Kinetics

Laser heating of solids to below the evaporation temperature can initiate chemical reactions on the surface or in the underlying heated layer. A general characteristic of such reactions (initiated by laser light) is the feedback arising in all cases when the optical properties of the reaction products differ from those of the primary substance [8]. The feedback may be positive (if the absorbance of the reaction products is higher than that of the primary compound) or negative (if the opposite is true). Rather than going into details of the kinetics of laser thermochemistry, considered in depth in Bunkin et al., Bauerle, Ibbs and Osgood [8–10], we emphasize here an important fact that plays a crucial role in the applications to be discussed. The essence of the matter is that the dynamics of the behavior of a system and the means of controlling this behavior are not so important as the possibility of controlling the dimensions and configuration of the chemical reaction zone, i.e., its spatial characteristics. Precisely this has become possible through the pulsed thermochemical action of the laser light. Under pulsed irradiation, the depth of the heated layer, $(a\tau)^{1/2}$ (a is the thermal diffusivity, τ is the pulse width), ranges from a fraction of a micrometer (at $\tau = 10^{-9}$ s) to several tens of micrometers (at $\tau = 10^{-3}$ s), whereas the thickness of the reacting layer is even smaller since the chemical reaction rate is often limited by the slower processes in which primary compounds are delivered (diffusion) and the reaction products are removed [11]. The configuration of the reacting layer differs very little from that of the optical image at small pulse durations. Thus, one can obtain a latent **(thermochemical) image** in the irradiated zone. The changes of the properties in the irradiated regions (in the heated zone) make it possible, in principle, to reveal the configuration (topology) of these regions by the widely-used microelectronic technology of selective etching, as well as by the selective (including epitaxial) growth. The thermochemical action of the laser light forms the basis of such processes [11].

7.2.1 Laser Oxidation of Metals

The existing approach to analysis of the laser oxidation of metals in the atmosphere at time intervals of 10^{-8} to 10^{-3} s is based on the concept of the oxidation as a heterogeneous process occurring under the solid-phase diffusion-kinetic control [12].

The rate at which the oxide layer thickness b increases as a function of temperature T is determined by the ratio between the rates of the aforementioned processes. Under isothermal heating, the oxidation kinetics is limited by the slowest stage, which is diffusion. In this case, the Wagner equation is valid:

$$\frac{db}{dt} = \frac{B}{b} \exp\left(-\frac{T_a}{T}\right),$$

(7.1)

where B is the constant of the parabolic oxidation law, T_a is the activation energy of the diffusion processes (in Kelvins), T is the temperature of the film and t is time.

To calculate the oxide layer thickness under non-isothermal laser oxidation, we use the approach developed in [12]. According to this approach, if the temperature of a film is substantially lower than T_a, the activation exponential sharply grows with increasing temperature, and the main contribution to the rise in the oxide film thickness occurs when the temperature T is close to the maximum value T_{max}. Hence, we can determine the thickness of the oxide layer formed under laser irradiation as the oxide layer thickness under isothermal heating for some equivalent time (shorter than the irradiation time) t_e:

$$b = \sqrt{2B \exp\left(-\frac{T_a}{T_{max}}\right) t_e} \quad, \tag{7.2}$$

The value of t_e is determined by the manner in which the temperature varies with time during the laser irradiation. In particular, if the maximum temperature is reached at an instant t_0 before the end of this process, t_e can be estimated as

$$t_e = \sqrt{\frac{2\pi T_{max}^2}{T_a \left|T_{tt}''(t_0)\right|}} \quad, \tag{7.3}$$

where $T_{tt}'' = \partial^2 T/\partial t^2$.

Thus, having found the time dependence of the temperature of a film at each of its points, we can determine the thickness of the oxide formed.

The subsequent etching of the film forms its topology via removal of its parts unprotected by a sufficiently thick oxide layer.

7.2.2 Laser Oxidation at Short Pulse Width

In the case of chromium and some other metals (Ni, Ti, Al, etc.), when a dense oxide film formed (Cr_2O_3, Ni_2O_3, TiO_2, Al_2O_3) protects the underlying layers from direct interaction with the atmosphere, this theory sufficiently well describes experiments at milli-, micro- and nanosecond laser exposure durations in a solid phase.

At the shorter pulse widths, the diffusion in the gas phase and adsorption of oxygen atoms may become a limiting stage of this heterogeneous reaction. The adsorption time (for a chemically clean metallic surface, when the adsorption activation energy is low enough), i.e., the time in which the surface layer of a metal is saturated with oxygen t_{ad}, is $t_{ad} = 4/nva$, where n is the oxygen concentration in the gaseous medium, v is the average thermal velocity of particles in the gas, and a is the lattice constant of a metal. For example, at normal pressure of atmospheric oxygen and at room temperature, $t_{ad} = 2 \cdot 10^8$ s for typical metals ($a^2 = 10^{-15}$ cm^{-2}). At a laser pulse width $\tau = 10^{-8} - 10^{-3}$, $t_{ad} \leq \tau$ and the adsorption is not the limiting stage of the oxidation reaction. This estimation is consistent with the results of laser oxidation experiments in a vacuum (residual pressure 0.1 Pa) and in air at a pulse width $\tau = 10^{-8}$ s. However, for shorter laser pulses or lower oxygen pressures in the presence of defects on the surface being oxidized, t_{ad} increases sufficiently for the gas diffusion to become the limiting stage [13, 14]. This situation has been observed, for example, in experiments on oxidation of diamond in the normal atmosphere [15].

7.2.3 Laser Oxidation with Exposure at a High Pulse Repetition Rate

If the resulting oxide is porous and cannot protect the underlying metal layers from further oxidation, the process kinetics will be determined by purely chemical (electronic) phenomena, which are much less sluggish than those related to the diffusion in solids and even the adsorption and diffusion of gases, and the apparent reaction rate must be determined by the temperature reached in the process and the exposure time in accordance with the Arrhenius kinetic equation. Analysis of the phase and chemical transformations occurring under laser heating in these metals and alloys suggests that they are generally governed by the laws of chemical thermodynamics. In this regard, the optimal modes are those with exposure at a high pulse repetition rate and spatial overlapping of the illuminated surface regions, when the temperature distribution is assumed to be quasi-stationary. In the case of a pulsed treatment, it becomes necessary to additionally use the kinetic approach, which makes it possible to estimate the time necessary for the occurrence of all the thermodynamically possible processes and to compare this time with the actual conditions [16].

7.3 Oxidation Lithography: Principles, Accuracy and Resolution

The laser-induced local oxidation of thin metal films was studied in detail in [17–20] and summarized in [11]. Thin chromium films widely used in the thin-film technology were chosen as objects of study.

The oxide layers formed upon laser exposure for even 10 ns are very stable and enhance the chemical stability of the irradiated film area (Fig. 7.2). The samples were placed in various etchants to examine the solubility of the metal films. It was found that the etching time for a film area irradiated even with a single pulse is substantially longer (by up to a factor of 50) than that for unirradiated films. This time depends on the experimental conditions (film thickness, pulse width, pulse energy, type and concentration of the etchant, etc.).

In general, the accuracy and the resolution of the laser-induced local oxidation method are preliminarily determined by the temperature field distribution (thermal image) within the irradiation zone. The distortion of the thermal image depends on the thermo-physical properties of the heated film and on the temporal parameters of the irradiation; it is due to the penetration of heat outside the IZ. The heat penetration length being given by $(a\tau)^{1/2}$, a radical way to enhance the accuracy of the method is to reduce the laser pulse width τ [11, 21].

The influence exerted by various parameters, such as the pulse width, pulse shape, light-flux density, size and shape of the IZ, on the accuracy and resolution of the lithographic process has been theoretically analyzed by means of a numerical simulation of the heat equation together with the oxidation kinetic equation [18]. The results of this simulation are shown in Fig. 7.3.

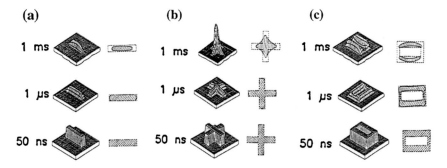

Fig. 7.3 2D images of the oxide layers in the irradiated zone for a *line* (**a**), a *cross* (**b**), and a *rectangular aperture* (**c**) and various pulse widths. (—: optical image to be recorded, ///: recorded image)

Three different types of features were chosen to characterize the following cases (Fig. 7.3): a line (a), an intersection of two lines with equal widths (b), and a "window" formed by two lines with different widths (c).

Figure 7.3 displays the calculated oxide-thickness distribution on the surface of a thin Cr film upon its exposure to 1 ms, 1 μs and 50 ns laser pulses. The same figure presents 1 nm oxide-thickness isolines. As mentioned above, the metal film will be etched in an appropriate etchant up to this level. Hence, these isolines represent the shape of the recorded image. It can be seen in Fig. 7.3 that the recorded image is distorted due to the interaction of the thermal fields from different parts of the optical image—IZ. This distortion is less pronounced for short pulses due to the shorter time for heat penetration beyond the irradiated zone. This circumstance leads to a distortion of the complex thermal field and to a distortion of the recorded image [22].

In the case of nanosecond laser pulses, the heat is concentrated in the IZ, the recorded complex pattern is nearly undistorted and the recording accuracy is independent of the pattern shape and size, in contrast to the cases of microsecond, and, especially, millisecond modes. The nanosecond mode is particularly suitable for the projection laser lithography, in which the recorded image consists of complex features of different shapes and sizes. The theoretical thermo-physical maximum resolution of the optical image in this case is about 1 μm for the 1.06 μm wavelength (Nd:glass, Nd:YAG laser) [23]. At the same time, the theoretical thermochemical resolution appears to be markedly higher due to the photo-thermo-chemical feedback and other phenomena (see below).

The microsecond mode is suitable for the scanning thermochemical image recording by a pulsed [24] or a continuous-wave (e.g., Ar-ion) laser source in a ring laser generator configuration [25] (see below). In this case, the irradiation time depends on the laser spot radius r_0 and scanning speed v_{sc} as $\tau = 2r_0/v_{sc}$. The accuracy, resolution and contrast of the recorded image also strongly depend on the type of etchant.

7.4 Applications of Oxidation Lithography: Thermochemical DOE Writing

The present-day diffractive optics requires development of techniques for repro-ducible formation of microstructures with feature sizes below 0.5 μm at a light field of 200–300 mm and even larger. The absolute accuracy of a pattern should be about 1/20 of the smallest diffraction zone size. In some cases, the diffractive structure should be produced on convex or concave optical surfaces with a large thickness [26]. The surface flatness of the thick and large optical substrate for the DOE should be better than λ/20. Substrates of this kind are very inconvenient for being coated with uniform and thin photoresist layers. These limitations can be overcome by fab-rication of the DOE by the above-described resistless thermochemical technique, which highly improves its accuracy and quality.

Another modern requirement is that a DOE with a specific ring structure should be generated. For example, the DOE designed to test and certify the aspheric wave front of a primary mirror in advanced telescopes must have circular diffractive structures with an accuracy of 10 nm at a minimal spacing less than 1 μm and overall sizes of several hundred millimeters [27]. These elements can only be fabricated with the desired accuracy and quality based on thermochemical method by means of a polar coordinate laser pattern generator or circular laser writing systems (CLWS).

7.4.1 Circular Laser Writing System for Direct DOE Fabrication

The laser thermochemical writing method was first applied to fabricate high-quality DOEs with amplitude transmission, [6, 28]. The irradiation was done with a focused laser beam under circular scanning of the substrate [25]. There are two main critical parameters of a machine used for this purpose: (1) laser power, which should be suf-ficient for heating a chromium film to the oxidation temperature, with consideration for the rotation speed and focused spot size, and (2) accuracy limit at the maximum rotation speed corresponding to the minimum exposure duration and size of the heat-affected zone. The writing laser power and sizes of written tracks are estimated in Sect. 7.4.2.

A CLWS [25, 29] is shown schematically in Fig. 7.4. The system can be subdivided into several main units:

Rotation unit. The substrate (diameter up to 300 mm and thickness up to 25 mm) with a deposited chromium film is fixed on the faceplate of the air-bearing spin-dle by a vacuum chuck. The angular encoder and the system of frequency mul-tiplication ensure a certain number of clock pulses per revolution (3–4 million, angular accuracy of about 1 arc sec.). The pulses are used to synchronize the laser writing beam modulation with the substrate rotation (rotation speed 300–800 rpm, instability 10^{-6}).

Radial displacement unit. The air-bearing linear table and the linear DC motor provide displacement of the optical writing head. The laser interferometer controls

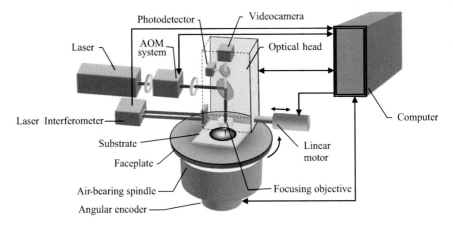

Fig. 7.4 Circular laser writing system

the displacement of the linear table with an accuracy of about 0.6 nm. The accuracy of positioning along the radial coordinate is better than 10–20 nm.

Writing beam power control. Two acousto-optic modulators (AOM system) control the beam power of the 2W laser operating at a wavelength of 532 nm. The use of two modulators allows us to extend the dynamic range of beam power control and to separate the beam modulation functions: compensation for the change in the linear scanning speed of the writing beam upon a radial coordinate displacement and modulation according to the microstructure design. The first AOM included in the feedback loop also suppresses fluctuations of the laser radiation power. The second AOM performs a binary or analog modulation (12 bits) of the laser beam, depending on the DOE type.

Optical writing head. The optical head is placed on the faceplate of air-bearing spindle and focuses the laser beam into a spot with radius r_0 of about 0.3 μm. The autofocus system holds the focal plane of the focusing objective lens (N.A. $= 0.65$) on the substrate surface during the writing process. A photodetector is used to measure the intensity of the beam reflected from a recording material.

7.4.2 Direct Thermochemical DOE Writing. DOE Writing Modes

The CLWS was used in experiments on irradiation of the chromium films [7, 25] at scanning velocities V_{sc} of 0.044–875 cm/s, which corresponds to writing radii of 0.007–140 mm at a rotation speed of 600 rpm and irradiation time τ ranging from ~100 ms in the central zones of DOE to ~50 ns at its periphery. The optimum power P at the writing spot for thermochemical writing was within 10–50 mW, which is sufficient for providing a power density of about $10^6 - 10^7$ W/cm^2 to heat the chromium film to its oxidation temperature. The exposed films were developed in an etchant composed of six parts of a 25 % $K_3Fe(CN)_6$ solution and one part of a

Fig. 7.5 Chromium test patterns (**a**) and dependence of recorded line width on the writing laser (**b**)

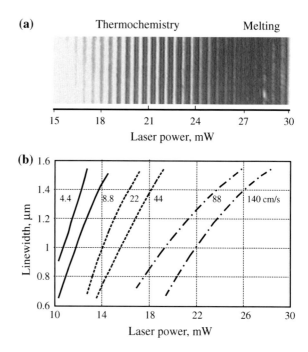

25 % NaOH solution. Cr-film-coated glass substrates (with a chromium thickness h_{Cr} of about 80 nm) were used in the experiments. Figure 7.5a shows the developed test pattern recorded at a scanning velocity of about 90 cm/s. The dependence of the recorded line width on the writing power is shown in Fig. 7.5b for different scanning velocities. According to these data, the laser power during the writing process is controlled so that the lines of equal width are written at any radius. Figure 7.5b demonstrates that a 40-fold change in the scanning velocity results in only a 3-fold variation of the laser power required for writing a 1-μm line. Typical ratios between the maximum and minimum powers (dynamic range) do not exceed 5 for writing of DOEs with diameters as large as 300 mm. So, the writing on Cr films appears to be a quite suitable method for production of high-quality amplitude DOEs and masks.

DOE writing accuracy. The writing process introduces errors into the written microstructure. The main source of errors is the incorrect choice of the writing power, radiation power and writing strategy [7]. We consider the influence of the chosen writing radiation power on the angular and radial sizes (width and length) of the recorded line (track) in the thermochemical writing. After the writing beam power is switched-on at the instant of time x_1, the film temperature grows linearly with increasing radiation intensity and exposure time. After the end of the radiation pulse, at the instant of time x_2, the film starts to cool down. The time of cooling of a chromium film with a thickness of about 80 nm to the temperature at which the thermochemical changes cease to occur ($T \sim 0.4 - 0.5 T_0$) is shorter than 0.05 μs [30] which approximately corresponds to the duration of the leading edge of the light pulse. If T_{tc} is the threshold temperature for the thermochemical changes in a film, then the coordinates of the points of beginning and end of the written track are x'_1 and x'_2, rather than the given x_1 and x_2. Thus, the thermochemical writing is characterized by errors in the coordinates of the beginning and end of the written track [31]. The extent of these distortions is primarily determined by the properties of chromium, size and motion speed of the focused beam, and threshold-surpassing coefficient. An angular shift of the track takes place, and this fact should be taken into account in fabrication of off-axis DOEs and optical angular scales by CLWS.

Evaluation of Performance Parameters. The resolution for linear structures written with a circular pattern generator depends on their orientation and radial position. Circular lines coinciding with the scanning direction can be written with an excellent quality and a minimum spacing of down to 0.6 μm. Figure 7.6a shows a SEM micrograph of a circular grating with a 1.2-μm spacing, written in a chromium film at a 28-mm radius. Arbitrarily oriented lines are usually wider and have rougher edges due to the radial raster pitch and angular quantization. Figure 7.6b shows a resolution test (test images with 2-, 1.6-, 1-, and 0.8-mm line spacing) for lines oriented at ±45° relative to the scanning direction, written with a 0.25-μm pitch at a radius of 20 mm.

The most presently interesting question is about what is the theoretical and experimental resolution limit for DOE structures produced by thermochemical technology [32], and we are going to make it clear in the following section.

Figure 7.6c shows a test pattern with a track width of about 0.3 μm (SEM micrograph). The track profile (furnished by AFM) is shown in Fig. 7.6d. It can be seen that the track profile has a peaked (0.3-μm wide) shape at the minimum writing power, whereas the writing spot has a nearly Gaussian distribution with a larger width (0.6 μm). This result is explained below and confirmed by a theoretical evaluation.

7.4.3 Application of DOE for Aspheric Optics Testing

To inspect the primary mirrors of telescopes with diameters of 6.5 and 8.4 m (MMT, Magellan, and LBT projects), DOEs have been developed and fabricated to control the null lenses for the visible (633 nm) and IR (10.6 μm) spectral ranges [26, 27]. The reflective DOEs were fabricated using CLWS by the thermochemical technology on a chromium film (about 60 nm thick) deposited on high-quality astro-glass-ceramic substrates. The DOEs are axial-amplitude reflective structures with a diffraction efficiency in the +1st order of about 5 %. The main parameters of the fabricated DOEs (diameter D, number N of circular zones, and the minimum size l of features) are listed in Table 7.1.

The writing errors of the CGH structure of these two DOEs were within 50 nm, as verified by the methods described in [7]. The maximum wavefront error was under $\lambda/15$–$\lambda/20$. Reflective axial Fresnel zone plates with the same D and l were simultaneously written and tested (using the Fizeau interferometer) for additional control over the CLWS writing accuracy. Because the writing time of DOEs was about 4 h for the MMT test and more than 5 h for LBT, the coordinate drift was

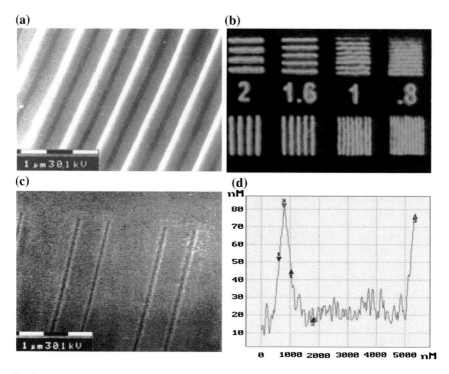

Fig. 7.6 Resolution tests: SEM micrograph of a circular diffractive element with 1.2-μm spacing (**a**), test images with 2, 1.6, 1, and 0.8-μm line spacing (**b**), SEM micrograph of 0.3-μm lines (**c**), and their profile (**d**)

Table 7.1 The main parameters of the fabricated DOEs

Telescope	Primary mirror	DOE specifications (633 nm)		
		D (mm)	N	l (μm)
MMT	6.5 m, f/1.25	136	32000	0.8
LBT	8.4 m, f/1.14	210	64000	0.6

minimized by the method of periodic correction of the origin of coordinates and of the trajectory of the CLWS spindle rotation [33].

An error ($\Delta n \sim 10^{-5}$) was revealed in the refractive factor of the lens glass in an inspection of the null lens for the MMT and Magellan mirrors; the error resulted in a spherical aberration by about λ.

Figure 7.7a presents DOE (diffractive structure 210 mm in diameter) fabricated for testing the 8.4 m LBT telescope mirror. Figure 7.7b, c show an interference pattern and a map of the surface of the 6.5 m Magellan primary mirror, obtained with the corrected null lens. The RMS errors were below 14 nm. The measurements were done with a Twyman-Green interferometer with phase modulation and computer processing. The general view of a large binocular telescope (LBT) and a new space

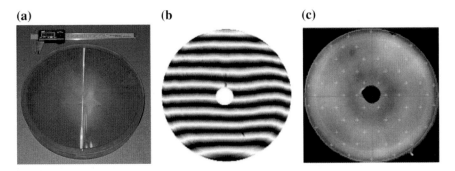

Fig. 7.7 DOE for testing the null lens for the primary mirror of LBT telescope (**a**), interference pattern of the Magellan telescope mirror (**b**), and phase map of the LBT null lens wavefront tested with DOE (**c**)

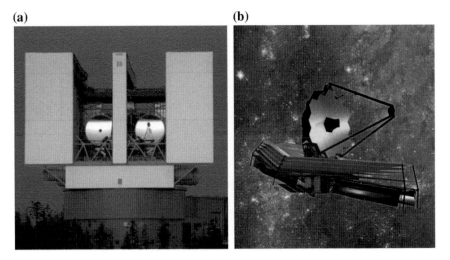

Fig. 7.8 Large telescopes with aspheric mirrors tested with thermochemically written DOEs: **a** large binocular telescope, **b** new James Webb Space Telescope

J. Webb observatory with a 6.5 m telescope having aspheric mirrors tested by the thermochemically written DOEs [34, 35] are shown in Fig. 7.8.

Let us note at the end of this section that the experimental resolution of DOE produced by CLWS ring pattern generators always exceeded that predicted by the simple thermal approach described above. We are going to make an attempt to explain the reasons for this behavior and to understand the limitations on the thermochemical writing resolution in the next section.

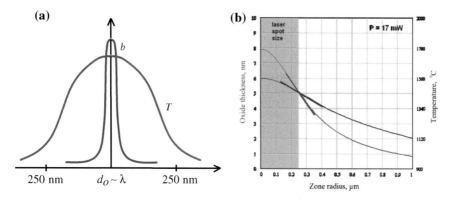

Fig. 7.9 Illustration of the thermochemical peaking effect: **a** the oxide thickness distribution is substantially narrower than that of the temperature distribution, **b** calculation of the thermochemical peaking: oxide-thickness and temperature distributions in a zone 1 μm in size

7.5 Super Resolution at the Thermochemical Writing

7.5.1 Thermochemical "Peaking"

One of the reasons for the high resolution of the thermochemical writing (the so far obtained minimum feature size d_{min} is 200 nm) consists in the so-called thermochemical "peaking" of the 'thermal' image [36]. The thermochemical peaking is based on the very sharp dependence of the chemical oxidation rate on temperature (7.1), which results in a strong restriction of the chemical reaction area in the range close to the maximum T_M (proportional to the intensity distribution peak).

It is apparent that a thermochemical image with the minimum size can be obtained for a non-uniform laser intensity distribution in a focal spot, with a maximum at the center. For the ordinary case of a Gaussian distribution (Fig. 7.9), the ratio between the characteristic size r of the image element formed in the film and the irradiated area radius $r_0(r < r_0)$ can be found from expressions (7.2) and (7.3) on the assumption that $T(r)/T(r=0) \approx \exp\left[-(r/r_0)^2\right]$ as [32]:

$$\frac{r}{r_0} = \sqrt{2\frac{T_0}{T_a} \ln \frac{b_0}{b_r}}, \tag{7.4}$$

where $T_0 \approx 1220$ K is the temperature sufficient (according to the experimental data) for a protective oxide layer to be formed on a chromium film.

Thus, the minimum size of the image element formed in the film by laser oxidation followed by etching depends on the size of the irradiated area and is determined by the properties of the oxide and by the etching conditions.

In particular, at $b_r/b_0 = 0.8$, the minimum element size r_{min} is $\sim 0.1r_0$. This means that the element size can be in principle markedly smaller than the focal spot radius and, correspondingly, substantially smaller than the radiation wavelength.

In general, the resolution $\delta = 1/2d_{min}$ of a thermochemical image exceeds that of a temperature image due to the strong dependence of the oxide growth rate on temperature. The gain in the resolution, roughly given by the ratio between the first derivatives with $1/b |\partial b/\partial x|$ and $1/T |\partial T/\partial x|$, reaches a value of 20 in our case, in which always grad $T <$ grad b, (Fig. 7.9b). A thorough theoretical analysis of this phenomenon was made separately in [32, 36].

7.5.2 Positive Feedback at the Thermochemical Action

The true (experimental) resolution of the thermochemical writing is even higher, compared with that calculated in the preceding paragraph, due to the photo-thermochemical feedback (between the absorbed power density, oxide thickness, and temperature) accelerating the oxide formation in the "hottest" area. It is of interest in this context to underline the specific features of multiple irradiations of Cr films (at a lower laser power and higher scanning speed) to improve the quality and reliability of the continuous thermochemical recording of separate DOE lines.

After the first pass, an oxide layer is formed whose thickness may be insufficient for protection from selective etching; the second and the subsequent passes considerably enhance the absorption of radiation and make the oxide layer thicker. For this reason, this thin oxides layer serves as a positive feedback element under repeated exposures: an increase in the absorption capacity of the previously irradiated areas results in a rise in their temperature and thereby leads to formation of a thicker oxide layer (Fig. 7.10). As a consequence, the difference between the oxide thicknesses at the center and on the periphery of recorded lines of multiply irradiated areas increases sharply, which results in a more contrasting, more precise and more reliable recording of separate tracks and enables a higher recording resolution. Results of corresponding physical model simulation is shown at the Figs. 7.10 and 7.11 [37].

The increase in the oxide thickness after different numbers of passes is shown in Fig. 7.11. Furthermore, the formation of an oxide layer also changes the thermophysical properties of a film: its thermal conductivity decreases. Accordingly, the local increase in the oxide layer thickness upon every next pass reduces the heat diffusion and enhances the contrast of lines and the recording contrast. The possibility of raising the writing contrast and improving the resolution by multipass irradiation has been recently confirmed experimentally [37].

Let us note that both the above-mentioned factors, the thermochemical peaking and positive feedback, make possible to obtain, in principle, a thermochemical image with a resolution even better than the optical image resolution upon a multipass irradiation due to higher contrast of the thermochemical recording, compared with the laser intensity distribution.

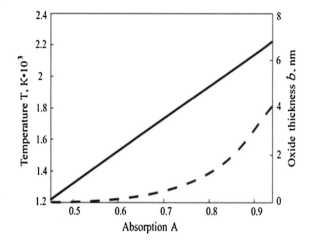

Fig. 7.10 Temperature of the irradiated area (*solid line*) and the oxide thickness (*dashed line*) versus the film absorption (for the center of the spot at a power $P = 3\,\mathrm{mW}$)

7.5.3 Effect of Short Pulse Lasers: Oxidation Plus Structural Modification

Another and, seemingly, apparent way to improve the thermochemical writing resolution is to use ultrashort laser pulses, because the shorter the pulse width, the smaller the differences between the "thermal" and optical images: actually, the size of the heat-affected zone for $\tau = 10^{-13}$ s is only 1 nm. At the same time, the situation appears not too promising upon a closer examination, because the usual rate-limiting step of the heterogeneous oxidation reaction at the chromium-atmosphere interface is a diffusion. It is a common knowledge that the heterogeneous diffusion is a rather slow process and has no sufficient time to occur in the cases of pico- and femtosecond pulses.

However, experiments have shown a selective solubility of the exposed and unexposed areas for all kinds of pico- and femtosecond lasers [38]. This fact indicates that chemical (or structural) changes occur in Cr films under the action of ultrashort laser pulses.

Oxidation. As it is clear from the aforesaid, the author's (see, e.g., [1–5]) and other researchers' [6–10] experience demonstrates that the main reason for the selectivity of chromium etching under the normal conditions is its oxidation to give the most stable oxide Cr_2O_3. Therefore, efforts were made to confirm the presence of this oxide presence in a study of the structure and composition of irradiated Cr films [38].

A micro-Raman spectroscopy served to test Cr films irradiated at three different energy levels (different numbers of pulses in the beam scanning technique described above). The Raman spectrum of an unexposed chromium film shows a weak broad asymmetric band peaked at $660\,\mathrm{cm}^{-1}$ (Fig. 7.12a) (the peak at $554\,\mathrm{cm}^{-1}$ is specific to Cr_2O_3 in a Raman spectrum). At the lowest irradiation energy (1st run), the intensity

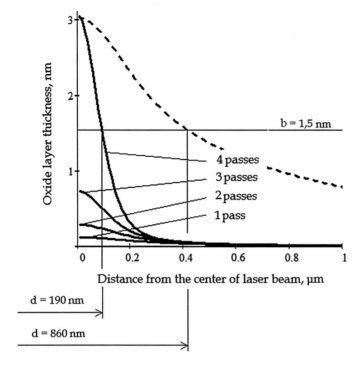

Fig. 7.11 Spatial distribution of the oxide thickness at different numbers of passes ($P = 3\,\text{mW}$, $r_0 = 0.2\,\mu\text{m}$, $h_{Cr} = 30\,\text{nm}$, $V_{sc} = 0.5\,\text{m/c}$). The *dashed line* corresponds to a single pass at $V_{sc} = 0.069\,\text{m/c}$ ($b = 1.5\,\text{nm}$ is the minimum thickness of the etching-resistant oxide film)

of the $660\,\text{cm}^{-1}$ peak grows, while its width decreases (Fig. 7.12b). The increase in the intensity shows that the amount of the oxidized substance (oxide thickness) grows, whereas the change in the peak width is indicative of the structural modifications of the material. At a higher exposure energy (2nd run), the micro-Raman study demonstrated a further rise in the intensity of the $660\,\text{cm}^{-1}$ peak in a Raman spectrum, along with that of the $554\,\text{cm}^{-1}$ peak (Fig. 7.12c), which is specific to Cr_2O_3 chromium oxide. Also, the intensity of the $554\,\text{cm}^{-1}$ peak becomes substantially higher than that of the $660\,\text{cm}^{-1}$ peak. For the highest exposure energy (3rd run), the Cr_2O_3 band dominates over the whole Raman spectrum of the sample, whereas the band at $660\,\text{cm}^{-1}$ is either very weak or is not observable at all (Fig. 7.12d).

To explain the origin of the $660\,\text{cm}^{-1}$ peak, an x-ray spectral microanalysis by scanning–electron microscopy was made. The results obtained demonstrate that the apparent ratio between Cr and O_2 best corresponds to CrO_2 oxide, which has not been observed in similar experiments before.

All the results presented here are well consistent with the constitutional diagram for the chromium-oxygen system (Fig. 7.13). It also will be recalled that CrO_2 (with a peak at $660\,\text{cm}^{-1}$ in the Raman spectrum) appears at the lowest exposure energies; with the exposure energy increasing together with temperature, a CrO_2 film grows and becomes ordered. With further increase in energy and temperature to above

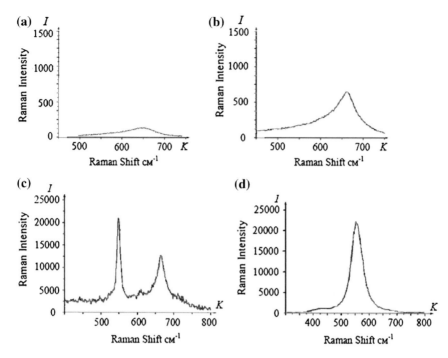

Fig. 7.12 Micro-Raman spectra of the sample under study: **a** chromium film before laser irradiation, **b** 1st run, **c** 2nd run, **d** 3rd run

510 °C, chromium dioxide (CrO_2) degrades to give Cr_2O_3, which is the most stable among the chromium oxides and is the only one appearing at high energies (up to the melting threshold) [39].

In summary, the band peaked at $660\,cm^{-1}$ in all Raman spectra for the chromium–oxygen system corresponds to a CrO_2 surface layer whose formation seems to be due to the interaction activation. There are chromium—oxygen bonds already present on the surface of a film exposed to air [40], whereas further changes involve a consolidation of the film, increase in its thickness, and its structural reordering due to the subsequent laser heating.

For samples irradiated in an ultrahigh vacuum (residual pressure $10^{-9}\,mm$ Hg), when no heterogeneous oxidation is possible, only a weak peak at $660\,cm^{-1}$, specific to unexposed chromium films, but sharpened drastically was recorded from the laser-irradiated area. The sharpening of the peak evidences that the substance undergoes structural modifications in the irradiated zone, [38]. The facts mentioned above suggest that not only oxidation processes occur under a short-pulse laser irradiation, but also some other process involving no oxidation. This seems to be a structural modification of chromium or chromium oxide films, which results from the increase in the etching selectivity ($k_{sel} > 1.2$) at low exposure energies.

Structural changes in thin chromium films under ultra short-pulse laser irradiation. The presence of the polycrystalline oxide CrO_2 was confirmed by X-ray

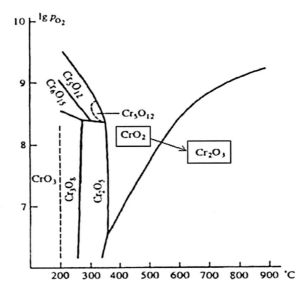

Fig. 7.13 Constitutional diagram of the chromium-oxygen system, with the stability zones of chromium oxides shown, P_{O_2} is the partial pressure of oxygen

diffraction analysis. The results of etching kinetics measurements for exposed and unexposed chromium films show that changes occur in the film bulk (Fig. 7.14). The photometric techniques from [7] were used for these measurements. It can be seen (Fig. 7.14) that there is a delay in the beginning of etching for both irradiated and unirradiated films (times t_1, t_2 and t_3, respectively), which is due to the presence of chromium oxides surface films. The thickness of the initial (atmospheric) oxide layer on the surface of an unirradiated film is about 1–2 nm, whereas during irradiation, its thickness increases up to 3–8 nm, judging by the increase in the etching time for this layer from 60 to 180 s. Meanwhile, the transmission of the film changes slightly, and only due to changes in a volume, as chromium oxides are sufficiently transparent. Structural rearrangements taking place in thin films during laser irradiation cause changes in their physical properties. In particular, grains grow in size, the number of crystal defects changes, bulk oxidation occurs, and, as the final result, the etching rate decreases. Figure 7.14 shows that, at the same film thickness, the etching times t_5 and t_6 of a bulk-exposed film are 3–8 times that of the original chromium film t_4.

It is especially important in this context to point to an appreciable difference between the slopes of the irradiated zone edges. With only the surface oxidation, it would be expected that the slope of curve (1) for the unirradiated sample is the same, but with t_2 or t_3 time delay [dotted curves (2a) and (3a)]. In fact, the slopes for irradiated films [curves (2) and (3)] strongly differ from that of curve (1), as it can be seen from Fig. 7.14. This difference ($t_4/t_{21} > t_5/t_2 > t_6/t_3$) means that not only a surface oxide layer appears, but also structural changes occur in the bulk of the film.

Scanning electron microscopy (SEM) was used for direct observation of structural modifications in chromium films. The surface structure images were obtained in the

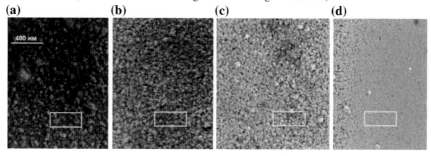

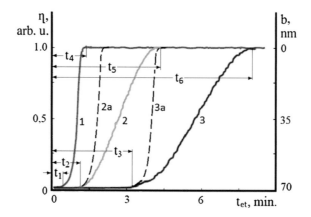

Fig. 7.14 Characteristic dependences of the transmission of 100-nm-thick chromium films unirradiated and irradiated under different power densities q on the time of etching t_{et} in $K_3Fe(CN)_6$ and NaOH solutions (measured in transmitted light at a wavelength of 633 nm)

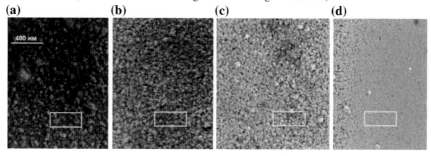

Fig. 7.15 Surface SEM images of irradiated Cr films. The number of crystallites per unit area grows due to the temperature rise (Gaussian distribution of the intensity in the laser focal spot) from 32 in the *left-hand panel* (**a**) to 450 (**d**) in the *right-hand panel*

range from unexposed chromium (Fig. 7.15a) to the middle of the line corresponding to the maximum temperature and CrO_2 formation (Fig. 7.15d). It appears that, as the temperature grows under laser heating, the degree of structural modification of exposed chromium increases. In particular, the number of crystallites per unit area becomes larger, which is apparently accompanied by the formation of a more strained structure, and the etching resistance increases when the oxide CrO_2 is consolidated.

The measurements demonstrated that the number of crystallite increases from 32 to 450 over a 300 × 150 nm area, i.e., the higher the laser annealing temperature, the smaller the size of the resulting crystallites. The minimum crystallite size obtained in the middle of a line is 10–15 nm (Fig. 7.15d), which suggests that the resolution of the method is limited by the grain size.

These results quite good agree with the observations of structural changes in transparent media under short laser pulses, described in Chaps. 11 and 12 of this book.

The above results can be consistently described with consideration for the fact that the time of thermal treatment under ultrashort pulse irradiation is apparently determined by the time of structure cooling, which mostly depends on the thermal characteristics of the ambient medium, rather than by the laser pulse duration proper.

7.6 Conclusion

The thermochemical action of a pulse laser radiation for patterning of metal films was discovered, studied, and introduced into experimental practice. The phenomenon was successfully applied for writing of diffractive structures in thin chromium films by scanning with a focused laser beam. The developed method was applied for direct writing of DOEs with a spatial resolution as good as $2000\,mm^{-1}$, a writing field of up to 300 mm, and a diffractive zone boundary accuracy of 25 nm. The method and the means for microstructure fabrication were used as a basis for fabricating and employing for the first time DOEs for high-accuracy testing of the shape of aspheric telescope mirrors 6.5 and 8.4 m in diameter.

The theoretical and experimental resolution limits depend on the minimum thickness b_{min} of an oxide layer resistant to etching. The smaller b_{min} and the laser spot size r_0 and the shorter the irradiation time τ, the higher the resolution. Properly the value of b_{min} is about several nanometers, which is substantially smaller than the optical wavelength.

The unique possibility of experimentally obtaining the highest resolution is furnished by the multipass irradiation of Cr films. Owing to the noticeable difference between the optical and thermal properties of chromium and its oxides and to the strong nonlinearities of the writing process, there appears a positive feedback between the oxide thickness and the film absorption. That is why every next pass makes more pronounced the difference between the oxide thicknesses at the center and periphery of laser tracks, which means the writing contrast strongly increases. This phenomenon allows, in principle, a thermochemical recording of optical images with details that cannot be resolved optically.

It should be noted that revealing the phase and chemical modifications of the surface under laser irradiation is a problem of primary importance for many other applications. These include laser patterning of silicon wafers, local oxidation of Ti films for photoelectric and new DOE applications, color laser marking, etc.

Acknowledgments Authors wish to thank collaborators who have contributed to our research efforts in this field and, in particular, Dr. Elena A. Shakchno, Michail V. Yarchuk, Dmitry A. Sinev from St-Petersburg National Research University of Information Technologies, Mechanics and Optics, and Voldemar P. Koronkevich, Viktor P. Korolkov, Evgeny G. Churin, Vadim Cherkashin, Anatoly I. Malyshev from Institute of Automation and Electrometry, Russian Academy of Sciences, Siberian Branch.
This study was supported by the Russian Federation Presidential Grant for leading scientific school SS–619.2012.2 and the Russian Foundation for Basic Research Grants 12–02–00974a, No 12–02–01118a and No 13–02–00971.

References

1. V.P. Veiko, M.N. Libenson, Vop. Radioelectroniki **3**(5), 99 (1964)
2. V.P. Veiko, G.A. Kotov, M.N. Libenson, M.N. Nikitin, in *Physico- Technical Basis of Laser Technology*, Proceedings School, Leningrad, USSR, 1979 (LDNTP. Leningrad, 1970, 1979), p. 43
3. V.P. Veiko, G.A. Kotov, M.N. Libenson, M.N. Nikitin, Dokl. Akad. Nauk. SSSR **208**,´587 (1973). [Engl. transl.: Sov. Phys.-Dokl 18, 83 (1973)]
4. V.P. Veiko, G.A. Kotov, M.N. Libenson, Elektronnaya Tekh. Microelectronica **3**(4), 48 (1973)
5. G.A. Kotov, M.N. Libenson, Elektronnaya Tekh. Microelectronica **3**(4), 56 (1973)
6. V.P. Koronkevich, A.G. Poleshchuk, E.G. Churin, Yu. Yurlov, Kvant. Elektron. **12**(4), 755–761 (1985). [Engl. Transl. Sov. J. Quant. Electr. 15, 497 (1985)]
7. V.V. Cherkashin, E.G. Churin, V.P. Korolkov, V.P. Koronkevich, A.A. Kharissov, A.G. Poleshchuk, J.H. Burge, Proc. SPIE. **3010**, 168 (1997)
8. F.V. Bunkin, N.A. Kirichenko, B.S. Luk'yanchuk, Usp. Fiz. Nauk **138**, 45(1982). [Engl. trasl.: Sov. Phys. - Usp. 138, (1982)]
9. D. Bauerle, *Chemical Processing with Lasers*, Springer Series in Material Science, vol. 1 (Springer, Heidelberg, 1986)
10. K.G. Ibbs, R.M. Osgood (eds.), *Laser Chemical Processing for Microelectronics* (Cambridge University Press, Cambridge, 1989)
11. V.P. Veiko, S.M. Metev, *Laser-Assisted Microtechnology*, Springer Series Materials Science, vol. 19 (Springer, Heidelberg, 1994, 1998)
12. A.M. Bonch-Bruevich, M.N. Libenson, Izvestiya Akad. Nauk SSSR, ser. Phys. **46**(6), 1104 (1982) (in Russian)
13. I.N. Goncharov, A.A. Gorbunov, V.I. Konov, A.S. Silenok, Yu.A. Skvortsov, V.N. Tokarev, N.I. Chapliev, preprint FIAN USSR No. 76 (1980)
14. V.A. Bobyrev, F.V. Bunkin, N.A. Kirichenko, B.S. Luk'yanchuk, A.V. Simakin, Kvant. Elektron. **9**, 45 (1982), [Engl. Transl. Sov. J. Quant. Electr. 12, 4 (1982)]
15. Konov V.I., Laser Photonics Reviews, 1–28, 2012/DOI 10.1002/lpor.201100030)
16. P. Veiko, A.A. Slobodov, G.V. Odintsova, Laser Phys. **23**(6), 066001 (2013)
17. S.M. Metev, K.V. Savchenko, K.V. Stamenov, V.P. Veiko, G.A. Kotov, G.D. Shandibina, IEEE J. Quant. Electr. **17**, 2004 (1981)
18. R.B. Gerassimov, S.M. Metev, K.V. Savchenko, J. Phys. D **17**, 1671 (1984)
19. V.P. Veiko, G.A. Kotov, V.A. Smirnov, G.D. Shandibina, E.B. Yakovlev, Kvant. Elektron. **7**, 2196 (1981). [English transl.: Sov. J. Quant. Electr. 11, 1338 (1981)]
20. V.P. Koronkevich, A.G. Poleshchuk, E.G. Churin, Yu. Yurlov, Pis'ma v "Zhurnal tehnicheskoi fiziki" **11**, 144–148. [Engl. Transl. (Sov. Technical Physics Letters) 11, (1985)]
21. V.P. Veiko, M.N. Libenson, Vop. Radioelectroniki **3**(4), 20 (1966)
22. V.P. Veiko, M.N. Libenson, Fiz. Khim. Obrab. Mater. **4**, 44 (1968)
23. V.P. Veiko, E.A. Touchkova, E.B. Yakovlev, Kvant. Elektron. **11**(4), 661 (1981). [English transl.: Sov. J. Quant. Electr. 11, (1981)]
24. V.P. Veiko, *Laser Treatment of Thin-Films for Microelectronics* (Mashinostroenie, Leningrad, 1986). (in Russian)
25. A.G. Poleshchuk, E.G. Churin, V.P. Koronkevich, V.P. Korolkov, A.A. Kharisov, V.A. Cherkashin, V.A. Kirianov, A.V. Kirianov, S.A. Kokarev, A.G. Verhoglad, Appl. Optics. **38**(8), 1295 (1999)
26. J.H. Burge, Proc. of SPIE. **2576**, 258 (1995)
27. J.H. Burge, L.R. Dettmann, S.C. West, OSA Trends Opt. Photonics **24**, 182 (1999)
28. V.P. Koronkevich, V.P. Kirianov, V.P. Kokoulin, I.G. Palchikova, A.G. Poleshchuk, A.G. Sedukhin, E.G. Churin, A.M. Sherbachenko, Yi. I. Yurlov. Optik **67**(3), 257 (1984)
29. V.P. Korolkov, A.G. Poleshchuk, Proc. SPIE **6732**, 67320X (2007)
30. U.C. Paek, A. Kestenbaum, J. Appl. Phys. **44**, 2260 (1973)
31. A.G. Poleshchuk, Optoelectr. Instrum. Data Process. **39**(6), 34 (2003)

32. V.P. Veyko, E.A. Shakhno, A.G. Poleshchuk, V.P. Korolkov, V. Matyzhonok, JLMN J. Laser Micro/Nanoeng. **3**(3), 201 (2008)
33. A.G. Poleshchuk, V.P. Korolkov, V.V. Cherkashin, S. Reichelt, J.H. Burge, Optoelectron. Instrum. Data Process. (Avtometriya) **38**(3), 3 (2003). (in Russian)
34. J.H. Burge, V.V. Cherkashin, V.P. Koronkevich, A.G. Poleshchuk, in *Proceedings of VI International conference "Applied Optics"*, vol. 3 (St. Petersburg, 2004), p. 203
35. G. Olczak, C. Wells, D. Fischer, M. Connoly, Proc. SPIE **8450**, 84500R (2012)
36. N.A. Kirichenko, B.S. Luk'yanchuk, Kvant. Elektron. **10**, 4 (1983). [Engl. Transl. Sov. J. Quant. Electr. **13**, 4 (1983)]
37. V. P. Veyko, E. A. Shakhno, D.A. Sinev., A. G. Poleshchuk, A.R.Sametov, A.G. Sedukhin, Comput. Optics. **36**(2(84)), 562 (2012). (in Russian)
38. V.P. Veyko, M.V. Yarchuk, M.V., A.I. Ivanov A.I., Laser Phys. **22**(8), 1310 (2012)
39. A.V. Baranov, K.V. Bogdanov, A.V. Fedorov, M.V. Yarchuk, A.I. Ivanov, V.P. Veiko, K. Berwick, J. Raman Spectrosc. **42**(10), 1780 (2011)
40. J. Benard, *Oxydation des Metaux* (Villars, Paris, 1964)

Chapter 8
Photophysics of Nanostructured Metal and Metal-Contained Composite Films

Nathalie Destouches, Frank Hubenthal and Tigran Vartanyan

Abstract In this chapter we give a brief introduction in the preparation and the optical properties of noble metal nanoparticles. We explain that their unique optical behavior is due to resonant absorption and scattering of light by the nanoparticles, i.e. by surface plasmon resonance's. We show the influence of various parameters on the optical properties and demonstrate how the optical properties of noble metal nanoparticles can be tailored for certain applications. Finally, we highlight four intriguing applications, based on the expertise of the authors.

8.1 Introduction

The optical properties of noble metal nanoparticles have been exploited since centuries, for example in stained glass to produce the magnificent colors of ancient windows, as in the Altenberg dome near Cologne in Germany or drink cups, such as the Lycurgus cup. However, the first explanation of the origin of the colors has been made by Michael Faraday. In his famous article published in 1857 he concluded: I think that in all these cases the ruby tint is due simply to the presence of

N. Destouches
Laboratory Hubert Curien, University of Lyon-University Jean Monnet, UMR 5516 CNRS,
18 Rue Pr. B. Lauras,F-42000 Saint-Etienne, France
e-mail: nathalie.destouches@univ-st-etienne.fr

F. Hubenthal
Institut für Physik and Center for Interdisciplinary Nanostructure Science and Technology,
CINSaT Universität Kassel, 40 Heinrich-Plett-Strasse, D-34132 Kassel, Germany
e-mail: hubentha@physik.uni-kassel.de

T. Vartanyan (✉)
Center of Information Optical Technologies,
St. Petersburg National Research University of Information Technologies, Mechanics and Optics,
49 Kronverksky pr., St. Petersburg 197101, Russia
e-mail: tigran@vartanyan.com

V. P. Veiko and V. I. Konov (eds.), *Fundamentals of Laser-Assisted*
Micro- and Nanotechnologies, Springer Series in Materials Science 195,
DOI: 10.1007/978-3-319-05987-7_8, © Springer International Publishing Switzerland 2014

finely-divided gold [1]. The first theoretical explanation has been made in 1908 by Gustav Mie, who applied the Maxwell equations in spherical coordinates to a small sphere in a homogeneous environment [2]. The aim of these calculations was to explain the experimental findings of his students. Up to now, the Mie application [3] is the only exact solution for the scattering and absorption problem of a small metal sphere in a matrix. With this outstanding work Mie could trace the origin of the colors back to a resonant absorption of light by metal nanospheres in a matrix. This resonant absorption is nowadays called localized surface plasmon polariton resonance (LSPPR) and is due to a collective excitation of the conduction band electrons in metal nanoparticles.

Since the Mie application is an exhausting numerical procedure, it became popular not before the invention of the computer. After computers have become available the previously time consuming calculations took now only seconds or milliseconds. Therefore, the Mie application has found increasing interest, which is reflected in thousands of citations. Mie's application has been afterwards extended, for example, to non-spherical nanoparticles, core-shell particles and to non-homogeneous environments. Several books have published computer programs for Mie applications [4, 5] and various Internet pages are available for the calculations.

The term plasmon has been introduced in 1956 by David Pines to describe the quantum of an elementary excitation associated with a high frequency collective motion [6]. Later on, this term has been adopted for surface plasmon resonances. The additional terms localized and polariton describe that a non propagating plasmon in a nanoparticle, excited by optical means is considered. These extensions are necessary to avoid a conceptual confusion with propagating surface plasmons in thin films and with plasmons excited by swift electrons either in a thin film or in a nanoparticle [3, 4].

Since decades metal nanoparticles, which bridge the gap between the properties of an atom and bulk material, have been studied in great detail. Metal nanoparticles exhibit unique optical properties that depend strongly on their shape, size, composition, and dielectric surrounding [4, 7–10]. They are promising candidates, e.g., as bio-sensors [11], catalysts [12, 13], data storage media [14, 15], for dichroitic glass [16, 17], or in thermal cancer therapy [18, 19]. Accompanied with the excitation of a LSPPR is a local field enhancement, which is the bases of many and extremely different applications, for example, in surface enhanced Raman spectroscopy [20, 21], surface enhanced fluorescence [22, 23], to enhance the efficiency of solar cells [24], or to structure surfaces [25, 26]. For most applications the morphology of the nanoparticles is a crucial parameter. Thus, besides the importance to know the electrical and optical properties of the nanoparticles as a function of their morphology, techniques are strongly required, which allow to tune or to change the morphology of the nanoparticles in a defined way. Such techniques have been developed in recent years. Based on the expertise of the authors, it will be shown, how the morphology of the nanoparticles can be precisely tailored by applying laser based processes. In addition, we will highlight several applications, in which the optical properties of designed nanostructured metal and metal-contained composite films are exploited.

8.2 Optical Properties of Noble Metal Nanoparticles

Noble metal nanoparticles exhibit unique shape and size dependent properties. In particular their extinction spectra are dominated by pronounced localized surface plasmon polariton resonances, which are attributed to coherent oscillations of the conduction band electrons excited by means of an electromagnetic field. In the general case, the field distribution in a nanoparticle is inhomogeneous, which leads to a complex optical response. To calculate this response, there is only one exact solution, which is known as Mie-application [3]. However, several approximations have been derived. The most common one is the quasistatic approximation that assumes the nanoparticle radius R to be very small compared to the incoming wavelength ($R < 0.06\,\lambda$) [4]. Thus, the conduction band electrons oscillate coherently against the ion core and the nanoparticle acts as a Hertzian dipole. The quasistatic approximation allows to derive the absorption (8.1) and the scattering (8.2) cross sections of light due to the nanoparticles, as demonstrated in references [3, 4]:

$$\sigma_{abs} = k \cdot \mathrm{Im}\,\{\alpha\,(\omega)\} = 4\pi k R^3 \,\mathrm{Im}\left\{\frac{\varepsilon - \varepsilon_m}{\varepsilon + 2\varepsilon_m}\right\}, \qquad (8.1)$$

$$\sigma_{sca} = \frac{k^4}{6\pi}\,|\alpha\,(\omega)|^2 = \frac{8\pi k R^6}{3}\left|\frac{\varepsilon - \varepsilon_m}{\varepsilon + 2\varepsilon_m}\right|^2, \qquad (8.2)$$

where k is the wavenumber and ω the angular frequency of the incident light, α is the polarizability, $\varepsilon = \varepsilon_1 + i\varepsilon_2$ and ε_m the dielectric permittivities of the nanoparticle and the surrounding medium, respectively. The extinction cross section, which is usually measured in the experiments, is given by the sum of absorption and scattering cross section:

$$\sigma_{ext} = \sigma_{abs} + \sigma_{sca} = 12R^3 \varepsilon_m^{3/2}\frac{\omega}{c}\frac{\varepsilon_2}{[\varepsilon_1 + 2\varepsilon_m]^2 + \varepsilon_2^2} \qquad (8.3)$$

Since absorption scales with the third and scattering with the sixth power of the nanoparticle radius to the extinction cross section, in the quasi static regime absorption dominates over scattering and one can write: $\sigma_{ext} \cong \sigma_{abs}$.

For metals whose optical properties are determined by the quasi-free conduction band electrons alone, i.e. Drude metals, the dielectric function $\varepsilon^{Drude}(\omega)$ is given by [4, 7, 27]:

$$\varepsilon^{Drude}(\omega) = 1 - \frac{\omega_p^2}{\omega(\omega + i\gamma)} \qquad (8.4)$$

with γ the damping constant of the metal, $\omega_p = \sqrt{\frac{ne^2}{m_{eff}\cdot\varepsilon_0}}$, e the charge of an electron, n the electron density, m_{eff} the electron effective mass and ε_0 the vacuum permittivity.

8.2.1 Damping of Plasmon Excitations Localized
in Metal Nanoparticles

The nanoparticle acts as an antenna, which strongly absorbs light at its resonance frequency $\Omega = \frac{\omega_p}{\sqrt{1+2\varepsilon_m}}$ obtained when $\varepsilon_1^{Drude} = -2\varepsilon_m$. The width of the resonance is given by the following equation:

$$\Gamma(\Omega) = \frac{2\hbar\varepsilon_2(\Omega)}{\sqrt{\left(\frac{d\varepsilon_1(\omega)}{d\omega}\Big|_{\omega=\Omega}\right)^2 + \left(\frac{d\varepsilon_2(\omega)}{d\omega}\Big|_{\omega=\Omega}\right)^2}} \qquad (8.5)$$

The homogeneous linewidth $\Gamma(\Omega)$ is an important parameter, because it reflects the strength of the damping of the plasmon resonance. As larger the homogeneous linewidth is as larger is the damping. The homogeneous linewidth is connected via the uncertainty relation with the dephasing time T_2:

$$T_2 = \frac{2\hbar}{\Gamma_{hom}} \qquad (8.6)$$

T_2 is the time, in which the coherent oscillating electrons lose their phase. The knowledge of T_2 is of tremendous importance [8, 28, 29], because it is directly proportional to the field enhancement, which is the base of a variety of applications.

For nanoparticles with reduced dimensions additional size dependent damping mechanisms arise, which cause a broadening of the homogeneous linewidth and, thus, a shortening of T_2. The corresponding processes have been widely discussed in literature [3, 4, 28].

Within the quasistatic approximation five mechanisms are relevant:

(a) *Direct emission of electrons* takes place within the first 1.5 to 4 fs after excitation. It is accompanied by a fragmentation of the plasmon resonance into nearby one-particle-one-hole states [30].
(b) *Surface scattering* is the inelastic scattering of the coherently oscillating electrons at the particle surface. It gets relevant as soon as the nanoparticle radius is smaller than the mean free path of the electrons at the Fermi edge. The damping is determined by the Fermi velocity divided by the nanoparticle radius [4].
(c) *Chemical interface damping* (CID) is caused by a dynamic charge transfer of electrons into and out of adsorbate or surface states. Due to the statistical nature of this process, the electrons lose their phase coherence, which results in an additional damping [28].
(d) *Increased Landau damping* describes the resonant generation of electron-hole-pairs [31] as an intraband transition. It is caused by a broadening of the discrete eigenstates of the holes and quasifree electrons, which are determined by the particle surface [32]. However, Landau damping has not been observed experimentally yet.

(e) *Band structure changes* are a relatively new proposed damping mechanism [33]. It has been observed for gold nanoparticles near the onset of the interband transition. It is explained by broadening of the band structure, which promotes interband transitions, by reducing the energy gap between the bands. In this case, the plasmon is treated as a two level system, because for an interband transition a correlated effect of the electrons is assumed, in which the energy of the plasmon is transferred to a single electron, which makes the transition.

Since the dephasing time T_2 depends on the nanoparticle size it cannot be used to define a size dependent dielectric function and a size independent parameter, the damping parameter A, is necessary [3, 8]. The A-parameter quantifies the influence of the increased damping due to size effects of the nanoparticles and has to be included in the dielectric function. Since the damping is proportional to $1/R$ [3, 4], it is taken into account, for example, by:

$$\frac{1}{T_2} = \frac{1}{T_{2,\infty}} + \frac{2A}{R}$$

(8.7)

with $T_{2,\infty}$ the dephasing time included in the dielectric function of the bulk. The $1/R$ dependence simply reflects the surface to volume ratio of a nanoparticle. This yields the following size dependent dielectric function:

$$\varepsilon\,(\omega, R) = \varepsilon_{\text{bound}} + 1 - \frac{\omega_{\text{p}}^2}{\omega\left[\omega + i\left(\frac{1}{T_{2,\infty}} + \frac{2A}{R}\right)\right]}$$

(8.8)

Although (8.7) is theoretically well understood, an exact knowledge of the A-parameter is still lacking. The reasons are that experimentally A has to be determined by size dependent measurements of the dephasing time for small nanoparticle sizes ($<R = 15\,\text{nm}$). Since most preparation techniques yield ensembles of nanoparticles whose plasmon resonance is inhomogeneously broadened, sophisticated experimental techniques are required to determine T_2. Single particle spectroscopy [8, 34] and persistent spectral hole burning [27, 33] have been successfully applied to determine precisely the dephasing time and the damping parameter. On the other hand, also theoretically A is hard to obtain, because extensive quantuum mechanical calculations have to be performed for a precise description. Only in few simple cases, easier semi classical descriptions are sufficient. Nevertheless, the knowledge of the damping parameter becomes more and more important due to the variety of applications of small noble metal nanoparticles. Only with a precise A-parameter the optical properties and the field enhancement of noble metal nanoparticles can be sufficiently exact calculated.

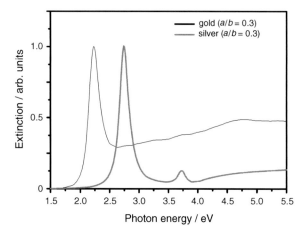

Fig. 8.1 Extinction spectra of single gold and silver nanoparticles with a radius of $r = 10$ nm and an axial ratio of $a/b = 0.3$

8.2.2 Optical Properties of Nanostructured Noble Metal Nanoparticles on Substrates

The energetic position of the plasmon resonance depends on the dielectric environment. A typical example for an inhomogeneous environment are nanoparticles at substrates prepared by Volmer-Weber growth (cf. Sect. 8.3.1). For such supported nanoparticles the environment is composed of the substrate and, for example, vacuum, air, or a solvent. In this case the inhomogeneous environment is taken into account by an effective dielectric function [3]. Furthermore, nanoparticles on substrates, in particular if prepared in Volmer-Weber mode, exhibit an oblate shape due to the growth kinetics. Since these nanoparticles can be approximated as rotational ellipsoids, two eigenmodes of the plasmon resonance can be excited. A high energetic (1,0) mode, which is correlated to an electron excitation along the short axis a of the nanoparticle and a doubly degenerated low energetic (1,1) mode, which is correlated to an excitation along the long axis b (it is degenerated, because the two axes b and c parallel to the substrate surface are equal). However, only for alkali metals both modes are nearly equally pronounced in the extinction spectrum. The high energetic (1,0) mode is for silver strongly and for gold totally damped by the interband transition, as shown in Fig. 8.1 for nanoparticles with an axial ratio of $a/b = 0.3$. Thus, for gold only the (1,1) mode appears in the extinction spectrum.

8.2.3 Reflection and Transmission of Supported Metal Nanostructures

When the nanoparticles are supported on a transparent substrate their optical properties are inferred through the measurement of the reflection and transmission coefficients of the whole sample. Hence, the connection between these values and the optical characteristics of individual particles is needed. There are several approaches to the solution of this very complicated problem that treat it at different levels of rigor [35]. The results of such calculations give invaluable insight in the nature of the electromagnetic interaction between the particles. Unfortunately they depend heavily on the detailed information about the system. As such information concerning the real systems is rarely available; an experimentalist is left without any practically useful tool for the analysis of the data. On the other hand in the most common case of self-organized granular metal films supported on transparent substrates, interaction between the particles is reduced, and the optical properties of the films may be connected with the optical properties of the individual particles in a simple and straightforward way [36].

Let f be the forward scattering amplitude of a nanoparticle that coincides with the backward scattering amplitude provided that the particle is small and the dipole scattering dominates. This quantity is defined by the volume, shape, and optical properties of the particle material as it was detailed in the previous sections. Consider, first a rarefied manifold of identical particles that form a plane with the surface number density c. Then the amplitude transmission and reflection coefficients of such a layer in vacuum are

$$t = 1 + \frac{2\pi i c f}{k} \text{ and } r = \frac{2\pi i c f}{k}, \tag{8.9}$$

where k is the wavenumber of the incident wave. Obviously, these expressions are unsuitable for the description of optical properties of dense films since the modulus of t and r according to (8.9) may easily exceed unity. On the other hand there is a well known formula for the reflection and transmission coefficients of a very thin continuous film with very large dielectric constant [37]. In this case the properties of the film are described by the complex constant $\alpha = \alpha' + i\alpha''$ that has the meaning of the product of the refractive index of the film by the phase difference of the electromagnetic wave during its passage through the film. Although the correctness of this approximation requires that the phase difference is much smaller than unity and the refractive index is considerably larger than unity, no restrictions are imposed on their product. If the film is at the interface between a transparent dielectric material with a refractive index n and vacuum, then, for a wave incident on the film from the side of the dielectric, the following formulas for the amplitude transmission and reflection coefficients are valid

$$t = \frac{2n}{n + 1 - i\alpha}, \quad r = \frac{n - 1 + i\alpha}{n + 1 - i\alpha} \tag{8.10}$$

Comparing (8.9) with the limiting values of (8.10) for $n = 1$ and $|\alpha| \ll 1$ we obtain $\alpha = 2\pi cf/k$. In the experiment one measures the reflection $R = |r|^2$ and transmission $T = |t|^2$ coefficients, while absorption is calculated according to $A = 1 - R - T$

$$R = \frac{(n - 1 - \alpha'')^2 + (\alpha')^2}{(n + 1 + \alpha'')^2 + (\alpha')^2}, \quad T = \frac{4n}{(n + 1 + \alpha'')^2 + (\alpha')^2},$$
$$A = \frac{4n\alpha''}{(n + 1 + \alpha'')^2 + (\alpha')^2}. \tag{8.11}$$

Then, α may be determined up to the sign of its real part through the measured characteristics of the film

$$\alpha'' = \frac{A}{T}, \quad |\alpha|^2 = \frac{4n - (n + 1)(2A + T(n + 1))}{T} \tag{8.12}$$

In most practical cases due to the inhomogeneous broadening f is to be understood as its weighted average over the distribution of particles shapes and volumes.

8.2.4 Mutual Modification of Silver Nanoparticle Plasmon Resonances and Absorptive Properties of Polymethine Dye Molecular Layers on a Sapphire Surface

Along with the widely known phenomenon of surface enhanced Raman scattering, in the near field of plasmon nanoparticles, there were also observed changes in the absorption and fluorescence of different molecules, including organic dyes [38, 39]. Granulated silver films were obtained by thermal evaporation in vacuum onto sapphire substrate. During the growth of the films, the substrates were maintained at room temperature. To shift the absorption band maximum, the island film was annealed in a vacuum chamber at 200 °C for 30 min [40–43]. A polymethine dye layer corresponding to about 11 monolayers of dye was deposited on the annealed film by spin coating.

Absorption of the hybrid material (dye on annealed silver nanoparticles) cannot be reduced to a simple combination of absorption of the metal nanoparticles and organic dye [44–46]. At the short wavelength side of the dye absorption band, absorption of the hybrid material appears to be even smaller than that of the metal film without dye. The difference spectrum shown in Fig. 8.2 is negative in the region of the wavelengths from 469 to 590 nm. At the longer wavelengths, extinction of the hybrid material noticeably exceeds that of the metal film without dye, in spite of the fact that the dye on the dielectric substrate in this region virtually does not show any absorption.

A decrease in the absorption of metal nanoparticles in the short wavelength region is explained by a change of their dielectric environment upon deposition of the dye

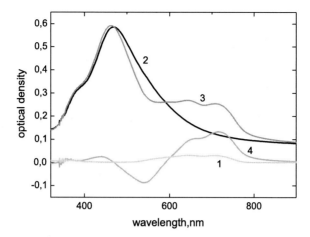

Fig. 8.2 Extinction spectra of a dye layer on the bare substrate (*1*), of silver nanoparticles after thermal annealing on sapphire surface (*2*), of the dye deposited on the annealed silver nanoparticles (*3*). (*4*) is the difference of the extinction spectra of silver nanoparticles coated and uncoated by the dye layer

solution. Indeed, the absorption band of the dye contains a region of anomalous dispersion, which causes a decrease in the permittivity of the layer in the vicinity of the short wavelength boundary of the dye absorption band. The decrease in the permittivity of the medium leads to an increase in the resonant frequency of plasmon oscillations [47–49]. Since the relative concentration of particles, in the spectral range under consideration, increases with increasing resonant frequency, the short wavelength shift of resonant frequencies of the plasmon oscillations leads to the observed decrease in the plasmon band absorption. One could expect that such a shift of resonant frequencies would lead also to a shift of the absorption band. Such a shift is indeed observed (Fig. 8.3), but it is rather small, because, while moving to shorter wavelengths, the plasmon resonance broadens and becomes weaker due to the effect of interband transitions in silver. The enhancement of the dye absorption in the main part of its absorption band can be naturally ascribed to the fact that dye molecules are located in the near field of nanoparticles, in which resonant plasmon oscillations are excited. Since the near field in the resonant nanoparticles is enhanced compared with that of the incident wave, the dye molecules deposited on the metal film absorb more strongly than the dye molecules adsorbed on the dielectric substrate without metal particles. As was noted above, the difference between the surface densities of dye molecules was insignificant and could not cause the observed enhancement of the hybrid material absorption compared with that of the dye on a dielectric substrate.

Two different mechanisms may be responsible for the absorption enhancement in the long wavelength region. First, similar to the short wavelength region, it may be caused by the frequency shift of plasmon oscillations in metal nanoparticles due to the anomalous dispersion of the dye. Second, the increased absorption outside the known bands of polymethine dye molecular components may be related to unusual spatial characteristics of the nanoparticle near field. Indeed, the characteristic spatial length within which the near field of the nanoparticle changes is determined by the size of the particle rather than by the light wavelength. Under these conditions, the probability of multipole (in particular, quadrupole) transitions in molecules (especially, in extended molecules like polymethines) increases and may become comparable with the probability of dipole transitions because the field noticeably changes within the length of the molecule [50]. Another aspect of the multipole transitions has been discussed in the Chap. 6 devoted to the anomalous light scattering by alkali metal nanoparticles. In that case the plasmon resonance is dominated by the quadrupole one even for very small particles. For noble metal nanoparticles discussed in this chapter anomalous light scattering is absent due to much stronger interband transitions.

8.3 Preparation and Defined Manipulation of Metal and Metal Contained Film

Metal nanoparticles can be prepared by various techniques, such as chemical or physical vapour deposition, wet chemical synthesis, pulsed laser deposition, gas phase aggregation, sol gel techniques, etc. However, most of the preparation techniques permit only minor possibilities to tailor the optical properties or the morphology of the nanoparticles. To overcome this drawback, the properties of metal nanoparticles on substrates or in a film can be manipulated during or after nanoparticle preparation, for example, by laser light or heat treatment. While a heat treatment leads to a non-selective manipulation of all nanoparticles in the ensemble, an irradiation with light allows a selective manipulation of the nanoparticles morphology and of their optical properties. In this section, physical vapour deposition, thermal and optical methods to form metal nanoparticles as well as different light driven manipulation techniques will be discussed.

8.3.1 Physical Vapour Deposition

A suitable method for a defined and highly reproducible generation of nanoparticles on substrates is physical vapour deposition under ultrahigh vacuum conditions. For this purpose a thermal metal atom beam is generated and directed onto a substrate. The atom deposition is followed by surface diffusion and nucleation, i.e. Volmer-Weber growth. A key feature of Volmer-Weber growth is a strong size and shape correlation of the generated nanoparticles. Small nanoparticles with $R_{eq} = 1$ nm are nearly spherical, but become more and more oblate for increasing size [10]. R_{eq} is the equivalent radius, i.e. a radius of a sphere with the same volume as the actual oblate nanoparticle. In the initial stages of the nanoparticle growth, where the

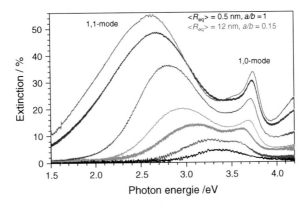

Fig. 8.3 Optical spectra of silver nanoparticles with different sizes grown in Volmer-Weber mode on a quartz substrate

nanoparticles are isolated, the flattening is due to the growth kinetics. Since most adatoms are deposited on the substrate and nucleate after surface diffusion at the edges of a nanoparticle, the lateral dimensions grow faster than the height of the nanoparticles. At higher coverage the flattening is intensified by coalescence, i.e. neighbouring nanoparticles merge together.

A consequence of the nanoparticle flattening during growth is that the plasmon resonances are shifted for increasing nanoparticle size. Due to the statistical growth behaviour the generated nanoparticle ensembles exhibit a broad Gaussian size and shape distribution and inhomogeneously broadened plasmon resonances. Note, the position and the width of the plasmon resonance depend on the size and shape of the nanoparticles. Consequently, the experimentally measured plasmon resonance of supported non-interacting nanoparticles with a certain size and shape distribution is a convolution of each individual nanoparticle plasmon resonance in the ensemble. Hence, it is inhomogeneously broadened.

To demonstrate the plasmon shift for increasing coverage, Fig. 8.3 depicts extinction spectra for silver nanoparticles on a quartz substrate with mean equivalent radii from $<R_{eq}> = 0.5$ nm up to $<R_{eq}> = 12$ nm. For silver both inhomogeneously broadened modes are shifted for increasing nanoparticle size. The high energetic (1,0)-mode is shifted to higher photon energies and the low energetic (1,1)-mode is shifted to lower photon energies. In addition, the extinction amplitudes of both modes increase, which reflects that the nanoparticles increase in size. Since the (1,0)-mode is strongly damped by the interband transition, both effects are less pronounced compared to the (1,1)-mode. However, the shift of the (1,1)-mode as a function of size is the base of size tailoring of the nanoparticles by laser light, as will be demonstrated in Sect. 8.3.3.

8.3.2 Laser-Induced Growth of Metal Nanoparticles in Glassy Matrix

Lasers have also been extensively used to generate metal nanoparticles in glass or in porous glassy matrix doped with ionic precursors. The growth mechanisms involve two steps: the reduction of metal cations and the migration of metal atoms to form nanoparticles. The size of the nanoparticles and their spatial distribution can be controlled by the conditions of the laser irradiation.

The precipitation of metal nanoparticles in glass results from the combination of a laser irradiation and a heat treatment [51]. Irradiation is often performed by means of femtosecond lasers that create active electrons and holes in the glass through multi-photon ionization, Joule heating and collisional ionization [52], and from plasma. As metal cations act as electron-trapping centers, the free electrons reduce them to metal atoms. A subsequent heat treatment is then required to increase the atomic mobility and aggregate the nuclei in the form of metal nanoparticles. Different studies have dealt with precipitating gold, silver and copper nanocrystals in glasses [53–56]. Interestingly, reversible mechanisms have also been demonstrated: gold nanoparticles, precipitated after femtosecond laser irradiation and annealing at 550 °C, were "dissolved" or broken into small-size particles or atoms by femtosecond laser irradiation and annealing at lower temperature (300 °C) [53].

Silver nanoparticles have been generated in silver-exchanged soda-lime glass by achieving simultaneously continuous wave ultraviolet laser exposure and high temperature annealing [57]. In such glasses, the growth of silver nanoparticles does not necessarily need high temperature annealing [58]. Actually, nanosecond laser exposures have been used successfully to form such nanoparticles, as they promote both electronic processes and thermal effects [59]. The latter are assumed to cause a solid–liquid phase transition; the high mobility of atoms in liquid phase and the segregation effects at the liquid–solid interface lead to a clustering of silver atoms at well-defined depths within the glass.

The mechanisms are similar in porous glassy matrix with interconnected pores but the precursors have a higher mobility than in dense glass and the nanoparticle growth occurs without any specific heat treatment. The important specific surface area also provides propitious conditions for the existence of a wide variety of defects, like oxygen vacancies or Si-OH dangling bonds [60], which provide electrons under laser irradiation. Metal nanoparticles have been generated with both femtosecond and continuous-wave laser irradiations in silica-based porous monoliths [61, 62]. A higher photoreduction efficiency under ultraviolet laser light has been obtained by using a hybrid mesostructured silica matrix rather than a porous silica one [63]. In this case, triblock copolymer polyethylene oxide-polypropylene oxide-polyethylene oxide (F127) added in the silica matrix acts as a sensitizer. The hydrophilic PEO part of the copolymer degrades under ultraviolet light and photochemically reduces Ag^+ to metallic Ag [64, 65]. A directed growth of silver nanoparticles has been achieved by producing 3D periodic patterns with critical dimensions smaller than 100 nm through multiple interferometric UV exposures (Fig. 8.4 from [66]).

Fig. 8.4 TEM picture on
the cross-section of a silver
containing silica-based
film after interferometric
illumination

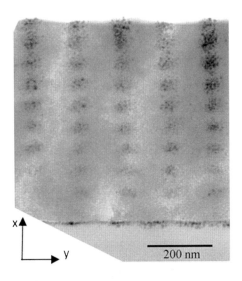

X

Y

200 nm

8.3.3 Laser-Induced Transformations of Supported and Embedded Metal Nanoparticles Ensembles

In this section the effects of post grown irradiation to noble metal nanoparticle ensembles on surfaces and in matrixes will be discussed. While under appropriate conditions a size tailoring of supported nanoparticles is achieved, irradiation of embedded nanoparticles leads to, for example, a shape change from spherical to prolate shape, as we will demonstrate in the following.

The aim of post grown irradiation of supported nanoparticles is to generate nanoparticle ensembles with a narrow size distribution. For nanoparticles which fulfil the relation $R < 0.06 \lambda$ the position of the plasmon resonance depends solely on the shape [3, 4]. In fact, size tailoring of supported nanoparticles has been initially developed for nanoparticles, which fulfil the relation $R < 0.06 \lambda$ and was later extended to larger nanoparticles. Due to the strong size and shape correlation of nanoparticles prepared by Volmer-Weber growth an effective size tailoring can be achieved, by addressing the plasmon resonances of nanoparticles with different shapes. Figure 8.5 depicts schematically the size tailoring process using two laser lines. For simplicity only the (1,1)-mode is drawn. First, the nanoparticle ensemble is irradiated with a photon energy $h\nu_1$, which excites the plasmon resonance of the large nanoparticles (Fig. 8.5a). These nanoparticles strongly absorb the light and convert it rapidly into heat. The temperature increase stimulates evaporation of surface atoms, which causes a size reduction solely of the large nanoparticles. The size reduction is accompanied by a shape change towards more spherical. Consequently, the plasmon resonance of the excited nanoparticles is shifted to higher photon energies as long as it does no longer interact with the laser light and the addressed nanoparticles have the desired size. Second, the small nanoparticles will be removed from the substrate

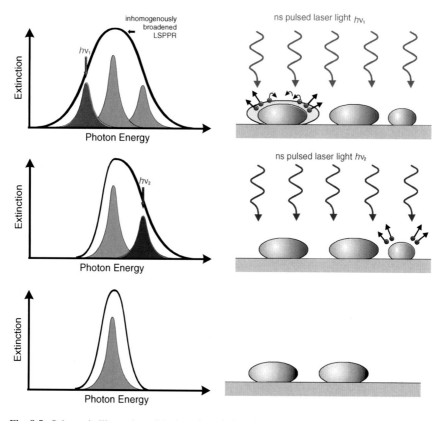

Fig. 8.5 Schematic illustration of the laser based size tailoring of metal nanoparticles. For clarity, only the (1,1)-mode is depicted

by irradiating the nanoparticle ensemble with a higher photon energy $h\nu_2$, which excites solely the small particles (Fig. 8.5b). A sufficiently high fluence leads to a strong temperature increase and causes a complete evaporation of the small nanoparticles. As a result, the standard deviation of the size and shape distribution is strongly reduced and the width of the plasmon resonance is significantly narrowed (Fig. 8.5c).

The effectiveness of size tailoring is demonstrated in Fig. 8.6. Figure 8.6a–c shows the optical spectrum, the AFM image, and the size distribution of a silver nanoparticle ensemble on a quartz substrate before irradiation. A broad plasmon resonance and size distribution is clearly observed. After irradiating the sample with a fluence of 150 mJ·cm^{-2}, applying first a photon energy of $h\nu_1 = 2.8$ eV and subsequently $h\nu_2 = 3.3$ eV, the size distribution is significantly improved, as immediately seen by comparing the AFM images before and after irradiation. A detailed analysis revealed that the width of the size distribution is narrowed from 27 to 14 % (Fig. 8.6c, f). This reduction in the size distribution is accompanied by a reduction of the plasmon resonance line width from 0.83 to 0.52 eV.

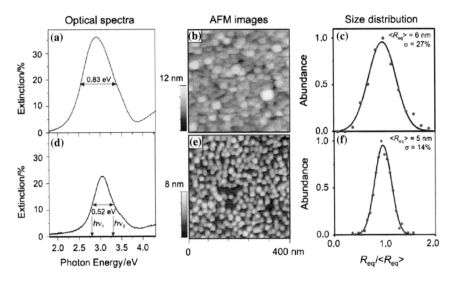

Fig. 8.6 Optical spectra, AFM images and relative size distributions of silver nanoparticles before (**a**)–(**c**) and after tailoring (**d**)–(**f**). The extinction spectra have been obtained with s-polarized light. Thus, the (1,0)-mode has not been excited. Reprinted with permission from [10]

While post irradiation of metal nanoparticles on substrates with ns-pulsed laser light aims at a narrowing of the size distribution, an irradiation of embedded nanoparticle ensembles with fs-pulsed laser light might cause an elongation of the nanoparticles shape. Extensive studies have been carried out on spherical nanoparticles produced in silver-exchanged soda-lime glass. The nanoparticles were shown to experience a persisting transformation into ellipsoidal shape when irradiated with intense femtosecond laser pulses at visible wavelengths. Actually different regimes exist depending on the laser intensity and the number of laser pulses [67]. Below an intensity threshold ($\sim$0.2 TW/cm^2) the laser has no effect on the nanoparticles.

At relatively low intensity (typically $<$2 TW/cm^2) the spheres are transformed to prolate spheroids (Fig. 8.7a) whose aspect ratio increases with the number of pulses per shot. This shape anisotropy can lead to a strong dichroism, due to the splitting of the plasmon band, useful for the production of micropolarizers [68]. At higher intensity, oblate spheroids are produced but only when single pulse irradiations are performed (Fig. 8.7b, d), otherwise the nanoparticles are partially destroyed (Fig. 8.7c). The shape changes occurring after only one pulse never lead to a strong dichroism. Whatever the produced spheroids the symmetry axis of the particles (i.e. the long axis for prolate and the short one for oblate) is always oriented along the laser polarization.

The growth of nanoparticles along the laser polarization that gives rise to prolate spheroids is likely to result from successive mechanisms, which can be described as follows. At wavelengths near the plasmon resonance of spherical nanoparticles, the near-field enhancement stimulates the emission of electrons from the surface

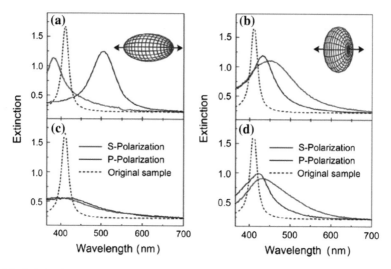

Fig. 8.7 Polarized extinction spectra of original and irradiated samples: **a** multi-shot regime (1,000 pulses per spot), peak pulse intensity $I_p = 0.6\,\text{TW/cm}^2$; **b** single shot regime, $I_p = 3\,\text{TW/cm}^2$; **c** multi-shot (5,000 pulses per spot), $I_p = 1.2\,\text{TW/cm}^2$; **d** singleshot, $I_p = 3.5\,\text{TW/cm}^2$ Reprinted with kind permission from [67]

of the metal particles. The emission process happens within a few femtoseconds and is maximal where the field is maximal i.e. for small nanospheres in the dipolar approximation, on each side of the diameter parallel to the incident laser polarization. The free electrons may then be trapped in the matrix and form color centers near the poles of the excited sphere. The ionized nanoparticles are likely to emit Ag^+ ions in statistical directions, in particular when after a few picoseconds electron thermalization and heat transfer to the silver lattice is finished. A recombination of silver ions and electrons can occur and give rise to a redeposition of Ag atoms on the nanoparticle preferentially near the poles where electrons were predominant. Silver ions can also reduce far away from the initial nanosphere and precipitate in the form of small clusters in a shell around the nanoparticle leading to a change in the local refractive index around the particle and to a red-shift of the plasmon bands [69]. Large aspect ratio nanoparticles exhibiting a strong dichroism can be achieved by successive or simultaneous irradiations with different visible and IR wavelengths (Fig. 8.8) [70, 71].

The described mechanisms require the presence of a rigid, ionic matrix and such a laser-induced transformation of metal nanoparticles has only been observed in glass nanocomposites to date. The maximum laser peak pulse intensity is more than one order of magnitude below the typical breakdown threshold of transparent glass [72]. It can be noted that heating the dichroic samples for at least several minutes above the glass transition temperature ($\sim 600\,°\text{C}$ in soda-lime glass) restores the original, non-dichroic optical extinction, due to a reshaping of the nanoparticles to spheres [73].

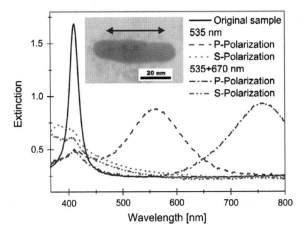

Fig. 8.8 Polarized extinction spectra of samples with Ag nanoparticles irradiated firstly at 535 nm and subsequently at 670 nm; 1,000 pulses per spot were applied, the peak pulse intensity was 1.5 TW·cm^{-2}. Inset: a TEM image of a deformed nanoparticle. Laser polarization is given as an arrow. Reprinted with kind permission from [70]

8.3.4 Reversibly Tuning the Size of Nanoparticles with Lasers

Low intensity continuous-wave lasers can also be used to tune in a reversible manner the size distribution of silver nanoparticle ensembles, provided that the latter are in contact with a semiconductor matrix and ambient atmosphere. The tuning of the nanoparticle size distribution results from a selective oxidation of silver nanoparticles when exposed to a monochromatic visible light and leads to a change in the material color [74, 75]. The laser-induced color changes, which characterize a photochromic behavior, depend on the incident wavelength and intensity; they are reversible and reproducible. Multicolor photochromism has been observed with few n type semiconductor matrixes exhibiting band gap energy in the UV range, but the most dramatic effects have been obtained with titanium dioxide. Numerous works have reported a photochromic behavior of silver nanoparticles either embedded in porous TiO$_2$ matrixes [76–81] or at the surface of TiO$_2$ single crystal plates [82–84]. In all cases, changes in the nanoparticle size result from charge transfers between the metal nanoparticles and TiO$_2$ under UV or visible photon excitation.

Under UV illumination, when the incident photon energy is larger than the band gap of TiO$_2$ (type 3.2 eV), the latter absorbs the incident radiation and releases electron-hole pairs. Electrons reduce Ag$^+$ ions, while the holes oxidize adsorbed organic compounds like water. Due to the high atomic and ionic mobility of silver in porous materials and on the surface of dense plates (especially in a humid atmosphere where water adsorbs on surfaces), reduced silver atoms coalesce to form nanoparticles, even upon low light intensity (of the order of mW·cm^{-2}). Increasing the light intensity from a few mW·cm^{-2} to a few kW·cm^{-2} tends to favour the growth of narrower size distributions of nanoparticles especially in mesoporous matrixes.

(a)

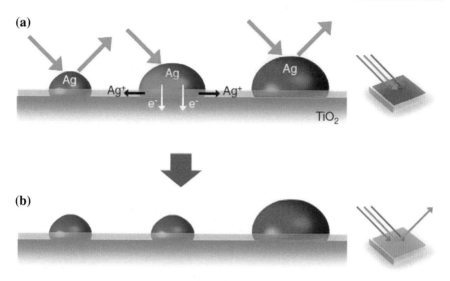

(b)

Fig. 8.9 Mechanisms of the multicolor photochromism. Particle colors stand for colors that the particles absorb. Reprinted with kind permission from [88]

Depending on the size distribution obtained, the materials have a more or less broad absorption band in the visible and a color varying from yellow brownish to greyish [77, 78, 85].

Exposing this nanocomposite material to visible light allows to excite electrons at the nanoparticle surface by means of the plasmon excitation. Due to a kind of Schottky junction existing between metal nanoparticles and the surrounding semiconductor, excited electrons transfer from the resonant nanoparticles to TiO_2 [80–88]. As a result, the resonant nanoparticles are gradually oxidized to Ag^+ ions. The transferred electrons are stabilized on adsorbed oxygen [74, 80] or reduce Ag^+ ions released from the resonant particles [82, 83]. Ag^+ ions diffuse or migrate from the initial particle [83], and form new small silver clusters or nanoparticles after recombination with free electrons, or deposit on other nonresonant Ag particles (Fig. 8.9). When using a monochromatic light, the resonant nanoparticles correspond to a range of sizes for which the plasmon band is centred on the incident wavelength. Only these nanoparticles are oxidized and a dip forms in the absorbance spectrum near the excitation wavelength leading to a resulting color that tends to the illuminant colour. Using white light leads to a bleaching i.e. the oxidation of all silver nanoparticles. Such a bleaching can also occur with a monochromatic light provided that the incident intensity is high enough (of the order of few $W \cdot cm^{-2}$) [77].

Oxidation of silver nanoparticles relies therefore on a visible light excitation, the presence of ambient oxygen in the vicinity of nanoparticles to stabilize the free electrons and a high ionic mobility. In order to prevent the spontaneous oxidation of silver nanoparticles in ambient conditions and preserve their color, one has to control the contact with oxygen or to reduce the ionic mobility. Subsequent chemical treatments with alkylthiol have been proposed to prevent contact of oxygen with silver [76]. However, embedding the nanoparticles in mesoporous matrixes with a well-controlled porosity seems to be a simpler way to get very stable color states [77, 89]. Actually controlling the porosity allows to limit the contact with oxygen and the ionic and atomic mobilities.

8.3.5 Optical Methods of Forming Metallic Nanostructures on the Surface of Insulating Materials

Non-thermal light-induced surface processes hold out hope for the development of new approaches to surface nanostructuring. Although they do not seem to be as flexible and universal as photolithography a number of niche applications are waiting for a specific cheap and easy process. The light induced atomic desorption is a reliable tool to control the surface number density of the adsorbed atoms in the course of the physical vapor deposition process. The strong enough illumination diminishes the number density of the adsorbed atoms below the threshold value needed for the beginning of the nucleation process. Hence, the deposition pattern reproduces the distribution of the illumination intensity over the surface.

Light-induced atomic desorption follows after the atom adsorbed on the surface absorbs a photon. From this point of view it is important to know the absorption bands of the atoms adsorbed on transparent substrates. This information is rather limited. One case studied in some detail deals with sodium atoms adsorbed on the sapphire surface [90]. Other examples include silver adsorbed onto AgCl single crystal surface and amorphous SiO_2 [91, 92].

Light-induced atomic desorption was observed for alkali [93–96] (Na, Rb and Cs) as well as for Zn and Sn [8.128]. To see if this process can lead to useful applications one needs to measure the photodesorption spectra as well as the cross section of photodesorption process. The only available data set belongs to sodium. According to [93, 94] the quantum yield of photodesorption is of the order of 10^{-3}. Hence, even at moderate levels of laser irradiation the photodesorption process outruns thermal desorption and makes a substantial contribution to the equilibrium between the gas phase atoms and the atoms adsorbed on the substrate. As a result of the competition between adsorption and desorption processes the surface number density of the adsorbed atoms in the illuminated region is reduced as compared to that in the dark regions. The buildup of a continuous metal film proceeds in this case via the Volmer-Weber growth mode. Obviously, it happens first in the dark regions where the adsorbed atoms do not desorb via photoinduced process. Keeping these conditions for a reasonable time interval one obtains metal deposits in dark regions while at the bright spots the substrate remains clean. The threshold intensity of a cw diode laser operated at the wavelength of 440 nm was found to be $1\ W \cdot cm^{-2}$ for the deposition rate of $0.02\ nm \cdot s^{-1}$. The microscopic images of the sodium deposits as well as the

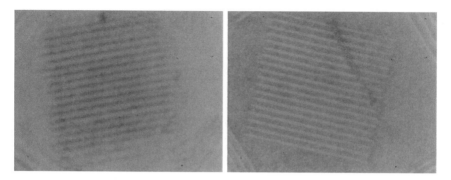

Fig. 8.10 *Left panel* microscopic image of the sodium deposits on sapphire substrate obtained via physical vapor deposition under simultaneous laser illumination trough the mire with a pitch of 10 μm depicted in the *right panel*

image of a mire are presented in Fig. 8.10. It is seen that the mire pattern is reproduced in the sodium deposits [97, 98].

8.4 Applications

Metal nanoparticles have a broad variety of applications, for example, in catalysis, surface enhanced Raman spectroscopy (SERS), bio-sensing, cancer therapy, wave guiding, and surface structuring. A description of all applications is not intended and behind the scope of this section. Instead, four examples based on the expertise of the authors will be highlighted.

8.4.1 Metal Nanoparticles as SERS Substrates

In environmental science the detection of toxic chemicals in low concentration is an important issue. For example, polycyclic aromatic hydrocarbons (PAHs) are toxic to biota and bioaccumulate in aquatic organisms. Furthermore, PAHs dissolve in water only in nanomolar concentrations due to their high octanol/water coefficient. Thus, a reliable routine technique is required, which yields a molecular fingerprint of the chemicals in water. Raman spectroscopy as a non-invasive vibrational technique is an ideal tool for the specific detection of molecules in aqueous environments.

However, Raman spectroscopy suffers from extremely low scattering cross sections, which limits its ability to detect low concentrations. To overcome this drawback, surface enhanced Raman spectroscopy (SERS) can be applied. SERS is based on the enhancement of the Raman signal of molecules in the vicinity of metal nanostructures.

Fig. 8.11 Detection limit of pyrene in distilled water as a function of plasmon wavelength of the silver nanoparticle ensembles, which serve as SERS substrates. Reprinted with permission from [99]

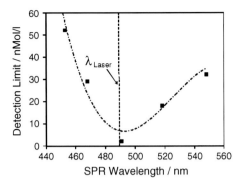

It has been shown that the SERS signal is in particular high, if the plasmon resonance is in the vicinity of the excitation wavelength for SERS [99]. Hence, tailoring the optical properties of metal nanoparticles is a crucial issue, as demonstrated in Fig. 8.11. It shows the detection limit of pyrene in water as a function of the plasmon resonance wavelength of the nanoparticle ensemble used for SERS. While for the nanoparticle ensemble whose plasmon resonance coincides with the excitation wavelength for SERS λ_{Laser}, the detection limit is 2 nMol/l only, it significantly increases by detuning the plasmon resonance wavelength. The achieved detection limit is below the allowed maximum concentration of pyrene in inland surface water (rivers or lakes) defined by the European Community. Thus, we have demonstrated that tailored noble metal nanoparticles are ideal candidates for SERS substrates, which are suited for an alarm sensor.

8.4.2 Exploiting Near Fields of Gold Nanoparticles for Surface Nanostructuring

Due to their great application potential, surface nanostructures with dimensions well below the diffraction limit have attracted considerable interest. A versatile possibility to generate such nanoscale structures is to focus fs-pulsed laser light by nanoantennas and obtain local fields, which are high enough to overcome the ablation threshold of the substrate. In particular, highly ordered triangular shaped nanoparticles prepared by nanosphere lithography have been the focus of interest. In this case the generated surface structures depend not only on the applied fluence, but also strongly on the polarisation direction of the laser light and on the size of the triangular nanoparticles [25, 26, 100, 101].

We have demonstrated [26, 100] that highly ordered small triangular shaped nanoparticles with a base of 74 nm are suitable to generate continuous nanostructures, such as nanochannels. The reason is that only such small triangular nanoparticles permit sufficient overlap of the enhanced local fields at the tips of a single particle

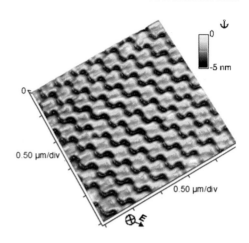

Fig. 8.12 Large scale AFM image of nanochannels generated by plasmon assisted local ablation. The channels are oriented along the polarization of the laser light (indicated by the *arrow*). Reprinted with permission from [26]

as well as of the two tips of neighboring nanoparticles to generate a homogeneous ablation depth along the channel. To achieve such continuous nanostructures, we have illuminated highly ordered triangular nanoparticles with fs-pulsed laser light. After irradiation and cleaning of the sample, micrometer long nanochannels in the fused silica substrate are obtained, as depicted in Fig. 8.12.

From a set of height profiles an average lateral width of the nanochannels of $<W> = (96 \pm 4)$ nm and an average depth of $<D> = (4.0 \pm 0.6)$ nm has been extracted. However, the length of the nanochannels was up to 10 μm [26]. With this work we have demonstrated that irradiation of triangular nanoparticle arrays with fs-pulsed laser light allows a fast and parallel surface nanostructuring of large areas. Most importantly, the dimensions of the nanostructures are well below the diffraction limit, which becomes feasible by exploiting the high and strongly localized near fields of small triangular nanoparticles.

8.4.3 Improvement of the Thermal Stability of Silver Films via UV Illumination

Annealing is known to greatly modify the granular metal films and change their optical properties. Fig. 8.13 plots the kinetics of the annealing process at the temperature of 230 °C followed up by recording extinction spectra of the silver granular film at different time intervals after deposition.

The temperature of annealing is too low to cause any significant loss of material through evaporation. Hence, the changes in the morphology of the film revealed by the optical extinction are due to diffusion. The general trend of the extinction spectra changes are rationalized in terms of the blue shift of the plasmon resonance due to the transformation of the nanoparticle shapes from oblate to more spherical form. Considerable reduction of the integral extinction in the course of annealing is due

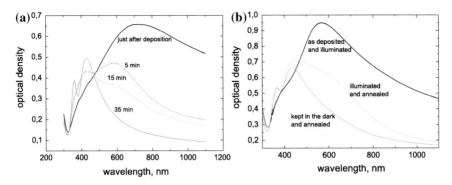

Fig. 8.13 a Annealing kinetics of the silver granular film on sapphire. The film was kept at the temperature of 230 °C. The time elapsed after the deposition is written at the curves. **b** Influence of UV illumination on the annealing of granular silver films. Illumination itself does not cause noticeable changes in the extinction spectrum (*black curve*). After annealing for 40 min at the temperature of 230 °C the difference in the extinction spectra of the illuminated film (*green curve*) and the film kept in the dark (*red curve*) becomes obvious

to the interband transitions in silver that lead to the deviations of the bulk silver optical properties from Drude model and damping of plasmon oscillation at shorter wavelengths.

A new and rather unexpected phenomenon was observed when the freshly prepared granular silver film was illuminated by UV light before annealing [102]. A mercury lamp radiation with a filter that transmits radiation with wavelengths shorter than 350 nm was used. The intensity of the illumination in this spectral range was 20 mW·cm^{-2}. The results are presented in Fig. 8.13. There are no noticeable changes in the extinction spectra after 3 hours of illumination. Then, the illuminated sample was annealed at the temperature of 230 °C for 40 min. The difference in the results of annealing for an ordinary film and the film subjected to UV illumination is drastic.

Optical density of the illuminated film is lager than that of the film kept in the dark by more than 0.2 in the range from 560 to 680 nm. This leads to the difference in the appearance of the corresponding parts of the film easily seen by eye as the illuminated area is more than 50 % darker. Illumination of granular metal film with longer wavelengths even with higher intensities does not produce any measurable effect on the annealing process.

8.4.4 Reversible or Permanent Laser-Induced Color Marking

Ag:TiO$_2$ nanostructured films have been shown to exhibit multicolour photochromism [74]. The reversible changes of color occurring under cw illumination relies on the control of the localized surface plasmon resonance of silver nanoparticles embedded in a nanoporous titania matrix (see Sect. 8.3.4). They result from the reversible tuning

Fig. 8.14 Printing/erasure
cycles on sol-gel films. Kept
off laser light, the printings
are stable for years [77]

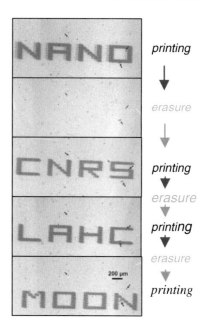

of the nanoparticle size/shape distribution through photo-activated redox reactions occurring specifically with the titania matrix [80]. Various colors have been obtained under monochromatic visible exposures resulting from a spectral hole burning in the film absorbance close to the excitation wavelength [74].

The porosity control of titania matrices through the elaboration of mesoporous films has allowed to report a high stability of the colored patterns without use of any additional chemical treatment (Fig. 8.14) [77]. Multicolour photochromism has been observed in such matrices with a wavelength-dependent selective oxidation of silver NP occurring at intensities of the order of $100\,\text{mW}\cdot\text{cm}^{-2}$ rather than few $\text{mW}\cdot\text{cm}^{-2}$ as reported on other kinds of $Ag:TiO_2$ films in the literature [74]. At higher intensity, silver nanoparticles can act as heat nanosources and crystallize the surrounding TiO_2 matrix leading to the formation of well-controlled craters [74, 103]. Thermal effects can also favor the growth of metal nanoparticles and result in a permanent color marking [104].

8.5 Conclusion

In this chapter, the optical properties of noble metal nanoparticles have been discussed. In the first part the fundamentals which lead to the unique optical behaviour and related effects have been explained. In the following we have introduced fundamental preparation techniques to generate defined nanostructured metal and metal-contained composite films. In addition, light driven processes to tailor the optical properties and the morphology of the nanoparticle ensembles have been explained. For example, we have shown that irradiation of metal nanoparticles in a matrix with fs-pulsed laser light causes significant shape changes of the nanoparticles. Depending on the laser parameters either prolate or oblate nanoparticles are generated. Based on the expertise of the authors some prominent applications of nanostructured metal and metal-contained composite films have been given in the Sect. 8.4. We have shown that metal nanoparticles have been useful and real life applications in the detection of toxic molecules, to generate sub-diffraction size nanostructures, as local heat sources and for reversible as well as permanent colour marking of metal-contained composite films. In addition, we have demonstrated that the stability of silver nanoparticles on substrates against heat induced reshaping can be drastically improved by illuminating the nanoparticles with UV light before a heat treatment. In conclusion, metal nanoparticles are not only of intriguing interest in fundamental science, they have already entered the way to real life applications.

Acknowledgments N.D. acknowledges the French National Research Agency (ANR) for financial support in the framework of project n° ANR-12-NANO-0006.
F.H. acknowledges the financial support of the Deutsche Forschungsgemeinschaft, the DAAD, and the European commission.
T.V. acknowledges the RFBR for financial support in the framework of research projects No. 11-02-01020-a and 12-02-00853-a and the European commission for support in the framework of grant PIRSES-GA-2013-612600—LIMACONA. His work was also partially financially supported by Government of Russian Federation, Grant 074-U01.

References

1. M. Faraday, Trans. Roy. Soc. (London) **147**, 145 (1857)
2. G. Mie, Annalen der Physik **25**, 377 (1908)
3. F. Hubenthal, in *Noble Metal Nanoparticles: Synthesis and Applications*, ed. by D.L. Andrews, G.D. Scholes, G.P. Wiederrecht. Comprehensive Nanoscience and Technology, vol 1 (Academic Press, Oxford, 2011), pp 375–435
4. U. Kreibig, M. Vollmer, *Optical Properties of Metal Clusters* (Springer, Berlin, 1995)
5. C.F. Bohren, D.R. Huffman, *Absorption and Scattering of Light by Small Particles* (Wiley, New York, 1983)
6. D. Pines, Rev. Mod. Phys. **28**, 184 (1956)
7. A. Pinchuk, G. von Plessen, U. Kreibig, J. Phys. D: Appl. Phys. **37**, 3133 (2004)
8. S. Berciaud, L. Cognet, P. Tamarat, B. Lounis, Nano Lett. **5**, 515 (2005)
9. P.K. Jain, K.S. Lee, I.H. El-Sayed, M.A. El-Sayed, J. Phys. Chem. **110**, 7238 (2006)
10. F. Hubenthal, Eur. J. Phys. **30**, S49 (2009)
11. A.J. Haes, S. Zou, G.C. Schatz, R.P. Van Duyne, J. Chem. Phys. B **108**, 109 (2004)
12. M. Valden, X. Lai, D.W. Goodman, Science **281**, 1647 (1998)
13. U. Heiz, E.L. Bullock, J. Mater. Chem. **14**, 564 (2004)
14. P. Zijlstra, J.W.M. Chong, M. Gu, Nature **459**, 410 (2009)
15. T.C. Chu, W.C. Liu, D.P. Tsai, Opt. Comm. **246**, 561 (2005)
16. L. Nadar, R. Sayah, F. Vocanson, N. Crespo-Monteiro, A. Boukenter, S. Sao Joao, N. Destouches, Photochem. Photobiol. Sci. **10**, 1810–1816 (2011)

17. N. Crespo-Monteiro, N. Destouches, L. Nadar, S. Reynaud, F. Vocanson, J.Y. Michalon, Appl. Phys. Lett. **99**, 173106 (2011)
18. S. Lal, S.E. Clare, N.J. Halas, Acc. Chem. Res. **41**, 1842 (2008)
19. E. Dickerson, E. Dreaden, X. Huang, I. El-Sayed, H. Chu, S. Pushpanketh, J. McDonald, M. El-Sayed, Cancer Lett. **269**, 57 (2008)
20. F. Hubenthal, D. Blázquez Sánchez, N. Borg, H. Schmidt, H.-D. Kronfeldt, F. Träger, Appl. Phys. B, **95**, 351 (2009)
21. E. Le Ru, J. Grand, N. Felidj, J. Aubard, G. Levi, A. Hohenau, J.R. Krenn, E. Blackie, P. Etchegoin, J. Phys. Chem. **112**, 8117 (2008)
22. J. Lakowicz, B. Shen, S. D'Auria, J. Malicka, J. Fang, Z. Gryczynski, I. Gryczynski, Anal. Biochem. **301**, 261 (2002)
23. A. Bek, R. Jansen, M. Ringler, S. Mayilo, T.A. Klar, J. Feldmann, Nano Lett. **8**, 485 (2008)
24. M. Westphalen, U. Kreibig, J. Rostalski, H. Lüth, D. Meissner, Sol. Energy Mater. Sol. Cells **61**, 97 (2000)
25. J. Boneberg, J. König-Birk, H.-J. Münzer, P. Leiderer, K.L. Shuford, G.C. Schatz, Appl. Phys. A **89**, 299 (2007)
26. F. Hubenthal, R. Morarescu, L. Englert, L. Haag, T. Baumert, F. Träger, Appl. Phys. Lett. **95**, 063101 (2009)
27. T. Vartanyan, J. Bosbach, F. Stietz, F. Träger, Appl. Phys. B **73**, 391 (2001)
28. U. Kreibig, Appl. Phys. B **93**, 79 (2008)
29. F. Hubenthal, C. Hendrich, F. Träger, Appl. Phys. B **100**, 225 (2010)
30. F. Calvayrac, P.-G. Reinhard, E. Suraud, C. Ullrich, Phys. Rep. **337**, 493 (2000)
31. P.-G. Reinhard, M. Brack, F. Calvayrac, C. Kohl, S. Kümmel, E. Suraud, C.A. Ullrich, Eur. Phys. J. D. **9**, 111 (1999)
32. C. Yannouleas, R. Broglia, Ann. Phys. **217**, 105 (1992)
33. F. Hubenthal, PLASMONICS. **8**(3), 1341–1349 (2013). doi:10.1007/s11468-013-9536-8
34. O.L. Muskens, P. Billaud, M. Broyer, N. Del Fatti, F. Vallée, Phys. Rev. B **78**, 205410 (2008)
35. A.N. Ponyavina, S.M. Kachan, in *Plasmonic Spectroscopy of 2D Densely Packed and Layered Metallic Nanostructures*, ed. by M.I. Mishchenko et al. Polarimetric Detection, Characterization and Remote Sensing (Springer, Berlin, 2011)
36. A.M. Bonch-Bruevich, T.A. Vartanyan, N.B. Leonov, S.G. Przhibel'skii, V.V. Khromov, Opt. Spectrosc. **91**, 779 (2001)
37. L.D. Landau, E.M. Lifshitz, *Electrodynamics of Continuous Media* (Pergamon, New York, 1984)
38. A.M. Glass, P.F. Liao, J.G. Bergman, D.H. Olson, Opt. Lett. **5**, 368 (1980)
39. H.G. Craighead, A.M. Glass, Opt. Lett. **6**, 248 (1981)
40. E.N. Kaliteevskaya, V.P. Krutyakova, T.K. Razumova, Opt. Spectrosc. **97**, 901 (2004)
41. T.A. Vartanyan, N.B. Leonov, S.G. Przhibel'skii, V.V. Khromov, Opt. Spectrosc. **106**, 697 (2009)
42. F. Stietz, J. Bosbach, T. Wenzel, T. Vartanyan, A. Goldmann, F. Träger, Phys. Rev. Lett. **84**, 5644 (2000)
43. J. Bosbach, C. Hendrich, F. Stietz, T. Vartanyan, F. Träger, Phys. Rev. Lett. **89**, 257404 (2002)
44. N.A. Toropov, E.N. Kaliteevskaya, N.B. Leonov, T.A. Vartanyan, Opt. Spec. **113**, 616 (2012)
45. I.I.S. Lim, F. Goroleski, D. Mott, N. Kariuki, W. Ip, J. Luo, C.J. Zhong, J. Phys. Chem. B **110**, 6673 (2006)
46. V.S. Lebedev, A.S. Medvedev, D.N. Vasil'ev, D.A. Chubich, A.G. Vitukhnovskii, Kvant. Elektron. **40**, 246 (2010)
47. C.F. Bohren, D.R. Huffman, *Absorption and Scattering of Light by Small Particles* (Wiley, New York, 1983)
48. N.V. Nikonorov, A.I. Sidorov, V.A. Tsekhomskii, K.E. Lazareva, Opt. Spectrosc. **107**, 705 (2009)
49. A.J. Haes, S. Zou, J. Zhao, G.C. Schatz, R.P. Van Duyne, J. Am. Chem. Soc. **128**, 10905 (2006)
50. V.V. Klimov, V.S. Letokhov, Phys. Rev. A **54**, 4408 (1996)

51. J. Qiu, M. Shirai, T. Nakaya, J. Si, X. Jiang, C. Zhu, K. Hirao, Appl. Phys. Lett. **81**, 3040 (2002)
52. B.C. Stuart, M.D. Feit, A.M. Rubenchik, B.M. Shore, M.D. Perry, Phys. Rev. Lett. **74**, 2248 (1995)
53. J. Qiu, X. Jiang, C. Zhu, M. Shirai, J. Si, N. Jiang, K. Hirao, Angew. Chem. Int. Ed. **43**, 2230 (2004)
54. Y. Teng, B. Qian, N. Jiang, Y. Liu, F. Luo, S. Ye, J. Zhou, B. Zhu, H. Zeng, J. Qiu, Chem. Phys. Lett. **485**, 91 (2010)
55. J. Shin, K. Jang, K.-S. Lim, I.-B. Sohn, Y.-C. Noh, J. Lee, Appl. Phys. **93**, 923 (2008)
56. J.M.P. Almeida, L. De Boni, W. Avansi, C. Ribeiro, E. Longo, A.C. Hernandes, C.R. Mendonca, Opt. Express **20**, 15106 (2012)
57. F. Goutaland, E. Marin, J.Y. Michalon, A. Boukenter, Appl. Phys. Lett. **94**, 181108 (2009)
58. J.-P. Blondeau, S. Pellerin, V. Vial, K. Dzierzega, N. Pellerin, C. Andreazza-Vignolle, J. Cryst. Growth **311**, 172 (2008)
59. A. Miotello, M. Bonelli, G. De Marchi, G. Mattei, P. Mazzoldi, C. Sada, F. Gonella, Appl. Phys. Lett. **79**, 2456 (2001)
60. C.J. Brinker, G.W. Scherer, *Sol-gel Science: The Physics and Chemistry of Sol-gel Processing* (Academic Press, San Diego, 1990), p. 620
61. H. El Hamzaoui, R. Bernard, A. Chahadih, F. Chassagneux, L. Bois, D. Jegouso, L. Hay, B. Capoen, M. Bouazaoui, Mater. Lett. **64**, 1279 (2010)
62. H. El Hamzaoui, R. Bernard, A. Chahadih, F. Chassagneux, L. Bois, B. Capoen, M. Bouazaoui, Mater. Res. Bull. **46**, 1530 (2011)
63. Y. Battie, N. Destouches, L. Bois, F. Chassagneux, A. Tishchenko, S. Parola, A. Boukenter, J. Phys. Chem. C **114**, 8679 (2010)
64. C. Luo, Y. Zhang, X. Zeng, Y. Zeng, Y. Wang, J. Colloid Interface Sci. **288**, 444 (2005)
65. J. Li, K. Kamata, T. Iyoda, Thin Solid Films **516**, 2577 (2008)
66. N. Destouches, Y. Battie, N. Crespo-Monteiro, F. Chassagneux, L. Bois, S. Bakhti, F. Vocanson, N. Toulhoat, N. Moncoffre, T. Epicier, J. Nanopart. Res. **15**, 1422 (2013)
67. A. Stalmashonak, A. Podlipensky, G. Seifert, H. Graener, Appl. Phys. B **94**, 459 (2009)
68. A. Stalmashonak, G. Seifert, A. Akin Unal, U. Skrzypczak, A. Podlipensky, A. Abdolvand, H. Graener, Appl. Opt. **48**, F37 (2009)
69. M. Kaempfe, G. Seifert, K.-J. Berg, H. Hofmeister, H. Graener, Eur. Phys. J. D. **16**, 240 (2001)
70. A. Stalmashonak, G. Seifert, H. Graener, J. Opt. A **11**, 065001 (2009)
71. A. Stalmashonak, C. Matyssek, O. Kiriyenko, W. Hergert, H. Graener, G. Seifert, Opt. Lett. **35**, 1671 (2010)
72. D. Du, X. Liu, G. Mourou, Appl. Phys. B **63**, 617 (1996)
73. A. Podlipensky, A. Abdolvand, G. Seifert, H. Graener, Appl. Phys. A **80**, 1647 (2005)
74. Y. Ohko, T. Tatsuma, T. Fujii, K. Naoi, C. Niwa, Y. Kubota, A. Fujishima, Nat. Mater. **2**, 29 (2003)
75. K. Naoi, Y. Ohko, T. Tatsuma, J. Am. Chem. Soc. **126**, 3664 (2004)
76. K. Naoi, Y. Ohko, T. Tatsuma, Chem. Commun. **10**, 1288 (2005)
77. N. Crespo-Monteiro, N. Destouches, L. Bois, F. Chassagneux, S. Reynaud, T. Fournel, Adv. Mater. **22**, 3166 (2010)
78. L. Nadar, R. Sayah, F. Vocanson, N. Crespo-Monteiro, A. Boukenter, S. Sao Joao, N. Destouches., Photochem. Photobiol. Sci. **10**, 1810 (2011)
79. J. Preclíková, F. Trojánek, P. Nemec, P. Malý, J. Phys. Status Solidi C **5**, 3496 (2008)
80. K. Kawahara, K. Suzuki, Y. Ohko, T. Tatsuma, Phys. Chem. Chem. Phys. **7**, 3851 (2005)
81. L. Bois, F. Chassagneux, Y. Battie, F. Bessueille, L. Mollet, S. Parola, N. Destouches, N. Toulhoat, N. Moncoffre, Langmuir **26**, 1199 (2010)
82. K. Matsubara, T. Tatsuma, Adv. Mater. **19**, 2802 (2007)
83. K. Matsubara, K.L. Kelly, N. Sakai, T. Tatsuma, Phys. Chem. Chem. Phys. **10**, 2263 (2008)
84. K. Matsubara, K.L. Kelly, N. Sakai, T. Tatsuma, J. Mater. Chem. **19**, 5526 (2009)
85. N. Crespo-Monteiro, N. Destouches, L. Nadar, S. Reynaud, F. Vocanson, J.-Y. Michalon, Appl. Phys. Lett. **99**, 173106 (2011)

86. Y. Tian, T. Tatsuma, Chem. Commun. 1810 (2004)
87. Y. Takahashi, T. Tatsuma, Nanoscale **2**, 1494 (2010)
88. T. Tatsuma, Bull. Chem. Soc. Jpn. **86**, 19 (2013)
89. N. Crespo-Monteiro, N. Destouches, T. Fournel, Appl. Phys. Express **5**, 075803 (2012)
90. A.M. Bonch-Bruevich, YuN Maksimov, V.V. Khromov, Opt. Spectr. **58**, 854 (1985)
91. A.N. Latyshev, O.V. Ovchinnikov, S.S. Okhotnikov, J. Appl. Spectr. **70**, 817 (2003)
92. J.M. Antonietti, M. Michalski, U. Heiz, H. Lones, K.H. Lim, N. Rösch, A.D. Vitto, G. Pacchioni, Phys. Rev. Lett. **94**, 213402 (2005)
93. I.N. Abramova, E.B. Aleksandrov, A.M. Bonch-Bruevich, V.V. Khromov, JETP Lett. **39**, 203 (1984)
94. A.M. Bonch-Bruevich, T.A. Vartanyan, A.V. Gorlanov, Yu.N. Maksimov, S.G. Przhibel'skii, V.V. Khromov, Sov. Phys. JETP **70**, 604 (1990)
95. A.M. Bonch-Bruevich, T.A. Vartanyan, YuN Maksimov, S.G. Przhibel'skii, V.V. Khromov, Surf. Rev. Lett. **5**, 331 (1998)
96. C. Marinelli, A. Burchianti, A. Bogi, F. Della Valle, G. Bevilacqua, E. Mariotti, S. Veronesi, L. Moi, Eur. Phys. J. D. **37**, 319 (2006)
97. T.A. Vartanyan, V.V. Khromov, N.B. Leonov, S.G. Przhibel'skii, Proc. SPIE **7996**, 79960H–1 (2011)
98. T.A. Vartanyan, V.V. Khromov, N.B. Leomov, S.G. Przhibel'skii, J. Opt. Technol. **78**, 505 (2011)
99. Y.-H. Kwon, R. Ossig, F. Hubenthal, H.-D. Kronfeldt, J. Raman Spectrosc. **43**, 1385 (2012)
100. R. Morarescu, L. Englert, B. Kolaric, P. Damman, R.A.L. Vallée, T. Baumert, F. Hubenthal, F. Träger, J. Mater. Chem. **21**, 4076 (2011)
101. A. Jamali, B. Witzigmann, R. Morarescu, T. Baumert, F. Träger, F. Hubenthal, Appl. Phys. A **110**, 743 (2013)
102. T.A. Vartanyan, N.B. Leonov, V.V. Khromov, S.G. Przhibel'skii, N.A. Toropov, E.N. Kaliteevskaya, Proc. SPIE **8414**, 841404–1 (2012)
103. N. Crespo-Monteiro, N. Destouches, L. Saviot, S. Reynaud, T. Epicier, E. Gamet, L. Bois, A. Boukenter, J. Phys. Chem. C **116**, 26857 (2012)
104. N. Destouches, N. Crespo-Monteiro, T. Epicier, Y. Lefkir, F. Vocanson, S. Reynaud, R. Charrière, M. Hëbert, Proc. SPIE 8609, Synthesis and Photonics of Nanoscale Materials X, 860905 (2013). doi:10.1117/12.2003178

Chapter 9
Selective Ablation of Thin Films by Pulsed Laser

Andreas Ostendorf, Evgeny L. Gurevich and Xiao Shizhou

Abstract Laser direct patterning of thin films with minimal substrate damage is receiving attention in many industrial applications, e.g., photovoltaic or flat displays. Substantial progress has been made in understanding of the laser-matter interactions and reveals that laser-induced thermal effects are significantly critical in most of laser ablation processes. The thermal penetration depth, determined by the optical absorption and subsequently the thermal diffusion length, are heavily dependent on the applied laser pulse duration. The ratios between the film thickness, the thermal and the optical penetration depths separate the ablation to be film-like or bulk-like behavior of the thin-film ablation.

9.1 Introduction

The interaction between laser radiation and solids has been investigated extensively for a large variety of materials, for different pulse duration and spectral domains. The spatio-temporal distribution of the laser-induced temperature profile plays a significant role in the initial stage of material ablation, although afterwards thermally induced stress, plasma and shockwave will be possibly involved [1, 2].[1] The laser energy deposition into the material starts from the photon-electron interaction and the electron temperature T_e increases. The following lattice temperature T_l rise is due to the elevation of the kinetic energy of the lattice through electron-lattice energy

---

[1] Physical model of the laser ablation and propagation of shock waves are discussed in the Chaps. 1 and 2 in the Part I of this book.

E. L. Gurevich (✉) · A. Ostendorf · X. Shizhou
Laser Applications Technology, Department of Mechanical Engineering, Ruhr-University Bochum, Universitätsstraße 150, ID 05/629, 44801 Bochum, Germany
e-mail: gurevich@lat.rub.de

V. P. Veiko and V. I. Konov (eds.), *Fundamentals of Laser-Assisted Micro- and Nanotechnologies*, Springer Series in Materials Science 195, DOI: 10.1007/978-3-319-05987-7_9, © Springer International Publishing Switzerland 2014

exchange [3, 4]. Usually the removal process (ablation) takes place as the material melts or evaporates.

The ablation geometry is highly dependent on the temperature distribution governed by the laser absorption profile and the heat diffusion length. The latter determines the heat affected zone (HAZ) surrounding the ablation spot. The HAZ is found to be strongly dependent on the applied laser pulse duration τ. A shorter pulse provides smaller heat affected zones and higher ablation accuracy. In the selective laser ablation of thin films [5], besides the lateral resolution, the precise in-depth profile is also critical. In order to avoid damage of the substrate, the ablation crater has to be restricted to the thin film thickness b. In this case, both optical penetration depth l_{opt} and heat diffusion length l_{th} have to be taken into consideration, because both affect the temperature distribution during the pulse irradiation. The depth to which the heat penetrates into the ablated material during a laser pulse is the effective thermal penetration depth l_{eff}.

In [6], the authors treated the effective thermal penetration depth by adding the optical penetration depth and thermal diffusion length. In [7, 8] the authors pointed out that for the effective optical penetration depth, both the optical penetration depth and the electron heat penetration depth have to be taken into consideration for ultrashort pulse laser ablation of metals. The effective thermal penetration depth is a superposition of l_{th} and l_{opt}, although it is difficult to give an explicit math expression from theoretical derivation. For strongly absorbing materials like metals where $l_{th} \gg l_{opt}$ is fulfilled, it is justified to treat the laser source as a surface heat source. Hence the thermal penetration depth is mainly determined by the heat diffusion length, which is proportional to the square root of the product of the thermal diffusivity α and the applied pulse length for nanosecond pulses [9]:

$$l_{eff} \simeq l_{th} = \sqrt{2\alpha\tau} \qquad (9.1)$$

In the ultrashort pulse regime, the situation is more complex because hot electron scattering plays a dominate role in the transportation of heat and phonon-phonon coupling could be neglected. In this case the thermal penetration depth is mainly caused by the electrons. In case when surface lattice melting occurs, it can be calculated for metals by using the TTM model [10]:

$$l_{eff} \simeq \left(\frac{128}{\pi}\right)^{1/8} \left(\frac{k^2 C_l}{C_0 T_m g^2}\right)^{1/4} \qquad (9.2)$$

Here $C_0 = C_e/T_e$ with C_e and C_l—the electron and lattice heat capacities, k—the thermal conductivity and g—the electron-phonon coupling constant.

However, for wide-bandgap semitransparent materials, which exhibit a large optical penetration depth, $l_{opt} \gg l_{th}$ is valid. Here the absorption of the laser energy has to be treated as a volume heat source, and the effective thermal penetration depends on both the optical penetration and the thermal diffusion. In this case it is difficult to give an explicit math expression from theoretical derivation.

Fig. 9.1 Schematically illustration of ablation threshold fluence for metals and semiconductors in different pulse duration regime. For short pulse ablation, the $\tau^{1/2}$ scaling rule is valid

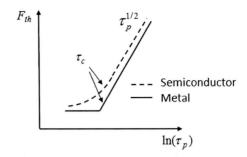

The relation of the thermal penetration depth and the pulse length can be determined by the damage threshold fluence q_* versus pulse length experiments. For short pulse $q_* \propto \tau^{1/2}$ for both metals and semiconductors. In ultrashort pulse regime, the ablation threshold keeps almost constant for metals, but a nonlinear feature is observed for semiconductors, schematically shown in Fig. 9.1 [11]. A characteristic time τ_c can be defined, which distinguishes the different dependence of the ablation threshold fluence on the pulse length and separates the short and ultrashort pulse regimes, as defined by the equation [10]:

$$\tau_c = \left(\frac{8}{\pi}\right)^{1/4} \left(\frac{C_l^3}{C_0 T_m g^2}\right)^{1/2}. \tag{9.3}$$

In bulk material ablation, the laser generated heat freely conducts in spatial dimensions in the target solid. The heat diffusion length is mainly determined by a critical temporal variable, such as the pulse width. However, when film-substrate system is concerned, the heat diffusion is affected by the film thickness also. The film thickness could affect the temperature distribution for certain pulse-width range. If the film thickness is smaller to the effective thermal penetration depth l_{eff}, the influence from the substrate has to be considered.

9.2 Thermal Penetration Depth in Laser Ablation of Films

The influence of film thickness on the ablation of metal films has been studied for different pulse length regimes [5, 12–15]. By nanosecond laser ablation of thin absorbing films the material is partly removed by two mechanisms: evaporation and melt flow out of the crater. The flow in this case is induced by reactive pressure of the evaporating material [16, 17]. The spots size by single picosecond laser pulse ablation of tungsten film on silicon substrate as the function of applied pulse energy for different film thicknesses are illustrated in Fig. 9.2a. A well-defined linear dependence in semi-log plot can be observed with almost a constant slope for all cases.

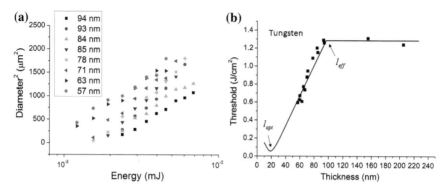

Fig. 9.2 a Dependence of the ablation spots size on the pulse energy; **b** dependence of ablation threshold fluence on the film thickness. Measured for tungsten film ablated by 10 ps laser pulse at 1,064 nm

The intersection of the interpolation of these lines with the energy-axis define the ablation thresholds. Figure 9.2b shows the dependence of ablation threshold on film thickness. The dependence trend of the ablation threshold fluence q_* on the film thickness b can be divided into two regimes. For films below a thickness of about 95 nm, the thresholds increase with the thickness. For thicker films, $b \gtrsim 95$ nm, the ablation threshold keeps constant equal to the bulk ablation threshold value. The effective thermal penetration length can be determined as $l_{eff} \approx 95$ nm, which separates the film and bulk ablation features for tungsten.

In most studies which deal with metal films, the influence of film thickness on the damage threshold exhibits two distinguished regimes. However, by taking very thin film into investigation, a third regime can be observed. In [18] femtosecond laser ablation of very thin gold films was investigated. If the film thickness is in the range of optical penetration depth, the ablation threshold fluence is found to decrease with the film thickness.

In summary, the threshold as a function of film thickness can divide into three regimes for metals. It decreases with the film thickness if $b \lesssim l_{opt}$. When the film thickness exceeds the optical penetration depth it starts increasing with the thickness and finally saturates at a constant value of the ablation threshold of the bulk material for $b \gtrsim l_{eff}$, see Fig. 9.2b.[2] Although the laser pulse parameters such as the pulse length or laser wavelength will affect the ablation threshold value, it always reveals these three distinguished regimes for metal films, where the condition $l_{th} \gg l_{opt}$ is fulfilled.

The ablation of transparent ITO films on glass substrate for different film thicknesses are investigated as well. Figure 9.3a shows the linear dependence of the ablation spots size on the applied single pulse energy in semi-log plot. More energy is required for thinner films ablation to obtain the same ablation spot size compared

---

[2] Ablation of transparent bulk materials is discussed in the Chaps. 1 and 11 in the Parts I and V of this book.

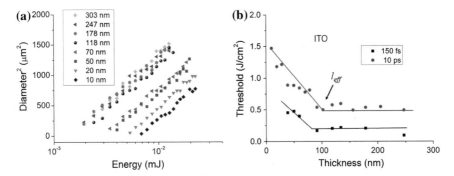

Fig. 9.3 **a** Ablation spots size increases as the rising of pulse energy for ITO films ablated by 10 ps pulses at 1,064 nm; **b** dependence of threshold fluence on the film thickness for ITO ablated by 10 ps pulses at 1,064 nm and 150 fs pulses at 800 nm

to thicker ones. Figure 9.3b illustrates how the ablation threshold fluence depends on the film thickness for femtosecond and picosecond laser ablation. The dependence of the ablation threshold fluence q_* on the film thickness b for ITO film demonstrates two regimes, in contrast to metals where three regimes can be distinguished. For thinner film, the ablation threshold fluence decreases as the film thickness increase, similar to the case of metals, which thickness is in the range of optical penetration depth [18]. The decreasing trend alters when the film thickness exceeds a certain value, where the effective thermal penetration depth of ITO is defined. Afterwards it tends to a constant value, which is equal to the ablation threshold of the bulk material. The same trend can be found for the low absorption polymer films [19]. The optical penetration depth of ITO is ~2 μm at 1,064 nm by taking linear absorption part ($\delta = l_{opt}^{-1} = 5.3 \times 10^5$ cm^{-1}) into account. Whereas the thermal diffusion length is ~15 nm be calculated by $\sqrt{2\alpha\tau}$ for 10 ps pulse length irradiation. The difference from metals is due to the fact that $l_{opt} \gg l_{th}$ for such low absorption ITO films.

In order to describe the dependence of the film thickness on the threshold fluence we assume a purely thermal origin of the observed ablation effects. Regarding the film temperature T reaching its melting point T_m as the criterion for ablation. Neglecting the lateral heat diffusion and only considering the depth profile due to the large beam size, the temperature increase in the heated volume due to the absorption of laser energy can be expressed as

$$\Delta T_m = \Delta Q / C, \tag{9.4}$$

where ΔQ is the total absorbed energy and C is the heat capacity at constant pressure. A uniform model for threshold fluence of melting can be obtained based on (9.4) written in the form [13]:

$$q_* \simeq \frac{\Delta T_m}{(1 - e^{-\delta b})(1 - R)} \left[\rho_f c_f - \left(\frac{l_{th,s}}{l_{th,f}}\right) \rho_s c s \right] L_f + \frac{\Delta T_m}{(1 - e^{-\delta b})(1 - R)} l_{th,s} \rho_s c_s. \tag{9.5}$$

Here ΔT_m is the temperature increase needed for melting, which is a constant for specific material. ρ, c are the mass density and specific heat, where the subscript f and s denote the film and substrate quantities, respectively. L_f is the minimum dimension of the heated volume, which can be the film thickness, the optical penetration depth or the effective thermal penetration length. From this equation the different regimes of ablation threshold variation trends for metal and transparent ITO films can be identified.

For metal films, $l_{opt} \ll l_{th}$, the ablation threshold dependence on film thickness can be characterized in three regimes:

- In the first regime I, $b \ll l_{opt} \ll l_{eff}$ and $L_f = b$ are fulfilled. For small b in the (9.5), the limit in the first item on the right hand is found to be constant: $\lim_{b \to 0} \frac{b}{1-e^{-\delta b}} = \frac{1}{\delta}$. However, for the second item which defines the influence of the substrate, the limit goes to infinity due to $(1 - e^{-\delta b}) \sim 0$ when $b \to 0$. This implies that as film becomes thinner, the substrate influence increases and more pulse energy is required to promote the film to its melting temperature, $q_* \sim 1/b$ is found.

- In the second regime II, $l_{opt} \ll b \ll l_{eff}$ and $L_f = b$ are fulfilled. In this case the pulse energy is regarded to be totally absorbed, where the item $(1 - e^{-\delta b}) \sim 1$. The influence from the substrate is limited and one can find that $q_* \sim b$.

- In the third regime III, $l_{opt} \ll l_{eff} \ll b$ and $L_f = l_{eff}$ are fulfilled. The laser energy is regarded to be totally absorbed and the heat volume is limited by l_{eff}. The influence from the substrate is neglected. Hence the ablation threshold q_* is no longer dependent on the film thickness, behaves as ablation of bulk material.

For transparent ITO films, $l_{opt} \gg l_{th}$, the ablation threshold dependence on film thickness can be characterized in two regimes:

- In the regime I, $b \ll l_{eff} \ll l_{opt}$ and $L_f = b$ are fulfilled. Similar to the case of metal films in the regime I, the substrate influence plays a dominate role on ablation threshold fluence. q_* decreases as the rising of film thickness.

- In the regime II, $l_{eff} \ll b \ll l_{opt}$ and $L_f = l_{eff}$ are fulfilled. Because of the large optical penetration depth l_{opt}, the changing of the film thickness b in this regime only leads to a very small variation of $(1 - e^{-\delta b})$ item, which can be regarded approximately as a constant. On the other hand, the heat volume is controlled by l_{eff} which is independent on the film thickness b. Therefore q_* keeps approximately as a constant.

These regimes are simulated according to the (9.5) in the Fig. 9.4 for metals and ITO.

9.3 Front- and Rear-Side Laser Ablation of Films

For selective laser ablation of thin films, weak absorption of the substrate and high absorption of the film are needed. Some experiments [20, 21] advise that rear-side ablation can achieve better ablation quality. Here we investigate the thermal penetration depth influence on the ablation of ITO films by front- and rear-side irradiation.

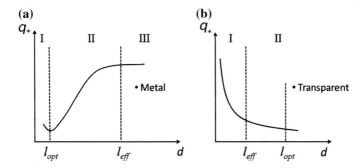

Fig. 9.4 Simulation result of threshold fluences dependence on the thickness of metal and transparent ITO films

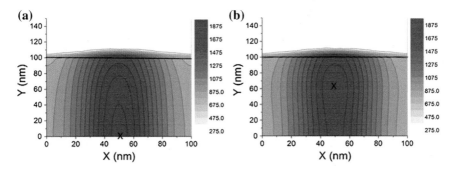

Fig. 9.5 Numerical simulation results of temperature distribution induced by laser irradiation from **a** front-side, and **b** rear-side. The location marked by "x" denotes the maximum temperature point

Previous works already demonstrated that the different ablation geometry could happen for laser ablation of metals by front- and rear-side irradiation, when the film thickness exceeds the thermal penetration depth [22, 23]. The film thickness is crucial in determining the amount of energy guided to the substrate: If it is smaller than the thermal penetration length, the temperature field can be regarded to be the same for both irradiation directions. On the other hand, if the film thickness is larger than the thermal penetration length, the influence of the substrate can be neglected for front-side ablation. Hence the temperature distribution induced by laser irradiation from the front- or rear-side ablation could make a strong difference and alter the surface ablation morphology. Figure 9.5 demonstrates the numerical simulation of the temperature distribution for the film thickness larger than the thermal penetration depth. One can detect a significant difference of the temperature distribution for front- and rear-side irradiation.

The SEM images in Fig. 9.6 are the single-pulse ablation results of transparent ITO films of 100 nm thickness on glass substrates, which are irradiated by 300 ms ($\lambda = 1,064$ nm), 10 ps ($\lambda = 1,064$ nm) and 150 fs ($\lambda = 800$ nm) laser pulses from front- and rear-side. The laser fluence is the same for front- and rear-side ablation.

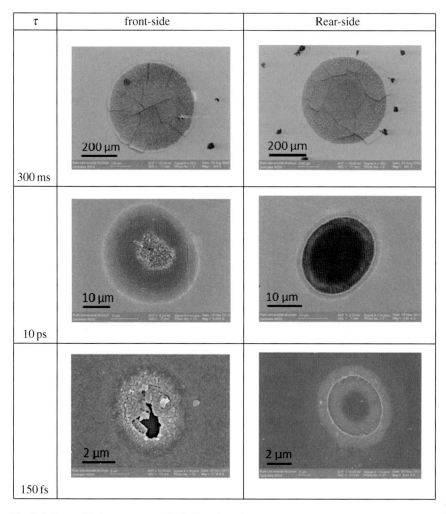

Fig. 9.6 Laser ablation geometry of ITO films from front- and rear-side single pulse irradiation by 300 ms pulse with $\lambda = 1{,}064$ nm, 10 ps pulse with $\lambda = 1{,}064$ nm and 150 fs pulse with $\lambda = 800$ nm

For the 300 ms pulse ablation, the film is seriously cracked for both thicknesses due to the large laser induced thermal stress. The ablation spot geometries do not show so much difference due to the large thermal penetration depth at this pulse duration. For 10 ps and 150 fs pulses ablation, the films are likely to be removed by thermal melting and vaporization without film cracking. For the thickness of 100 nm films, much cleaner and clearer ablation spots can be obtained by rear-side ablation. The images reveal that the ablation quality is different for those film thicknesses exceeding the thermal penetration depth, see Fig. 9.6 for $\tau = 10$ ps and $\tau = 150$ fs pulse ablation of 100 nm ITO films. However, no apparent difference in the ablation geometry for the front- and the rear-side ablation can be found for a thinner film with $b = 50$ nm.

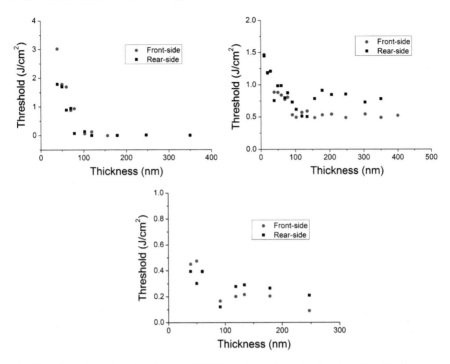

Fig. 9.7 The ablation threshold fluence of ITO films alternative as the function of the film thickness for different pulse length of 300 ms (*left*), 10 ps pulse (*left*) and 150 fs (*bottom*). Both front- and rear-side irradiations are presented

The corresponding thresholds for the front- and the rear-side ablation of the films of different thickness can be found in Fig. 9.7. The threshold fluences curves are different when the film thickness exceeds the thermal penetration depth. Here the rear-side ablation is slightly higher compared to the constant value for front-side ablation. In the case of long pulse irradiation, or for thin films, the ablation does not demonstrate so much difference in threshold fluence and spot morphology. Also the ablation spots show much better rim quality by the rear-side short and ultrashort pulse irradiation of the 100 nm ITO films.

The difference between metal and ITO materials is schematically shown in Fig. 9.8, where the the *film ablation* and the *bulk ablation* features can be separated by the thermal penetration depth.

9.4 Incubation Effect in Laser Ablation of Films

Laser-induced damage to a material surface under multi-pulse irradiation demonstrates an interesting phenomenon: the material surface becomes damaged at pulse fluence far below the single-shot ablation threshold, so called "N-on-one"

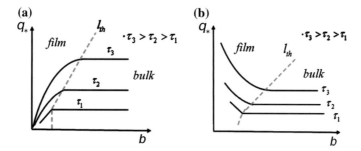

Fig. 9.8 Schematic dependence of laser ablation threshold as a function of film thickness for **a** metal and **b** ITO films for different pulse lengths. l_{th} indicates the thermal diffusion length which differentiates the film and bulk solid features for laser ablation. The results were obtained by laser front-side irradiation

accumulation effect. The incubation phenomena have been observed in various materials processing by pulsed lasers, including polymers [24, 25], metals [26, 27], semiconductors [28] and insulators [29].

The origin of the incubation is not yet fully revealed. For example, in uv laser processing of PMMA, some authors argued that the incubation was associated with the buildup of pressure due to the formation of polymer fragmentation [30–33]. On the contrary, Küper and Stuke showed that the incubation of PMMA with 248 nm laser was attributed to the photochemical degradation and mechanical stability reduction by means of spectroscopy studies [24, 25]. Graciela and co-workers suggested that the decomposition of PMMA by uv incubation pulses was the result of photoinduced formation of defect centers, which enhanced the absorption of uv light [34]. From their calculation, the ablation threshold is associated with the pulse number by

$$q_*(1) = q_*(N)[1 + kq_*(N)(N - 1)], \tag{9.6}$$

where $q_*(1)$ and $q_*(N)$ are the single-shot and N-shot damage thresholds, respectively. k reflects the incubation, which is the function of absorption cross section of PMMA (σ) and the absorption cross section of the induced defect center (β). Based on the (9.6) the authors successfully obtained the value of σ and β.

For metals, it was reported that for multi-pulse laser induced damage by 10 ns Nd:YAG laser pulses with 1,064 nm wavelength, the accumulation process was attributed to thermal stress-strain energy storage [26]. The authors argued that multi-pulse laser ablation on a site was similar to the bulk mechanical fatigue damage. On the analogy of the fatigue failure induced by the stress for N cycles, the authors derived a cumulative equation

$$q_*(N) = q_*(1)N^{S-1}, \tag{9.7}$$

where S is so called incubation coefficient which quantifies the degree of incubational behavior. The model gives a reasonable interpretation for the incubation behavior of metals for the reported experiments in many research works, see e.g., [15, 35]. It is also appropriate to fit the experimental data in pulsed laser ablation of some semiconductors [28]. It also appears in evaluation of the incubation effect in femto- and nanosecond laser ablation of doped PMMA conducted by Krüger et al. [36].

For dielectric or some semiconductor materials with "N-on-one" irradiation, the incubation effect can be attributed to laser induced defects. Multi-shot irradiation onto a site of dielectrics could induce the generation of point defects by multi-photon excitation, for example formation of color centers [37, 38]. The strength of such accumulation process is related to the excitation and generation of electrons initiated by combined multi-photon and avalanche ionization. The threshold fades due to the decrease of defect accumulation for increasing N until reaching a constant level. Therefore a different equation was proposed for this case [39]:

$$q_*(N) = q_*(\infty) + [q_*(1) - q_*(\infty)]e^{k(N-1)}, \qquad (9.8)$$

where $q_*(\infty)$ gives the maximum fluence at which the material could be permanently irradiated without causing damage, k is the factor characterizing the incubation degree. In this case the ablation threshold drops dramatically at low pulse numbers and stays constant for a large number of pulses. This dependency is mainly used in laser processing of dielectric materials [40].

Equation (9.7) is only valid for a limited number of pulses since $q_* \neq 0$ [41]; on the other hand it is also not applicable to all experiments with metals. An exception has been described by Kern et al. for femtosecond laser-induced damage of thin Au films [42]. They adopted (9.8) to fit the multi-shot damage threshold versus the pulse numbers and argued that surface modification is responsible for incubation for films significantly thinner than the characteristic penetration depth [27].

Here we investigate the incubation behavior of the wide band-gap semiconductor ITO material when exposed to picosecond laser pulses. Structuring of ITO films has a great impetus from industrial applications. Accordingly, laser patterning of ITO thin films on glass substrates has been extensively studied with various pulsed laser sources [43–51]. However, a detailed study of the incubation behavior of ITO films and its influence on laser patterning has not yet been reported. Here we will show the decline of ablation thresholds as a function of the pulse number and investigate the incubation coefficient as well as its dependence on film thickness and its influence on laser line patterning of thin ITO films.

Figure 9.9 illustrates the typical surface damage morphologies of ITO films irradiated by different number of laser shots recorded by SEM. The laser fluence was set to $1.33\,\text{J/cm}^2$, slightly above the single-shot threshold. By increasing the number of incident laser pulses the damage site area grows and the rims are getting sharper. Figure 9.10 demonstrates the surface damage morphologies induced by different laser fluences with a fixed number of laser shots ($N = 10$). Increasing the laser fluence will enlarge the ablation area, which is basically the same effect as to increasing the number of laser pulses.

To determine the damage thresholds of ITO films as a function of the incident pulse number as well as its dependence on film thickness, a series of experiments

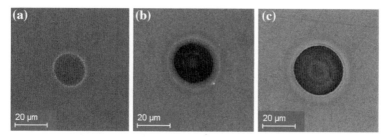

Fig. 9.9 Laser induced damage micro morphologies of the ITO films irradiated by different pulse numbers recorded by SEM, where **a** 1 pulse, **b** 10 pulses and **c** 100 pulses. The applied laser fluence was 0.83 J/cm^2. The film thickness was 100 nm for all cases

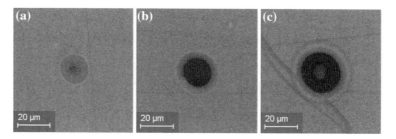

Fig. 9.10 Laser induced damage micro morphologies of the ITO films irradiated by different pulse energies recorded by SEM, where **a** 0.53 J/cm^2, **b** 0.88 J/cm^2 and **c** 1.33 J/cm^2. The applied pulse number was 10. The film thickness was 100 nm for all cases

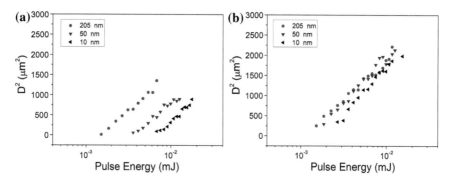

Fig. 9.11 The measured squared diameter of ablated spots plotted against the applied pulse energy, **a** 1 shot, **b** 1,000 shots

was performed with different laser fluence levels on samples with thicknesses of 10, 50 and 205 nm. Figure 9.11 presents the relationships of measured squared diameter plotted against the logarithm of the applied pulse energy for $N = 1$ and 1,000 laser pulses, respectively. From the data plots a well-defined linear dependence in semi-log plot can be observed with almost a constant slope for all cases which yields a beam radius of about 20 μm. In all plots, for the same pulse energy, the ablation site is

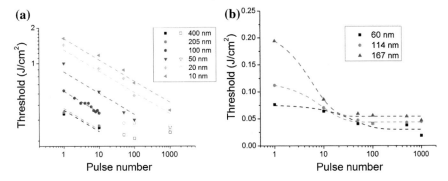

Fig. 9.12 **a** Log plot of the ablation thresholds in dependence on the number of pulses for ITO. The lines are a fit to (9.7) with only the filled data points are used in the fit. The incubation coefficient can be estimated as $S = 0.82$. **b** Ablation threshold fluence of various Au layer thicknesses in dependence of pulse number per damage site. The lines are a fit to (9.8), the fit parameters for an infinite number of shots on a layer of given thickness are presented in the graph

bigger for thicker films, especially for lower pulse shots. As the pulse number has been further increased up to 1,000, the ablation site area changes almost similar for the three film thicknesses. This effect can be reflected from the calculated damage thresholds. For a 1 pulse shot, the damage thresholds are 1.48, 0.89 and 0.35 J/cm² for 10, 50 and 205 nm films, respectively. On the contrary, for 1,000 pulse shots, the threshold values become 0.36, 0.26 and 0.25 J/cm². In this case the differences between the three thicknesses are getting much smaller in terms of ablation thresholds. This implies for a large pulse number that the damage threshold is nearly independent on the film thickness. Another result one can observe is that the thinner ITO film with lower pulse shots yields higher damage thresholds, which however, drops faster with an increasing number of laser shots. This is contrary to the metal films where the ablation thresholds decrease when the film thicknesses getting thinner if it is in the range of the thermal penetration penetration length [52].

It has to be mentioned that for multi-pulse shots, the delay between two neighboring pulses in our experiments is 5 μs, which is much longer than all relaxation times of electrons and phonons, in the range below 10 ps [53].

The threshold fluence versus the number of applied pulses is shown in Fig. 9.12a. It is obvious that the threshold fluence drops strongly when the surface is exposed more than once with otherwise identical laser parameters. Especially for thinner films, the damage threshold drops dramatically when increasing the number of pulses. For thicker films, a less pronounced drop can be observed, only apparent for a low number of laser pulses. For example, at 1 shot and 1,000 shots, for a thickness of only 10 nm the ablation threshold drops by 75 % but only by 30 % in the case of a 205 nm film thickness. These observations indicate that for very thin films and only a few pulses, a strong cumulative effect occurs. As also shown in Fig. 9.12a, the thresholds will drop with an increasing number of incident pulses.

However, beyond a specific number of pulses the fluence threshold stays constant. The threshold value turned to its final value with less pulses for thick films, e.g. for a thickness of 205 nm the characteristic value for laser pulses is about 10, whereas for 50 nm thick films the threshold falls within the first 100 pulses. Therefore, it is reasonable to exclude the points in Fig. 9.12a which have reached the final threshold value for high pulse numbers [41]. The relationship between threshold and pulse number can be quantitatively expressed by (9.7), as shown in Fig. 9.12a by the dotted lines. It is interesting to see that the linear fitting lines yield the same slope values of −0.18, although there is a high uncertainty for the 205 nm thickness due to a lack of enough experimental data. This is justified to conclude that the incubation effect is attributed to the surface stress-strain energy storage by the linear relationship in the logarithmic plot of the damage thresholds against the incident pulses number, yielded by (9.7). The incubation coefficient can be calculated to $S = 0.82$.

The dependency of the threshold fluences on the film thickness is different for metals and ITO: For metal films, the ablation threshold fluence is found to increase as the rising of film thickness in the range of thermal penetration depth, when same pulse number is applied. For ITO films, the inverted trends can be defined, where the ablation threshold is found to decrease as the rising of thickness. The incubation effect depends differently on the film thickness for the both types of materials.

Usually the storage of strain-stress energy induced by multipulse irradiation is regarding to the reason of the incubation effect when talking about the ablation of thick metals. Quite a few people come to this point of view where (9.7) is found to be fitted quit well with their experimental data [26, 27]. However, when the film becomes thinner, the dependency of the ablation threshold fluence on the pulse number changes. As shown in Fig. 9.12b obtained in [42], the multi-shot damage thresholds fit quit well with (9.8) instead of (9.7) when the film thickness is in the range of thermal penetration depth. From their experiment, the largest differences between single and multi shot of laser induced damage occur for the thickest layers, which gets higher cumulative effect. For thinner films that are significantly smaller than the thermal penetration depth, the strengthen of the surface modification is regarded to be responsible for incubation increases, rather than in multipulse ablation the storage of strain-stress energy as the bulk metal samples.

For wide bandgap semiconductors, the multi pulses irradiation on a bulk solid with fluence below single pulse ablation threshold is more likely to cause the defect of the material's band structure, for example form color center [46]. This is considered to be the reason for the incubation effect here. Hence the (9.8) is more suitable to describe the incubation effect rather than the strain-stress deformation model by (9.8), which can be found in several publications [46]. However, as we found in our experiment on ablation of ITO films, the strain-stress deformation model described by (9.7) gives a better fitting to the experimental data, shown in Fig. 9.12a. Different to metal films, the largest differences between single and multi shot of laser induced damage occur for the thinner layers. It can be regarded that the behavior of the storage strain-stress energy induced by multipulse irradiation is responsible for the incubation effect for ITO film.

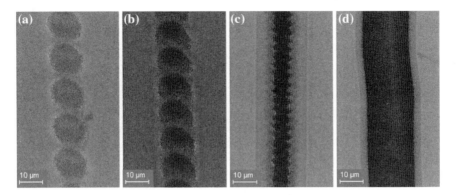

Fig. 9.13 Laser induced patterning line morphologies on the ITO films irradiated by different overlapping rates recorded by SEM, where **a** shows the results for $R_{ov} = 0$, **b** $R_{ov} = 30\%$, **c** $R_{ov} = 75\%$ and **d** $R_{ov} = 95\%$. The applied laser fluence has been 0.88 J/cm². The film thickness was 100 nm for all cases

The discussion above was completely referred to multi-pulse shots on a single site, based on the "N-on-one" experiment. In most practical applications, however, laser patterning of lines or areas is the result of a laser pulse train with a fixed overlapped area R_{ov} usually determined by the laser scanning speed v, laser repetition rate f and the spot diameter D_1. If $0 < v/f \leq D_1$, the overlapped area can be expressed as $R_{ov} = 1 - v/(fD_1)$.

Assuming there is no incubation effect, the width of the patterning line would be always equal to the diameter of single-shot site, the line width $L = D_N = D_1$, no matter how much overlapping rate has been used. Figure 9.13 illustrates the patterning lines with different overlapping rates at the laser fluence of 0.88 J/cm². As one can see the line width increases with higher overlapping rate and stays no longer equal to the diameter of a single-shot site. This indicates the significant incubation effect on the width of the patterning lines.

For further calculations we replaced the pulse number N in "N-on-one" experiment by $N = fD_1/v = (1 - R_{ov})^{-1}$, which can be considered as the effective pulse number overlapped within the single-spot site. Combining this with the (9.7), one can obtain

$$L = D_1\sqrt{1 + \frac{(S-1)}{ln[q/q_*(1)]}ln(1 - R_{ov})}, \qquad (9.9)$$

with L representing the ultimate line width at a given laser fluence q and q_* is the single-shot ablation threshold. It is obvious, that for an overlapping rate $R_{ov} = 0$, i.e. no overlapped area between two succeeding pulses, incubation effects disappear and the width of the line equals to D_1. As the overlapping rate increases, the width of the patterning line increases, as well. Assuming the other extreme case $R_{ov} = 100\%$, the width theoretically approaches infinity which is equal to the "N-on-one" case

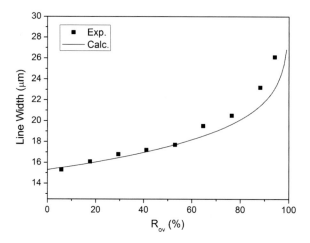

Fig. 9.14 Width of patterning lines versus different overlapping rate. The applied laser fluence has been 0.88 J/cm². The *solid line* is fitted according to (9.9)

with an infinitely number of pulses. However, in this situation (9.7) is not valid any longer.

In order to verify (9.9), we conducted a series of experiments with 100 nm ITO films applying different overlapping rates, as plotted in Fig. 9.14. The incubation coefficient was evaluated in Fig. 9.12, where it comes to $S = 0.82$. The single-shot ablation threshold is 0.59 J/cm² and the applied laser fluence $q = 0.88$ J/cm² induces a damage site diameter of $D_1 = 15.3 \, \mu\text{m}$. Substituting these values for the parameters in (9.9), the result is plotted in solid line shown in Fig. 9.14.

The plot reveals that the (9.9) agrees quite well with the experimental data. A slight deviation appears at higher overlapping rates, which is equivalent to an extremely high number of incident pulses on one site and results from the failure of (9.7) for high values of N. Taking the course of the development into account, it is reasonable to generalize (9.9) to the laser patterning of material with an incubation behavior which can be described by (9.7), e.g. most of metals. In case where (9.8) should be used to calculate $q_*(N)$ for multi-pulse ablation, the similar derivation course could be used to model the width of the patterning line, e.g. for most of dielectrics.

9.5 Conclusion

The thermal penetration depth for both metals and transparent ITO films are discussed. For metals, the thermal penetration depth can be analytically calculated by treating the laser as surface heat source. Due to the complicated nonlinear absorption of laser pulses for transparent semiconductors, a theoretical prediction of thermal penetration depth is difficult. By experimentally measuring the relation of the ablation threshold fluence dependent on the film thickness, the effective thermal penetration depth for both metal and ITO films can be obtained. When the film thickness is in the range of thermal penetration depth, the influence from the substrate can not be neglected and the ablation features are heavily dependent on the film thickness, demonstrating a different behaviors to the ablation of bulk material, such as the ablation threshold fluence for front- and rear-side ablation, for multi-pulse ablation.

References

1. Samuel S. Mao, Xianglei Mao, Ralph Greif, Richard E. Russo, Influence of preformed shock wave on the development of picosecond laser ablation plasma. J. Appl. Phys. **89**(7), 4096–4098 (2001)
2. M. Hauer, D.J. Funk, T. Lippert, A. Wokaun, Time resolved study of the laser ablation induced shockwave. Thin Solid Films **453–454**, 584–588 (2004)
3. C. Kittel, H. Kroemer, *Thermal Physics* (W.H Freeman, San Francisco, 1980)
4. D. von der Linde, K. Sokolowski-Tinten, J. Bialkowski, Laser-solid interaction in the femtosecond time regime. Appl. Surf. Sci. **109–110**, 1–10 (1997)
5. S.M. Metev, V.P. Veiko, *Laser-Assisted Microtechnology* (Springer, Berlin, 1998)
6. Friedrich Dausinger, Femtosecond technology for precision manufacturing: fundamental and technical aspects. RIKEN Rev. **50**, 1–10 (2003)
7. C. Momma, S. Nolte, B.N. Chichkov, F.V. Alvensleben, A. Tünnermann, Precise laser ablation with ultrashort pulses. Appl. Surf. Sci. **109–110**, 15–19 (1997)
8. Jianjun Yang, Youbo Zhao, Nan Zhang, Yanmei Liang, Mingwei Wang, Ablation of metallic targets by high-intensity ultrashort laser pulses. Phys. Rev. B **76**, 165430 (2007)
9. S.K. Lau, D.P. Almond, P.M. Patel, Transient thermal wave techniques for the evaluation of surface coatings. J. Phys. D **24**(3), 428 (1991)
10. P.B. Corkum, F. Brunel, N.K. Sherman, T. Srinivasan-Rao, Thermal response of metals to ultrashort-pulse laser excitation. Phys. Rev. Lett. **61**, 2886–2889 (1988)
11. B.C. Stuart, M.D. Feit, S. Herman, A.M. Rubenchik, B.W. Shore, M.D. Perry, Optical ablation by high-power short-pulse lasers. J. Opt. Soc. Am. B **13**(2), 459–468 (1996)
12. A. Rosenfeld, E.E.B. Campbell, Picosecond UV-laser ablation of Au and Ni films. Appl. Surf. Sci. **96–98**, 439–442 (1996)
13. E. Matthias, M. Reichling, J. Siegel, O.W. Käding, S. Petzoldt, H. Skurk, P. Bizenberger, E. Neskc, The influence of thermal diffusion on laser ablation of metal films. Appl. Phys. A **58**, 129–136 (1994)
14. S.-S. Wellershoff, J. Hohlfeld, J. Güdde, E. Matthias, The role of electron-phonon coupling in femtosecond laser damage of metals. Appl. Phys. A **69**, S99–S107 (1999)
15. J. Krüger, D. Dufft, R. Koter, A. Hertwig, Femtosecond laser-induced damage of gold films. Appl. Surf. Sci. **253**(19), 7815–7819 (2007)
16. V.P. Veiko, S.M. Metev, A.I. Kaidanov, M.N. Libenson, E.B. Jakovlev, Two-phase mechanism of laser-induced removal of thin absorbing films. I. Theory. J. Phys. D: Appl. Phys. **13**, 1565 (1980)
17. V.P. Veiko, S.M. Metev, K.V. Stamenov, H.A. Kalev, B.M. Jurkevitch, I.M. Karpman, Two-phase mechanism of laser-induced removal of thin absorbing films. II. Experiment. J. Phys. D: Appl. Phys. **13**, 1571 (1980)
18. S.J. Henley, J.D. Carey, S.R.P. Silva, Pulsed-laser-induced nanoscale island formation in thin metal-on-oxide films. Phys. Rev. B **72**, 195408 (2005)
19. S. Xiao, S. Fernandes, C. Esen, A. Ostendorf, Picosecond laser direct patterning of poly(3,4-ethylene dioxythiophene)-poly(styrene sulfonate) (pedot:pss) thin films. JLMN **6**(3), 249–254 (2011)

20. J. Bovatsek, A. Tamhankar, R.S. Patel, N.M. Bulgakova, J. Bonse, Thin film removal mechanisms in ns-laser processing of photovoltaic materials. Thin Solid Films **518**(10), 2897–2904 (2010)
21. H.P. Huber, M. Englmaier, Ch. Hellwig, A. Heiss, T. Kuznicki, M. Kemnitzer, H. Vogt, R. Brenning, J. Palm, High speed structuring of cis thin-film solar cells with picosecond laser ablation. Proc. SPIE **7203**(10), 72030R (2009)
22. G. Heise, M. Englmaier, Ch. Hellwig, T. Kuznicki, S. Sarrach, H.P. Huber, Laser ablation of thin molybdenum films on transparent substrates at low fluences. Appl. Phys. A **102**, 173–178 (2011)
23. K. Yung, Zh Cai, H. Choy, Selective patterning and scribing of ti thin film on glass substrate by 532 nm picosecond laser. Appl. Phys. A **107**, 351–355 (2012)
24. S. Küper, M. Stuke, Femtosecond UV excimer laser ablation. Appl. Phys. B **44**, 199–204 (1987)
25. S. Küper, M. Stuke, UV-excimer-laser ablation of polymethylmethacrylate at 248 nm: characterization of incubation sites with fourier transform IR- and UV-spectroscopy. Appl. Phys. A **49**, 211–215 (1989)
26. Y. Jee, M. Becker, R. Walser, Laser-induced damage on single-crystal metal surfaces. J. Opt. Soc. Am. B **5**, 648–659 (1988)
27. J. Güdde, J. Hohlfeld, J.G. Müller, E. Matthias, Damage threshold dependence on electron-phonon coupling in au and ni films. Appl. Surf. Sci. **127–129**, 40–45 (1998)
28. J. Bonse, J.M. Wrobel, J. Krüger, W. Kautek, Ultrashort-pulse laser ablation of indium phosphide in air. Appl. Phys. A **72**, 89–94 (2001)
29. A. Rosenfeld, M. Lorenz, R. Stoian, D. Ashkenasi, Ultrashort-laser-pulse damage threshold of transparent materials and the role of incubation. Appl. Phys. A **69**, S373–S376 (1999)
30. R. Srinivasan, V. Mayne-Banton, Self-developing photoetching of poly(ethylene terephthalate) films by far-ultraviolet excimer laser radiation. Appl. Phys. Lett. **41**, 576–578 (1982)
31. H.H.G. Jellinek, R. Srinivasan, Theory of etching of polymers by far-ultraviolet high-intensity pulsed laser- and long-term irradiation. J. Phys. Chem. **88**(14), 3048–3051 (1984)
32. R. Srinivasan, B. Braren, D.E. Seeger, R.W. Dreyfus, Photochemical cleavage of a polymeric solid: details of the ultraviolet laser ablation of poly(methyl methacrylate) at 193 and 248 nm. Macromolecules **19**(3), 916–921 (1986)
33. R. Srinivasan, B. Braren, K.G. Casey, Nature of "incubation pulses" in the ultraviolet laser ablation of polymethyl methacrylate. J. Appl. Phys. **68**(4), 1842–1847 (1990)
34. G.B. Blanchet, P. Cotts, C.R. Fincher Jr., Incubation: subthreshold ablation of poly-(methyl methacrylate) and the nature of the decomposition pathways. J. Appl. Phys. **88**(5), 2975–2978 (2000)
35. P.T. Mannion, J. Magee, E. Coyne, G.M. O'Connor, T.J. Glynn, The effect of damage accumulation behaviour on ablation thresholds and damage morphology in ultrafast laser micromachining of common metals in air. Appl. Surf. Sci. **233**, 275–287 (2004)
36. J. Krüger, S. Martin, H. Mädebach, L. Urech, T. Lippert, A. Wokaun, W. Kautek, Femto- and nanosecond laser treatment of doped polymethylmethacrylate. Appl. Surf. Sci. **247**(1–4), 406–411 (2005)
37. R.T. Williams, Optically generated lattice defects in halide crystals. Opt. Eng. **28**, 1024–1033 (1989)
38. N. Itoh, K. Tanimura, Effects of photoexcitation of self-trapped excitons in insulators. Opt. Eng. **28**, 1034–1038 (1989)
39. D. Ashkenasi, M. Lorenz, R. Stoian, A. Rosenfeld, Surface damage threshold and structuring of dielectrics using femtosecond laser pulses: the role of incubation. Appl. Surf. Sci. **150**(1–4), 101–106 (1999)
40. F. Liang, R. Vallee, D. Gingras, S. Chin, Role of ablation and incubation processes on surface nanograting formation. Opt. Mater. Express **1**, 1244–1250 (2011)
41. J.B. Nielsen, J. Savolainen, M.S. Christensen, P. Balling, Ultra-short pulse laser ablation of metals: threshold fluence, incubation coefficient and ablation rates. Appl. Phys. A **101**, 97–101 (2010)

42. C. Kern, M. Zürch, J. Petschulat, T. Pertsch, B. Kley, T. Käsebier, U. Hübner, C. Spielmann, Comparison of femtosecond laser-induced damage on unstructured versus nano-structured Au-targets. Appl. Phys. A **104**, 15–21 (2011)
43. M. Inoue, T. Matsuoka, Y. Fujita, A. Abe, Patterning characteristics of ito thin films. Jap. J. Appl. Phys. **28**, 274–278 (1989)
44. T. Szörényi, Z. Kotor, L. Laude, Atypical characteristics of KrF excimer laser ablation of indium-tin oxide films. Appl. Surf. Sci. **86**, 219–222 (1995)
45. M. Takai, D. Bollmann, K. Haberger, Maskless patterning of indium tin oxide layer for flat panel displays by diode pumped Nd-YLF laser irradiation. Appl. Phys. Lett. **64**, 2560 (1994)
46. D. Ashkenasi, G. Mueller, Fundamentals and advantages of ultrafast micro-structuring of transparent materials. Appl. Phys. A **77**, 223–228 (2003)
47. C. Molpeceres, S. Lauzurica, J.L. Oca na, J.J. Gandía, L. Urbina, J. Cárabe, Microprocessing of ito and a-Si thin films using ns laser sources. J. Micromech. Microeng **15**, 1271 (2005)
48. M. Xu, J. Li, L. Lilge, P. Herman, F2-laser patterning of indium tin oxide (ito) thin film on glass substrate. Appl. Phys. A **85**, 7–10 (2006)
49. G. Raciukaitis, M. Brikas, M. Gedvilas, T. Rakickas, Patterning of indium tin oxide on glass with picosecond lasers. Appl. Surf. Sci. **253**, 6570–6574 (2007)
50. M. Chen, Y. Ho, W. Hsiao, K. Huang, Y. Chen, Analysis of thermal effect on transparent conductive oxide thin films ablated by uv laser. Thin Solid Films **518**, 1067–1071 (2009)
51. A. Risch, R. Hellmann, Picosecond laser patterning of ito thin films. Phys. Procedia **12**(Part B), 133–140 (2011)
52. S.S. Wellershoff, J. Hohlfeld, J. Güdde, E. Matthias, The role of electron-phonon coupling in femtosecond laser damage of metals. Appl. Phys. A **69**, S99–S107 (1999)
53. S.K. Sundaram, E. Mazur, Inducing and probing non-thermal transitions in semiconductors using femtosecond laser pulses. Nat. Mater. **1**, 217–224 (2002)

Part IV
Bulk Micro Structuring
of Transparent Materials

Chapter 10
Reversible Laser-Induced Transformations in Chalcogenide- and Silicate-Based Optical Materials

Alexander V. Kolobov, Junji Tominaga and Vadim P. Veiko

Abstract When exposed to laser light, materials such as chalcogenide and oxide glasses undergo various structural transformations. Often, the effect of light is to heat the material, which may lead to processes such as crystallization. However, there are numerous examples when the electronic excitation plays a very important role as well. In this chapter, we describe various light-induced phenomena in chalcogenide and silicate-based optical materials, putting major accent—where possible—on non-thermal processes. The chapter starts with chalcogenides, where such effects are more strongly pronounced. This is followed by laser-induced processes in silicates. Current applications of photo-induced phenomena in chalcogenides and silicates conclude the chapter. Readers are also invited to see the related chapter by Kononenko and Konov in this volume.

10.1 Chalcogenide Glasses and Phase-Change Alloys

Chalcogenides are materials that contain group VI elements such as sulphur, selenium and tellurium. While oxygen is also a chalcogen, for historical reasons oxides are usually treated as an independent class of materials. Chalcogenide optical materials can be divided into two major classes: chalcogenide glasses and phase-change alloys. The former class includes sulphides and selenides, the latter one is usually

A. V. Kolobov (✉) · J. Tominaga
Nanoelectronics Research Institute, National Institute of Advanced Industrial Science and Technology, Tsukuba Central 4, 1-1-1 Higashi, Tsukuba, Ibaraki 305-8562, Japan
e-mail: a.kolobov@aist.go.jp

V. P. Veiko
Chair of Laser Technologies and Applied Ecology,
St.Petersburg National Research University of Information Technologies,
49 Kronverksky pr., St. Petersburg 197101, Russia
e-mail: veiko@lastech.ifmo.ru

V. P. Veiko and V. I. Konov (eds.), *Fundamentals of Laser-Assisted*
Micro- and Nanotechnologies, Springer Series in Materials Science 195,
DOI: 10.1007/978-3-319-05987-7_10, © Springer International Publishing Switzerland 2014

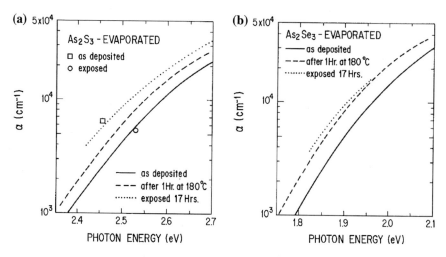

Fig. 10.1 Reversible shift of the absorption edge of typical chalcogenide glasses, As$_2$S$_3$ and As$_2$Se$_3$, under photoillumination and thermal annealing. Reprinted from [5] with permission from Elsevier

limited to tellurides. Intrinsically metastable, chalcogenide glasses can undergo various structural transformations when exposed to external stimuli such as light. Some of these processes leave the material in the amorphous (glassy) phase while come others can lead to crystallisation, which in often reversible. Phase-change alloys, on the other hand, switch between the crystalline and amorphous states. This section sets in with a description of reversible light-induced phenomena that take place within the glassy phase. We subsequently proceed to crystallisation-amorphisation phenomena in chalcogenide glasses. Finally, reversible phase-change processes in tellurides are discussed.

10.1.1 Reversible Photostructural Changes and Photo-Induced Anisotropy

The most known reversible photo-induced process in chalcogenide glasses is photodarkening. The first reports of photodarkening date back to the late 1960s-early 1970s [1–3]. The effect was first observed in thin films but later was reproduced for melt-quenched bulk glasses [4]. The effect manifests itself as a reversible change of optical absorption induced by the band-gap light and is represented in Fig. 10.1 for the typical chalcogenide glasses As$_2$S$_3$ and As$_2$Se$_3$ [5]. Upon exposure to the band gap light, the absorption edge of an as-deposited film (solid curve) shifts to lower energies, or longer wave lengths (dotted curve). Since the absorption at a fixed photon energy is increased, this effect is called photodarkening. Subsequent annealing near the glass-transition temperature results in a recovery of the initial parameters,

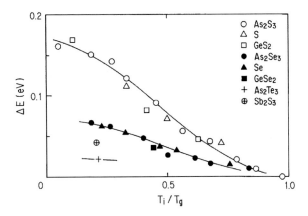

Fig. 10.2 Photodarkening value as a function of temperature, Reprinted from [6] with permission from Elsevier

which, however, is incomplete: the curve corresponding to the annealed state occupies an intermediate position between those of the as-deposited and irradiated films. Should the annealed film be irradiated by the same light, one observes *a completely reversible behaviour during the irradiation-annealing cycle*. The annealing temperature strongly correlates with the glass transition temperature (Fig. 10.2) suggesting that bond-rearrangement is a crucial requirement for the observed change.

In As-rich selenides and in some sulfides, photoirradiation of previously darkened films results in partial bleaching [7, 8], provided that the second illumination is performed at a higher temperature than the initial one. This is illustrated in Fig. 10.3. Should the film photodarkened to saturation at room temperature (A → B) be heated in the dark to a higher temperature (G) and exposed to the same light as that which caused the photodarkening, its transmission increases (G → F). The steady state value of the absorption established on illumination of a previously photodarkened film (F) is the same as that achieved by illumination of the annealed sample by the light with the same intensity and wavelength and at the same temperature (E → F). This result demonstrates that the final state of the illuminated film is fully determined by the parameters (in this case, temperature) of the latest treatment and does not depend on the film's earlier history.

Reversible changes in the optical absorption are accompanied by (reversible) changes in the refractive index [9], electrical and photoelectric properties [10–13], volume [4], microhardness [14], glass-transition temperature [15, 16], dissolution rate in various solvents [17], etc. The totality of these changes has led investigators to the conclusion that the photoinduced changes in optical absorption are caused by changes in the structure. The first direct proof of the reversible structural changes was given by Tanaka [18], who demonstrated that the first peak (the so-called first sharp diffraction peak) in the X-ray diffraction pattern underwent reversible changes on illumination and subsequent annealing.

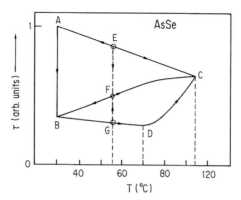

Fig. 10.3 Exposure of certain chalcogenides (As$_{50}$Se$_{50}$ case is shown) can lead to both photodarkening and photobleaching depending on the state of the film prior to the latest illumination. The final steady-state value is fully determined by temperature and the parameters of light during the latest illumination. Reprinted from [7] with permission from Wiley-VCH Verlag GmBH

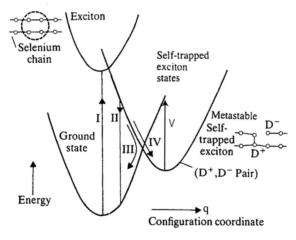

Fig. 10.4 Schematic illustration of the model for the self-trapping of excitons in chalcogenide glasses. Optical excitation (*I*) to an exciton state can be followed by two non-radiative decay channels [19], either directly back to the ground state (*III*) or to the metastable self-trapped exciton state (*IV*). Modified after [19]

Chalcogenide glasses are characterised by a strong electron-phonon coupling. As a result, the system in the excited state undergoes a significant lattice relaxation. The newly stabilised structure may be metastable and persist after cessation of the electronic excitation. This metastable state has been described as a self-trapped exciton [19]. A phenomenological description of the photostructural changes has been given using a configuration-coordinate diagram approach (Fig. 10.4), originally developed for description of defects in solids.

Any of a large number of atomic displacements may influence an atom in a solid. Often, a very specific combination of displacements is especially important and usually a combination of different normal modes are needed to make up the configurational coordinate Q. For a detailed description the interested reader is referred to a monograph by Itoh and Stoneham [20].

Optical reversibility of the process can be easily understood if in addition to the transitions initiating from the ground state shown as 'I', one also considers a similar transition starting from the metastable state corresponding to the self-trapped exciton (shown as 'V' in the Figure). The equilibrium between the two processes determines the resulting structural state of the material. In the general case, when the photon energy is smaller than the energy separation between the adiabatic potential curves corresponding to the two states, thermal excitation of the system is required, the latter being different for the transitions from the ground state to the self-trapped exciton state and vice versa, which determines different equilibrium conditions for different temperatures.

The microscopic origin of the observed structural change has long remained a subject of debate due an intrinsic difficulty of structural studies of amorphous materials. Extended x-ray absorption fine structure (EXAFS) is one of the very few available methods. Using this technique, one can obtain structural information on the short-range order such as the number of the nearest neighbours, the interatomic distances and the chemical nature of the chemical species around the absorbing atom. EXAFS studies were performed for both elemental chalcogens (selenium) [21] and binary chalcogenides [22]. The results for an elemental material are easier to interpret and we discuss them below.

In films of amorphous selenium under photoexcitation both an increase in the coordination number and the mean-square relative displacement (MSRD) were observed. After cessation of the exposure, the coordination number reverted back to the original value, while the increased disorder remained (Fig. 10.5, left).

The observed light-induced increase in the coordination number indicates the formation of higher-than-twofold coordinated sites. This may look very unusual because normally one would expect the occurrence of photoinduced bond breaking, and hence a decrease in the average coordination. This unusual behavior observed in a-Se has been interpreted as follows.

Under photoirradiation, electrons from the top of the valence band, formed by lone-pair electrons, are excited into the conduction band, leaving behind unpaired electrons in the former lone-pair orbitals. If the excited atom finds itself close to the neighbouring chain, its unpaired electron in the formerly lone-pair orbital interacts with lone-pair electrons of the neighbouring chain, creating an interchain bond (Fig. 10.5, right). Comparison of total energies for the photoexcited atom close to a neighbouring chain and for two cross-linked chains, shown at the bottom of the same figure, clearly demonstrates that such a process is energetically favourable [23]. A pair of three-fold neutral defect sites, $2C_3^0$, is therefore formed, introducing local distortion of atoms observed as an increase in MSRD.

In the processes described above, polarization of the light has not been considered. It was subsequently found that in addition to a change in absorption (scalar effect),

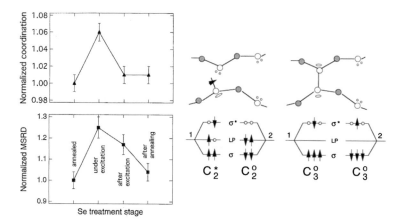

Fig. 10.5 *Left* Changes in the coordination and mean-square relative displacement of Se under different condition (reprinted with permission from A.V. Kolobov et al. [21]. copyright 1997 by the American Physical Society). *Right* Schematics of the formation of an interchain Se-Se bond following photoexcitation (from [23] with permission from Elsevier)

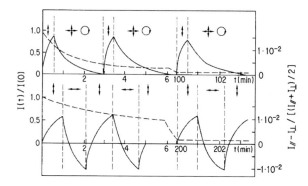

Fig. 10.6 Kinetics of photodarkening (*dashed line*) and photo-induced dichroism (*solid line*) in an $As_{50}Se_{50}$ film. *Arrows* represent linear polarisation and circles refer to circular polarisation. Reprinted from [25] with permission from Elsevier

dichroism and birefringence (vectoral effects) can be additionally induced in chalcogenide glasses, when exposed to linearly polarised light. An important finding was that photo-induced dichroism can be reoriented by changing the polarisation of the incident light. An identical dichroism is created afterwards in an orthogonal direction, i.e. reversal of photodichroism is possible. Should circularly polarised or unpolarised light be used in the second irradiation step, the induced dichroism is destroyed [24]. This behaviour is demonstrated in Fig. 10.6. Such reorientation can be performed repeatedly without any sign of a decrease in the effect. Interestingly, this light-induced reorientation or erasure of photo-induced dichroism is not associated with any change in the scalar photodarkening that continues monotonically with exposure time (Fig. 10.6).

10.1.2 Photocrystallisation of Selenium

Amorphous selenium is known to crystallise readily. Several works have been reported on the photocrystallisation of a-Se. Dressner and Stringfellow [26] reported that the radial growth rates of spherulitic crystals in a-Se films are sharply enhanced, by an order of magnitude, upon irradiation with light that has a wavelength shorter than a critical value. Based on the observation that the crystal growth rate was controlled by the production of electron-hole pairs and not by the density of the absorbed flux, they have concluded that this behavior reflects primarily the photo-, rather than the thermal, effect of the irradiation. Anisotropic crystallisation of a-Se under the action of linearly polarised light observed independently by several groups [27–31] is another strong argument in favour of the electronic nature of this process.

10.1.3 Photo-Induced Loss of Long-Range Order

10.1.3.1 Athermal Photoamorphisation

Light can also induced an opposite process, i.e. transform a crystallized chalcogenide glass into an amorphous phase. The first such observation was reported for $As_{50}Se_{50}$. An $As_{50}Se_{50}$ film crystallised on a silica glass substrate becomes amorphous again if irradiated by continuous low-intensity light [32–35]. XRD patterns of thermally crystallised and illuminated amorphous films are shown in Fig. 10.5. This change is athermal because the light-induced heating is negligible.

The positions of crystallised films correspond to the XRD powder pattern of As_4Se_4. Realgar-like As_4Se_4 is a molecular crystal [37]. The molecular structure of the crystallised film is further confirmed by Raman scattering data [34] where the spectrum of the crystal exhibits numerous narrow peaks characteristic of a molecular solid. Interestingly, after photocrystallisation no trace of molecular modes is left: the Raman spectrum of the photoamorphised material exhibits a single broad peak characteristic of a cross-linked amorphous solid.

Each As_4Se_4 molecule (Fig. 10.7, right) has two intramolecular homopolar As-As bonds; the shortest intermolecular separation (3.42 Å) is between an As atom in one molecule and a Se atom in the other. The As-As bond within the As_4Se_4 molecule (2.57 Å) is significantly longer than in, say, a-As (2.49 Å) and the As-As-Se bond angle subtended at either one of the As atoms comprising the As-As bonds (101.2°) is also larger than the average bond angle in a-As (98°). The authors conclude, that such intramolecular bonds are rather strained and, as a result, the electronic states associated with these bonds are likely to be at, or just above, the top of the valence band. Optical illumination might be expected preferentially to involve the excitation of such states, leading to bond scission; homopolar As-As bond formation could then be formed between closest atoms of neighbouring molecules, leading to a more cross-linked, and amorphous, structure. In addition, intramolecular As-Se bonds may also be broken upon illumination, giving rise to intermolecular As-Se bonds formation.

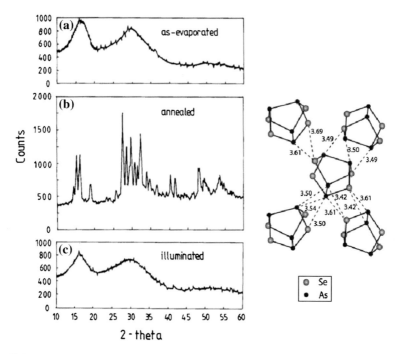

Fig. 10.7 *Left panel* X-ray diffraction patterns (Cu K radiation) of (**a**) as-evaporated, (**b**) crystallised, and (**c**) photo-vitrified $As_{50}Se_{50}$ film on a silica substrate. Reprinted from [36] with permission from Elsevier. *Right panel* crystal structure of As_4Se_4, after [37] with permission from IUCr

Photo-induced amorphisaton has also been observed in As_2S_3 [38].

The fact that photoinduced athermal amorphisation has been observed in several different chalcogenide alloys indicates that this phenomenon may be rather general. Later in this volume, photo-assisted amorphisation of GST225 is also described.

10.1.3.2 Photomelting of Selenium

It has also been found that selenium, in both amorphous and crystalline forms, can be melted by light at temperatures as low as 77 K [39]. The results of the Raman scattering study are shown in Fig. 10.8. The growing peak at $252\,cm^{-1}$ corresponds to the increased fraction of the disordered phase. Special care was taken to exclude the possibility that this effect is caused by heating due to light absorption. In particular, comparison of the Stokes and anti-Stokes peak intensities demonstrated that the temperature rise in this process did not exceed $10°$.

It was further found that the "amorphous" peak at $252\,cm^{-1}$ was strongly polarised (Fig. 10.8, bottom panel). This result indicates that quasi-free chains produced as a result of photo illumination are strongly oriented with respect to the polarisation of the excitation light. The authors concluded that photomelting of selenium is

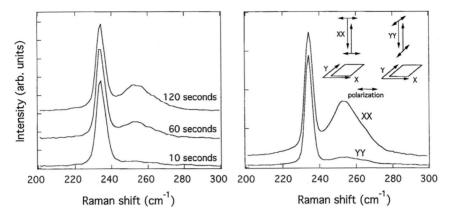

Fig. 10.8 *Left* Evolution of Raman spectra of bulk polycrystalline t-Se at 77 K during illumination; *right* XX and YY configurations at 77 K after illumination with a light polarised parallel to the X axis. Reprinted with permission from V. Poborchii et al. [39]. Copyright 1999, American Institute of Physics

associated with breaking of interchain bonds between Se chains oriented parallel to the polarisation plane of the inducing light [39].

Athermal photomelting has also been studied by computer simulations and the obtained results suggested that "the occupation of low-lying conduction states gives already weakly bonded atoms additional freedom to diffuse in the network and in extreme conditions leads to photomelting, while conventional thermal diffusion depends on the lattice dynamics of atoms" [40].

10.1.3.3 Athermal Amorphization of GST Memory Alloy

Unlike EXAFS where oscillations quickly damp with temperature, XANES is insensitive to temperature. At the same time, it is highly sensitive to the three-dimensional structure present, in particular, the white-line intensity of gst is significantly different among the crystalline, amorphous, and liquid states [42, 43], which can serve as a signature of the structure present. A three-dimensional area map of the fluorescence intensity at the x-ray excitation energy corresponding to the white line position of Ge is shown in Fig. 10.9 and demonstrates the usefulness of such an approach.

Figure 10.9 (right panel) shows the time evolution of the white-line intensity of GST as a function of delay after the excitation pulse [41]. One can clearly see from the figure that following exposure to the laser pulse the white-line intensity first monotonically decreases and reaches a minimum value in about 1 ns. This fast initial decrease in white-line intensity is followed by its partial recovery and within ~2 ns a new saturation value is reached. This final value coincides with the value of the white-line intensity in the amorphous state realised after exposure to a laser pulse.

It is of particular interest that the minimum value of the white-line intensity during the amorphisation process is *significantly higher* than that corresponding to the static

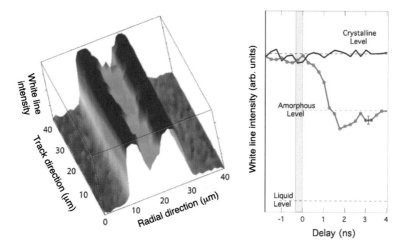

Fig. 10.9 *Left* a three-dimensional X-ray intensity map showing the crystallised band (*blue*), small amorphous marks (*green*), and as-deposited amorphous background (*red*). A ca. 1 μm diameter X-ray probe was used. *Right* Time evolution of the fluorescence intensity at the *white line*. A decrease in the white-line intensity following the excitation process is followed by a slower partial recovery. While the steady state white-line intensity reaches exactly the static amorphous value, the minimum intensity acquired during this process is *significantly higher* than the corresponding white-line intensity value for the liquid phase demonstrating that gst *does not melt* in a conventional sense on its way to the amorphous phase. The *yellow rectangle* indicates the duration and relative position of the excitation pulse and the cross bar indicates the estimated error on both axes. Reprinted with permission from Fons et al. [41]. Copyright 2010 by the American Physical Society

liquid state [43] clearly demonstrating that GST does not melt in a conventional sense upon its transformation from the crystalline to the amorphous phase [41]. This result demonstrates that the amorphisation process in the sub-nanosecond excitation regime *does not* consist of conventional melting and subsequent quenching of the melt into the amorphous phase and may be a solid-solid process. Ultrafast amorphisation of Ge-Sb-Te alloys induced by femtosecond laser pulses [44–46] provides further support for this conclusion. At the same time, for cases when the change is induced by longer laser or current pulses thermal effect may play the dominant role.

10.2 Silicate Phase-Changing Materials

10.2.1 Glasses and Glass-Ceramics: Different Sides of the Same Coin

According to the general notion, the glass is a supercooled melt of an amorphous inorganic material, obtainable by melt-quenching of a glass-forming substance. The glass-forming substances can be identified as inorganic substances which do not crystallize in melt-quenching and are solidified as an amorphous phase.

The glassy state of a substance is non-equilibrium and has larger values of entropy, enthalpy and, in general, specific volume than the crystalline state. The structure and properties of glasses depend on the chemical composition of the mixture of base substances, conditions of their synthesis, rate of supercooling within the glass-transition range, and additional thermal treatment processes.

In contrast to glasses, glass-ceramics (GCs) are vitro-crystalline (micro- or nanocrystalline) materials produced by planar (catalyzed) crystallization of special-composition glasses. Basically, GCs are polycrystalline materials prepared from glass containing crystal-forming substances (nucleating agents) by heat treatment for 6–10 h. The crystallisation process begins with the formation of billions of sub-microscopic nuclei per cubic millimeter, each serving as a growth centre. Owing to the great diversity of glass compositions and heat-treatment procedures, GCs offer a broad variety of properties [47].

As opposed to a conventional glass, whose properties are determined primarily by its chemical composition, in the case of GCs a key role is played by their microcrystalline structure and phase composition. In addition to being similar in manufacturing process, various GCs contain silica as a key component and consist of both glassy and crystalline phases. Characteristically, GCs combine high hardness and mechanical strength with excellent insulating properties, high softening temperature, thermal and chemical stability and other features, due to their fine-grained, uniform semicrystalline structure.

There is one specific group of advanced photostructurable or photosensitive glasses (PG). These materials have been attracting significant attention due to the unique opportunity to control structural phase transitions in these materials by photoactivation. Traditional technology consists from three steps including UV photoactivation (10–15 min) and two stages of heat treatment with total duration of more than 3 h with subsequent inertial cooling [48]. In this way, the writing process becomes indirect and highly time-consuming. For example, in the case of lithium aluminosilicate photostructurable glass after a first photochemical stage colloidal silver clusters with submicrometer dimensions appear. It leads to light yellow coloration of material without pronounced changes in absorbance spectra. After the second stage, there is a first heat treatment at 400–500 °C, lithium metasilicate (LMS) Li_2SiO_3 crystalline phase precipitates onto silver colloidal particles. Polycrystalline phase growth and formation of all new material properties (including drastic fall of optical transparency due to absorption and scattering of visible light) are the results of the third stage which is a second heat treatment at 550–650 °C.

As mentioned above, glass crystallisation in the GC fabrication process involves several prolonged heating/cooling steps [47]. At the same time, experiments [49] have shown that, under CO_2 laser irradiation, the structural and phase changes involved occur very rapidly, on the order of several seconds. It is also important to note that typical crystallization process of PG seems to be nonreversible because of the complicated character of photochemical and thermophysical processes. At the same time reversible crystallization-amorphization cycle of PG glasses was demonstrated under the CO_2 laser action.

The mechanisms of these rapid and reversible structural changes are not yet fully clear. A detailed understanding of these mechanisms would allow one to create advanced optical materials with tailored properties and to substantially extend the application area of GCs.

The objective of this work is to review the recent developments in experimental studies of laser amorphisation and crystallisation in GCs and to analyse and interpret the results obtained.

The key features of structural modification to GCs will be illustrated by the amorphisation and crystallization of:

(i) titanium aluminosilicate glass-ceramics (ST-50-1 type), which consists of rutile (α-TiO_2) which mainly works as the nucleation agent, cordierite ($2MgO \cdot 2Al_2O_3 \cdot 5SiO_2$) and an amorphous phase (60.5 SiO2, 13.5 Al_2O_3, 8.5 CaO, 7.5 MgO, 10 TiO2 in wt %), and

(ii) lithium aluminosilicate photostructurable glass (Foturan^{TM} type) that consists primarily of silica (75–85 SiO2), along with various stabilizing oxide admixtures, such as Li_2O (7–11 wt %), K_2O and Al_2O_3 (3–6 wt %), Na_2O (1–2 wt %), ZnO (<2 wt %), and Sb_2O_3 (0.2–0.4 wt %), [58]. Trivalent cerium (Ce_2O_3) is the photosensitizer, and silver (Ag_2O_3) serves as the nucleation agent.

10.2.2 Physical Processes of Laser-Induced Phase-Structure Modifications of Oxide Glass-Ceramics

10.2.2.1 Titanium Aluminosilicate Glass-Ceramics (TAGC) Laser Modification Under CO_2-Laser Irradiation

The experimental results have been presented in recent publications, so here we only briefly outline them. For more details, we refer the reader to previous reports [50, 51]. The phase composition and structural state of TAGCs were identified by standard X-ray diffraction (XRD) techniques. Figure 10.10 shows the XRD pattern of the unprocessed GC. The peaks at 1.688 and 2.495 Å are due to rutile (α-TiO_2), the halo around 3.505 Å is characteristic of amorphous SiO_2, and the other peaks arise from cordierite ($2MgO \cdot 2Al_2O_3 \cdot 5SiO_2$). The relative area of the grey regions represents the amorphous content of the GC. Bleaching occurs when crystals, which have strongly scattering boundaries, disappear (Fig. 10.11), and the amorphous content rises.

As a result of the amorphisation process, the transmission band of the material shifts from the mid-IR (with a maximum at 5.5 μm) to the visible/near-IR region, 0.3–2.8 μm (Fig. 10.11), extending up to 5 μm. At incident power densities and heating rates below those needed for amorphisation, TAGC crystallises, as evidenced by XRD results (Fig. 10.10c), optical spectra of recrystallised GCs (Fig. 10.11) and the visual appearance of unprocessed, amorphised and recrystallised GCs. Figure 10.12 shows typical time dependences of the surface temperature over one heating/cooling

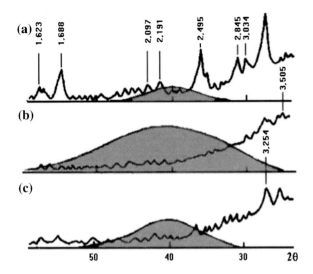

Fig. 10.10 X-ray diffractograms: angular light scattering intensity dependence of GC (ST-50-1) and corresponding amorphous phase (*blue region*): **a** initial polycrystalline structure; **b** after action of laser irradiation (amorphisated GC); **c** after the next action of laser irradiation (partial crystallized GC)

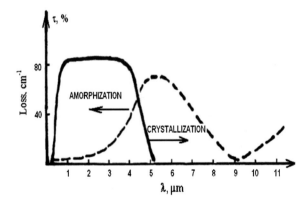

Fig. 10.11 Transmission spectrum of glass-ceramic ST501 and corresponding glass before (*1*) and after (*2*) laser amorphisation

cycle for a 0.6 mm-thick sample under CO_2 laser irradiation at a laser spot diameter of 1–2 mm and various incident power densities.

Laser amorphisation occurs at heating rates $v_{heat} \approx$ 30–80 K/s and cooling rates $v_{cool} \approx$ 50–100 K/s 1. In this process, the temperature of the zone being heated reaches the melting point T_m, and the rate of subsequent cooling is high enough for melt freezing and small enough for the formation of a stable, fracture-free amorphous material. The formation of amorphised and recrystallised zones in GCs during laser irradiation is illustrated in Fig. 10.12.

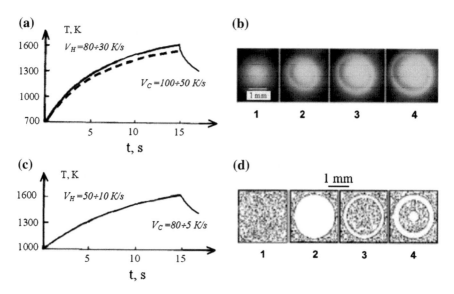

Fig. 10.12 Surface temperature of TAGC under laser irradiation and corresponding visually observed phase-structural variations. **a** amorphisation kinetic (incident power P = 1–3 W, power density $q = 3 \cdot 10^5 - 10^6$ W/m^2, initial temperature $T_0 = 700$ K), **c** reverse crystallisation kinetic ($P \leq 0.3$ W, $q \leq 10^5$ W/m^2, $T_0 = 1000$ K) (*solid curves* experimental, *dashed one* theoretical, **b** formation of amorphisation zone at laser action on the sample: video images obtained at 6 s (*1*), 6.5 s (*2*), 7 s (*3*), and 8 s (*4*) after the action begins; **d** photographs of the sample after a second exposure at different beam diameters. The unprocessed and secondary recrystallised zones appear *grey*, and the amorphised zone appears *white*. All images and pictures are obtained in transmitted light. Sample thickness $h = 0.6$ mm, laser spot diameter d = 1–2 mm. Irradiance characteristics: laser power P = 1–3 W, power density $q = 3 \cdot 10^5 - 1 \cdot 10^6$ W/m^2, initial temperature $T_0 = 700$ K; laser power P = 0.3 W, power density $q \leq 10^5$ W/m^2, initial temperature $T_0 = 1000$ K

10.2.2.2 Titanium Aluminosilicate Glass-Ceramics Laser Modification Under ND-YAG-Laser Irradiation

The situation is substantially different when TAGC is exposed to visible and near-IR radiation. In the case of cw Nd : YAG laser irradiation the appearance of oscillations of transmission in the irradiated region was observed (Fig. 10.13)—bleaching waves [52]. Such a situation were observed in 0.6 mm-thick ST-50-1 plates irradiated by the 6×10^3 W cm^{-2} Nd: YAG laser beam of diameter 300 μm. Samples were heated up to 450 °C before irradiation.

Irradiation by the cw Nd:YAG laser causes self-induced bleaching of a glass ceramic plate. First radiation is absorbed in the surface layer of the plate and produces its heating. The surface layer becomes transparent for incident radiation after melting. The region absorbing radiation moves from the surface to the polycrystalline part of glass ceramics behind the melting front. After melting of the glass ceramic plate over the entire depth, it becomes transparent for incident radiation and heating ceases. Then crystalline structures begin to form in the glass ceramic melt due to the

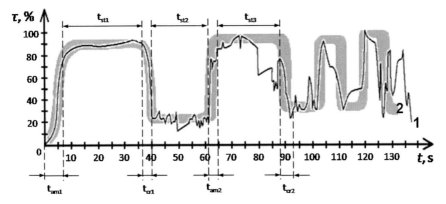

Fig. 10.13 Change in the transmission of a ST-50-1 devitrified glass plate irradiated by a CW Nd: YAG laser (*1*) and the corresponding smoother dependence (*2*)

gradual radial outflow of heat from the heated (melted) region. In this case, absorption drastically increases and transparency decreases. Glass ceramics again begins to absorb radiation and the process is repeated. Thus, the transmission of the plate oscillates and the phase-structure transitions becomes reversible.

Such a phenomenon doesn't appears under CO_2 laser action because $10.6 \mu m$ radiation is always absorbed, both in the polycrystalline and amorphous phase in a thin surface layer (factor Bouguer $\alpha = 8 \cdot 10^3 cm^{-1}$), and the melting front moves inside glass ceramics due to heating caused by the heat conduction.

The main properties of the process in this case are determined by nonlinear heating due to interrelation between the absorption ability and temperature of glass ceramics. The modeling of this phenomena have shown that in the certain range of acting parameters the abrupt change in the absorption ability of the plate leads to oscillations of transmission. The kinetics of this process, i.e. the duration of absorption-transmission cycles is determined to a great degree by the radiation power, the scattering efficiency, and the residual transparency observed after the first crystallisation. The duration of the cycles varies. For experimental parameters used in calculations, it decreases from cycle to cycle.

10.2.2.3 Lithium Aluminosilicate Photosensitive Glass (LAPG) Laser Modification Procedure

The above mentioned results on laser modification of titanium aluminosilicate glass-ceramics led us to the supposition that IR CO_2 laser heating is able in the same manner act on to other silicate glasses.

The authors idea was not only to receive a local crystallization of LAPG but also to reduce drastically a heating stage time, and more of that to make an attempt to realize a reverse amorphization of the crystalline structure what wasn't observed

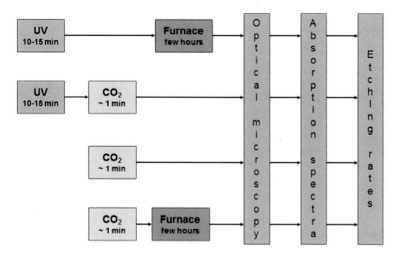

Fig. 10.14 The overall layout of the experiments for crystallization process: methods of crystallization and comparison of a nascent crystal phases. The time between consequent processing steps could vary from minutes to days. The penetration depth of radiation at $\lambda = 10.6\,\mu$m in the glass is about 1.25 μm

before. With this purpose the series of experiments on CO_2 laser action on lithium aluminosilicate photosensitive glass were carried out [3, 7]:

1. The basic experiment including traditional steps: CW UV He-Cd-laser irradiation and standard double-stage heat treatment (UV + T). It was shown for photosensitive glass that heating by means of a CO_2-laser replaces heat treatment in a furnace and results in the formation of crystal structures which, however, haven't been investigated completely;
2. The next step was the CW UV He-Cd-laser irradiation (or, for example, pulse N_2-laser) followed by CO_2-laser action instead of heat treatment in a furnace (UV + CO_2);
3. The subsequent development of the previous experiment, which was positive in sense of crystal phase formation also, became CO_2-laser irradiation only without the use of UV and heat treatment (CO_2);
4. The next experiment was the control and combined CO_2-laser action and traditional heat treatment (CO_2 + T).

The general layout of the experiments for crystallization process is presented in Fig. 10.14.

Each of the processing regimes mentioned above have led to formation of polycrystalline (glass ceramic) nontransparent structure. All experiments on crystallization were followed by successful CO_2-laser induced amorphisation. All resulting structures, both polycrystalline and amorphous, were investigated under a microscope, as well as by the X-ray diffraction (XRD), optical spectroscopy and Raman scattering spectroscopy.

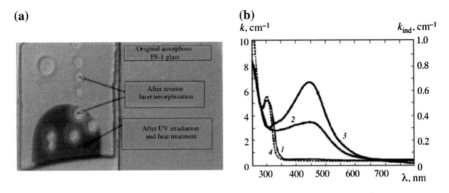

Fig. 10.15 External view of Foturan glass before and after laser-induced crystallization and secondary amorphisation (**a**), and corresponding absorption spectra (**b**): the original (*1*) and secondary amorphisated (*4*) glass; induced spectrum after UV action and the first stage of heat treatment at °C (*2*) and after the second stage of heat treatment at 600 °C (*3*), where k is the absorption coefficient, k_{ind} is the induced absorption coefficient

In all processing regimes CO_2-laser exposure parameters were the same: the power density is $q \approx 3.0 \times 10^5$ W/m^2, the preheating time is 20–40 s approximately, and the exposure duration is 40–240 s. With preheating to $T_0 = 400$ °C (it is necessary to avoid cracking of the sample and after temperature reaches this value, the general processing begins) the measured value of temperature T_{cr} at which crystallization processes begin at the surface of irradiation spot with diameter 5mm was $T_{cr} = 950$–1150 °C.

The external view of FS-1 material before and after CO_2-laser processing is shown in Fig. 10.15. After laser amorphization the yellow coloration and the strong light scattering disappear. Irradiated area becomes transparent for visible light (Fig. 10.15). In other words IR laser-induced treatment results in the transition to the initial glass. The measured value of temperature at the beginning of amorphisation process was $T_{am} = 1200$–1300 °C, while the threshold power density $q \approx 1.0 \times 10^7$ W/m^2. The preheating temperature is the same as for crystallization procedure. The heating time by laser is 70–100 s.

Radical changes in the optical transparency of the material can be observed by the naked eye in the form of the simultaneous (with a material irradiation by a laser beam) appearance of an opaque region in the transparent material. Absorbance spectra of areas crystallized under different methods are presented on Fig. 10.15a. Predictably material absorption characteristics have changed after laser action. For the other details on the crystal structures in LAPG defined by XRD and micro Raman spectroscopy let referee to the publication [53].

10.2.2.4 Laser-Induced Modification Results and Discussion

The amorphisation process we associate with the laser melting of GCs followed by vitrification as a result of rapid cooling. Note first of all that we do not assume that rutile and cordierite microcrystals, typically present in TAGC, or lithium metasilicate in LAPG undergo complete melting at the short laser action. Their presence in the amorphous matrix is evidenced by characteristic XRD peaks (Fig. 10.10b, c). It is these microcrystals which we believe to be responsible for the residual absorption in the GCs (20 %).

Bleaching kinetics are governed by the rate at which the melting front advances into the bulk of the material. Estimations of the heating and cooling rates made in [54] based on thermophysical models correlate with the experiments good enough. So there is no doubt in the thermal nature of an amorphisation process. Whether the solidified melt is amorphous or crystalline depends on the cooling rate (and his prehistory) and also on the nature of the melt. It can be shown theoretically that any melt can be amorphised by sufficiently fast cooling [54].

The crystallization process. As mentioned above, the melt obtained by laser exposure is considered to be in a nonequilibrium state, with residual traits of the parent crystalline phase. The short-range order of ST-50-1 and probably the structure of the main types of crystals in the parent material continue to manifest themselves after melting and amorphisation; in particular, the crystallite size decreases, as evidenced by small-angle X-ray scattering data (Fig. 10.10b) and the residual absorption in the amorphised material (Fig. 10.11). In this situation, density and structure fluctuations may be of importance. In the case of local laser heating, such fluctuations are inherently due to the extremely sharp temperature gradients about 100–300 K/s near the melting isotherm in our case and more. Indirect evidence for the important role of the temperature gradient is provided by Fig. 10.16, which illustrates GC crystallisation during CO_2 laser irradiation. The photograph shows clearly that crystallisation occurs along the temperature gradient.

10.2.2.5 Reversibility of Phase-Structure Transitions

In the case of photo-thermo-induced crystallization (UV + T) the given mechanism is well-known (see for example [48, 55]) and is based on the photochemical reaction of cerium ions ionization by UV radiation with the generation of free electrons, the deoxidation of silver atoms and the subsequent formation of crystallization centers, where on thermal growth of the crystalline phase occurs (clustered crystallization).

It seems the mechanism of single step IR laser-induced crystallization is the following. Nonequilibrium, non-stationary and nonhomogeneous material heating occurs because of fundamental absorption at vibrational-rotational transitions of silicon-oxygen bonds and results in lattice temperature increase. Temperature fields with strong thermal gradient ($\sim$500 K/s) appear and cause nonequilibrium crystallization (with the limited diffusion and drift of ions) at density fluctuations.

Fig. 10.16 Photo of irradiated
area of a plate at the stage of
partial crystallization (the
numbers in the picture are
in mm)

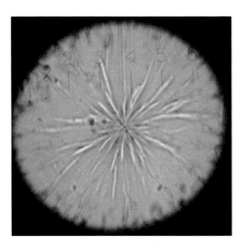

Microstructures of nascent crystals have a strong dependence on time and heating-cooling rates [54].

In turn the secondary amorphisation mechanism tentatively must be a result of the following phenomena:

1) melting of Li_2SiO_3 microcrystals with interruption and release of long-range order bonds in liquid melt:

$$Li_2O - SiO_2(crystal) \xrightarrow{melting} Li_2O - SiO_2(liquid\ phase) \qquad (10.1)$$

2) dissolution of silver clusters in the liquid melt with interruption of one weak external bond and valence electrons release

$$(Ag)_n \rightarrow nAg^+ + ne^- \qquad (10.2)$$

3) capture of released electrons by Ce^{4+} ions with their partial reduction:

$$Ce^{4+} + e^- \rightarrow Ce^{3+} \qquad (10.3)$$

The overall forms as well as identical absorption spectra are the experimental confirmation of the mechanism described above: the spectrum of glass after secondary amorphisation [53] provides an evidence of FS-1 original structure recovery with UV photosensitivity restoration having an absorption peak at $\lambda = 313$ nm as a consequence of regeneration of Ce^{3+} ions and it can be seen that the absorption band of colloidal silver ($\lambda = 420$ nm) completely disappears. It is possible to name such transition as the reverse structural modification, since the initial glass becomes a polycrystalline material and then it becomes a secondary glass with spectral characteristics closed to initial, and such transitions could be repetitive.

10.2.3 *Conclusions*

Research [47–51], made on titanium aluminosilicate glass-ceramics and lithium aluminosilicate photostructurable glass are shown, that local (focused in the spot about
100 μm) action of CO_2-laser radiation lead to its local amorphisation (or crystallization, depends on initial state of GC). Following irradiation with the different
intensities (and consequently with the others heating and cooling rates) allow to
realize the reverse crystallization (or amorphisation) of GC.

1. The capability of the IR CO_2-laser radiation with wavelength 10.6 μm to
   induce the local crystallization (C) of titanium aluminosilicate glass as well as
   photosensitive lithium-aluminosilicate glass has been proved and experimentally
   confirmed. The IR action leads to nucleation centers formation based on density
   fluctuations with subsequent growth of crystals with limited diffusion and drift
   of elements causing "freezing" of any given metastable state of liquid melt of a
   basic material due to high rates of heating-cooling.

   Rapid laser crystallisation of amorphous (glassy) materials can only be
   explained by taking into account the key features of their behaviour in a nonequilibrium state. To adequately account for this process, it is necessary to assume
   that the melt resulting from rapid laser heating contains residual quasi-ordered
   polycrystalline structures, which play a key role in the crystallisation process.
   Above given data provide evidence of such structures.
2. It is possible to have secondary amorphisation (and crystallization) of the structures, obtained by the local laser crystallization (amorphisation). Laser amorphisation the titanium aluminosilicate glass-ceramics and the kinetics of this process
   can be fully understood in terms of the proposed thermophysical model (heating,
   melting, cooling) for the interaction of laser radiation with materials in an quasi-
   equilibrium state.

   In the case of photosensitive lithium-aluminosilicate glass the mechanism of
   laser amorphisation is more complicated and looks like disordering and melting
   of polycrystals following by the dissociation of silver molecules with separation
   of valence electrons and their subsequent trapping by Ce^{4+} ions and formation
   of the amorphous phase of the glass.
3. Repeated reversible phase-structure transitions in the silicate glass-ceramics of
   the type crystallization—amorphization—crystallization (C-A-C) and amorphization—crystallization—amorphization (A-C-A) are implemented

   In addition to that:

   – the CO_2-laser radiation with wavelength 10.6 μm can be used as a thermal probe
     discovering metastable phases fully in accordance with the Ostwald principle
     because of laser processing conditions, it is short-term and non-stationary heat-
     ing, and
   – a laser with a different wavelength with sufficient degree of absorption is also
     suitable for material structure modification, which allows a volume modification
     of the structure for transparent materials.

Fig. 10.17 A snapshot of a HDTV video recorded using a super-resolution disc. Reprinted with permission from Japan Society of Applied Physics from Nakai et al. [56]

10.3 Applications of Laser-Induced Transformations

Light-induced processes described in this chapter form the basis of several technological processes. Some of the best known examples are briefly described below.

10.3.1 Chalcogenides

Reversible photostructural changes and the accompanying volume expansion can be used to fabricate gratings and microlenses; the process has a distinct advantage since it does not include etching as a technological step.

The best known example of practical use of photo-induced phenomena in phase-change alloys is their use in optical memory devices such as re-writable optical discs (DVD) and recently they have also been used for Blu-Ray discs. Use of non-linear properties of chalcogenide phase-change alloys have enabled the development of super-resolution technology, when marks with sizes significantly smaller than optical diffraction limit have been successfully recorded and read out using super-resolution optical discs with a storage capacity of 50 GB (Blu-ray $\times$ 2).

10.3.2 Silicates

Local modification of GCs, resulting in optical transparency switching and influencing their surface morphology by virtue of the modification-induced variation of specific volume, makes them candidate materials for micro- and nanophotonic applications [52]. Owing to the large difference in solubility and ion-exchange properties

Fig. 10.18 ST-50-1 glassce-
ramics plate with microlens
array produced by laser amorphization (lens size of 0.9 mm,
high optical quality, full
processing time ~1 min)
(in *reflected light*)

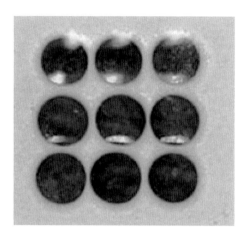

between their amorphous and crystalline components, GCs can be used equally successfully to produce both surface and bulk structures for lab-on-a-chip devices [57],
so-called nanosatellites [55], diffraction gratings [58] and other applications (see,
e.g. Fig. 10.18) [51].

This feature enables a 2D or 3D microstructuring of photosensitive glasses by photolithography and chemical etching. In photosensitive glass-ceramics fabricated from
photosensitive glasses, nontransparent white or colored 3D images can be formed.
The different solubilities of the crystalline and transparent glassy phases propose
an opportunity to form protruding images and manufacture technical products from
photosensitive glass-ceramics with a matrix of precisely produced openings with any
profiles. The physical and chemical properties that can be controlled by photoexcitation include the optical transmission, microhardness of a material, and its resistance
to chemical etching. Some more details on physics and technology of laser action on
to GS are considered in the recent review [58].

Acknowledgments The authors wish to thank all colleagues and collaborators who have contributed to our research efforts in this field and, in particular, Dr. Eduard I. Ageev from St-Petersburg
National Research University of Information Technologies, Mechanics and Optics. A part of this
work (on silicates) was supported by the Russian Federation Presidential Grant for leading scientific
school SS-1364.2014.2 and the Russian Foundation for Basic Research Grant 13-02-00033.

References

1. R. Chang, Mater. Res. Bull. **2**, 145 (1967)
2. J.S. Berkes, S.W. Ing, W.J. Hillegas, J. Appl. Phys. **42**(12), 4908 (1971)
3. S. Keneman, Appl. Phys. Lett. **19**(6), 205 (1971)
4. H. Hamanaka, K. Tanaka, S. Iizima, Solid State Commun. **23**(1), 63 (1977)
5. J.P. De Neufville, S.C. Moss, S.R. Ovshinsky, J. Non-Cryst. Solids **13**(2), 191 (1974)
6. K. Tanaka, J. Non-Cryst. Solids **59**, 925 (1983)

7. V.L. Averyanov, A.V. Kolobov, B.T. Kolomiets, V.M. Lyubin, Phys. Stat. Sol. (a) **57**(1), 81 (1980)
8. H. Hamanaka, K. Tanaka, K. Tsui, S. Minomura, J. de Phys. Paris **42**, C4 (1981)
9. S.B. Gurevich, N.N. Ilyashenko, B.T. Kolomiets, V.M. Lyubin, V.P. Shilo, Phys. Stat. Sol. (a) **26**(2), K127 (1974)
10. V.L. Averyanov, B.T. Kolomiets, V.M. Lyubin, M.A. Taguirdzhanov, in *Proceedings of 7th International Conference on Amorphous and Liquid Semiconductors*, ed. by W. Spear (CICL, University of Edinburgh, Edinburgh, 1977), p. 802
11. K. Shimakawa, S. Inami, S.R. Elliott, Phys. Rev. B **42**(18), 11857 (1990)
12. K. Shimakawa, S.R. Elliott, Phys. Rev. B **38**(17), 12479 (1988)
13. H. Naito, T. Teramine, M. Okuda, T. Matsushita, J. Non-Cryst. Solids **97**, 1231 (1987)
14. B.T. Kolomiets, V.M. Lyubin, Mat. Res. Bull. **13**(12), 1343 (1978)
15. B.T. Kolomiets, S.S. Lantratova, V.M. Lyubin, V.P. Shilo, Sov. Phys. Solid State **21**, 594 (1979)
16. H. Koseki, A. Odajima, Jpn. J. Appl. Phys. **22**, 542 (1983)
17. B.T. Kolomiets, V.M. Lyubin, V.P. Shilo, Fiz. Khim. Stekla (in Russian) **4**, 351 (1978)
18. K. Tanaka, Appl. Phys. Lett. **26**(5), 243 (1975)
19. R.A. Street, Solid State Commun. **24**(5), 363 (1977)
20. N. Itoh, M. Stoneham, *Materials Modification by Electronic Excitation* (Cambridge University Press , UK, 2001)
21. A. Kolobov, H. Oyanagi, K. Tanaka, K. Tanaka, Phys. Rev. B **55**(2), 726 (1997)
22. G. Chen, H. Jain, S. Khalid, J. Li, D. Drabold, S. Elliott, Solid State Commun. **120**(4), 149 (2001)
23. A.V. Kolobov, H. Oyanagi, A. Roy, K. Tanaka, J. Non-Cryst. Solids **232/234**, 80 (1998)
24. T. Kosa, I. Janossy, Philos. Mag. B **64**, 355 (1991)
25. V.M. Lyubin, V.K. Tikhomirov, J. Non-Cryst. Solids **114**, 133 (1989)
26. J. Dresner, G. Stringfellow, J. Phys. Chem. Solids **29**(2), 303 (1968)
27. V. Tikhomirov, P. Hertogen, C. Glorieux, G. Adriaenssens, Phys. Stat. Sol. (a) **162**(2), R1 (1997)
28. K. Ishida, K. Tanaka, Phys. Rev. B **56**(1), 206 (1997)
29. V. Lyubin, M. Klebanov, M. Mitkova, T. Petkova, J. Non-Cryst. Solids **227**, 739 (1998)
30. V. Lyubin, M. Klebanov, M. Mitkova, T. Petkova, Appl. Phys. Lett. **71**(15), 2118 (1997)
31. V. Lyubin, M. Klebanov, M. Mitkova, Appl. Surf. Sci. **154**, 135 (2000)
32. A.V. Kolobov, S.R. Elliott, Philos. Mag. **71**, 1 (1995)
33. A.V. Kolobov, V.A. Bershtein, S.R. Elliott, J. Non-Cryst. Solids **150**(1–3), 116 (1992)
34. A.V. Kolobov, S.R. Elliott, J. Non-Cryst. Solids **189**(3), 297 (1995)
35. R. Prieto-Alcon, E. Marquez, J.M. Gonzalez-Leal, R. Jimenez-Garay, A.V. Kolobov, M. Frumar, Appl. Phys. A **68**(6), 653 (1999)
36. S.R. Elliott, A.V. Kolobov, J. Non-Cryst. Solids **128**(2), 216 (1991)
37. P. Goldstein, A. Paton, Acta Cryst. B **30**(4), 915 (1974)
38. M. Frumar, A.P. Firth, A.E. Owen, J. Non-Cryst. Solids **192**(193), 447 (1995)
39. V. Poborchii, A. Kolobov, K. Tanaka, Appl. Phys. Lett. **74**, 215 (1999)
40. J. Li, D.A. Drabold, Phys. Rev. Lett. **85**(13), 2785 (2000)
41. P. Fons, H. Osawa, A.V. Kolobov, T. Fukaya, M. Suzuki, T. Uruga, N. Kawamura, H. Tanida, J. Tominaga, Phys. Rev. B **82**(4), 041203 (2010)
42. A. Kolobov, Nat. Mater. **7**, 351 (2008)
43. A.V. Kolobov, P. Fons, M. Krbal, R.E. Simpson, S. Hosokawa, T. Uruga, H. Tanida, J. Tominaga, Appl. Phys. Lett. **95**, 241902 (2009)
44. T. Ohta, J. Optoelectron. Adv. Mater. **3**(3), 609 (2001)
45. M. Konishi, H. Santo, Y. Hongo, K. Tajima, M. Hosoi, T. Saiki, Appl. Opt. **49**(18), 3470 (2010)
46. H. Santoh, Y. Hongo, K. Tajima, M. Konishi, T. Saiki, in *Proceedings of the 2009 EPCOS Meeting*, (Aachen, Germany, 2009)
47. A.L. Berezhnoi, *Glass-ceramics and Photo-sitalls* (Plenum Press, NY, 1970)
48. S.D. Stookey, Ind. Eng. Chem. **41**(4), 856 (1949)

49. V. Veiko, G. Kostyuk, N. Nikonorov, A. Rachinskaya, E. Yakovlev, D. Orlov, in *Advanced Laser Technologies 2006* (International Society for Optics and Photonics, Bellingham, 2007), pp. 66,060Q–66,060Q
50. P.A. Skiba, V.P. Volkov, K.G. Predko, V.P. Veiko, Opt. Eng. **33**(11), 3572 (1994)
51. V. Veiko, K. Kieu, Quant. Electron. **37**(1), 92 (2007)
52. V. Veiko, B.Y. Novikov, E. Shakhno, E. Yakovlev, Quant. Electron. **39**(1), 59 (2009)
53. V.P. Veiko, E. Ageev, M.M. Sergeev, A.A. Petrov, M.A. Doubenskaia, J. Laser Micro/Nanoeng (to be published)
54. V. Veiko, E. Yakovlev, E. Shakhno, Quant. Electron. **39**(2), 185 (2009)
55. F.E. Livingston, H. Helvajian, *Photophysical Processes that Lead to Ablation-free Microfabrication in Glass-ceramic Materials* (Wiley, Weinheim, 2006)
56. K. Nakai, M. Ohmaki, N. Takeshita, M. Shinoda, I. Hwang, Y. Lee, H. Zhao, J. Kim, B. Hyot, B. Andre, L. Poupinet, N.T. Shima, T., J. Tominaga, Jpn. J. App. Phys. **49**, 08KE02 (2010)
57. K. Sugioka, Y. Cheng, K. Midorikawa, Appl. Phys. A **81**(1), 1 (2005)
58. V.P. Veiko, E. Ageev, A.V. Kolobov, J. Tominaga, J. Optoelectron. Adv. Mater. **15**(5–6), 371 (2013)

Chapter 11
Fs Laser Induced Reversible and Irreversible Processes in Transparent Bulk Material

V. V. Kononenko and V. I. Konov

Abstract Laser-induced bulk modification of initially transparent solids is currently one of the hot topics in laser–matter interaction studies. The realization of these effects requires the application of focused intense beams of ultrashort pulsed (pulse durations <1 ps) lasers and the non-linear absorbtion of light in the focal volume. This chapter will describe various types of irreversible (permanent) changes inside amorphous and crystalline materials induced by sequences of fs laser pulses. To understand the mechanisms of fast material transformations, investigations of transient processes in the laser activated zone are strongly needed. The pump-probe optical technique combined with high temporal (fs) and spatial (um) resolution allows a better understanding of the problem. We shall focus on a number of transient phenomena: non-linear material optical response (Kerr effect), beam self-focusing, interband transitions and material ionization, carriers, and heat relaxation.

11.1 Introduction

One of the most important aspects of the light–matter interaction is the possibility of realizing irreversible structural transformations in band gap materials [1–3]. This fundamental capability of modifying the solid structure is attracting tremendous scientific and practical interest. Indeed, under certain circumstances these structural changes can considerably affect the essential physical and optical properties of the laser-treated material. Nowadays, the non-linear interaction of tightly focused fs radiation with transparent medium in such a way as to induce photomodification is a rapidly developing field in science [4]. Besides other things, such an approach allows producing a local structural modification in the bulk of a solid, as the material is transparent to intensive radiation everywhere other than the beam waist.

V. V. Kononenko (✉) · V. I. Konov
A. M. Prokhorov General Physics Institute of RAS, Moscow, Russia
e-mail: vitali.kononenko@icloud.ru

V. P. Veiko and V. I. Konov (eds.), *Fundamentals of Laser-Assisted Micro- and Nanotechnologies*, Springer Series in Materials Science 195, DOI: 10.1007/978-3-319-05987-7_11, © Springer International Publishing Switzerland 2014

This chapter is devoted to the state of the art of investigations in the domain of local modifications stimulated by fs laser irradiation of solid bulk material. The ultrashort radiative influence on particular materials will be described to illustrate all the main kinds of structural modifications in crystalline and amorphous wide gap semiconductors: the formation of point defects in Sect. 11.4.1, material densification in Sect. 11.4.2, and allotropic transformation in Sect. 11.4.3, i.e. a total change of the bond coordination geometry. It should be noted that the data presented are based largely on experiments made lately at the General Physics Institute, Russian Academy of Sciences, Moscow. In particular, the kinetics of the activation of the color centers in a bismuth-doped boro-alumino-phosphate glass irradiated by a fs laser will be discussed [5]. The experimental data on laser modification in pure fused silica will be presented, as well [6]. Finally, recent achievements in the technique of local graphitization in the bulk of diamond monocrystals will be reported and discussed [7].

A very important scientific point here is that the processes involved are intrinsically non-equilibrium. This, in turn, makes their nature rather puzzling. Actually, the photoexcitation of the binding electrons causes not only an increase in their energy but also perturbs the corresponding atomic bonds. This perturbation is not small enough to be negligible, and the lattice photoexcitation and subsequent relaxation could result in the reconfiguration of the disturbed molecular clusters undergoing the action of non-compensated internal Coulomb forces. The probability of this photochemical reaction in a solid depends strongly on the irradiation conditions and there still remains a big challenge in predicting exactly the eventual result of a laser treatment. That is the reason why the experimental study of the transient processes in band gap materials is so important for the comprehension of their radiative modification mechanisms.

The way this challenge can be met is the pump-probe optical technique, allowing the measurement of the transient light absorption with a fs time resolution in the affected zone. Its interferometric variant will be described in detail. This technique was found to be quite powerful, because any change of the state of the matter, including both permanent (irreversible) and transient (reversible) changes, leads to well-defined positive or negative variations in the refractive index. The difficulty lies in extracting the information contained in the data, but if this succeeds, fundamental conclusions about the light–matter interaction can be drawn. It will be demonstrated in this chapter that many of processes in the bulk of a solid can be investigated thoroughly in such a way. We will focus on a number of reversible laser-induced phenomena:

- The laser pulse energy deposition through non-linear absorption and the subsequent electronic subsystem excitation—interband transitions of the carriers and the formation of an electron–hole (e–h) plasma (Sects. 11.3.1 and 11.3.2);
- The non-linear optical response of the media—the so-called Kerr effect (Sect. 11.3.1), which on the one hand could disturb substantially the laser pulse and lead eventually to strong beam filamentation. On the other hand, the

spatial-temporal characteristics of the propagation of the laser pulse in the bulk of the material could be controlled by watching the Kerr-enhanced permittivity;
- The both localized and non-localized channels of energy dissipation: the formation of localized electronic states (Sect. 11.3.3) and lattice heating (Sect. 11.3.4), respectively.

11.2 Experimental Techniques

11.2.1 Laser Bulk Treatment Conditions: Sample Scanning and Focusing

Laser treatment is a very flexible and scalable technique, which allows, in principle, producing very complicated pre-given reliefs on the surfaces of solids and structures in their bulk. The perfect control of the laser spot characteristics and the development of advanced scanning methods are the main challenges that have to be tackled to apply lasers to a particular domain.

The second task can be solved using either laser beam or sample scanning. The first way is a lot faster and is applied presumably in the processing industry. The second one provides high precision and regularity of structuring, being used mainly in scientific investigations. Likewise, in the majority of the experiments described in this chapter, a sample placed on a computer-driven $X - Y$ stage was translated controllably, so a selected region was irradiated by a certain number of laser shots.

As regards the control of the light intensity inside the beam waist, its difficulty strongly depends on the extent of the inherent optical non-linearity of the particular material. If it is relatively small (i.e. for surface processing), the utilization of focusing optics (lenses and mirrors) with attenuating elements provides an easy and appropriate manipulation of a Gaussian laser beam. Otherwise, in the case of material bulk treatment with a high laser intensity, there could be strong non-linear effects. These effects (self-phase modulation, non-linear dispersion, self-focusing, conical emission, etc.) result in remarkable distortions of the original Gaussian beam. The weaker is the applied focusing, the more the original light path is affected by the non-linearity of the material. Under weak focusing, for instance, plasma filaments tend to appear in the bulk of the material [8]. This behavior sometimes is and sometimes is not useful (Sect. 11.3.1) in terms of controllable bulk structuring.

The experiments described here were performed using a laser system composed of a $Ti : Al_2O_3$ oscillator (Tsunami, Spectra Physics) and regenerative amplifier (Spitfire, Spectra Physics) emitting at a wavelength of $\lambda = 800$ nm with pulse duration at half-width of $\tau \approx 100$ fs and energy up to $E = 1$ mJ. In the case of $c : Si$ treatment, a parametric amplifier (OPA-800, Spectra Physics) was employed to ensure the penetration of the radiation into the bulk of the sample.

The pulse energy was varied using attenuation filters. The beam was focused by aspherical lenses with focal lengths in the range of 4–16 mm (Fig. 11.1) inside

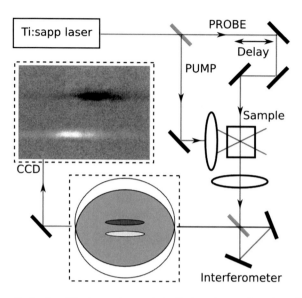

Fig. 11.1 The optical setup. The *insets* demonstrate (i) the scheme of overlap of the interfering beams (each of them contains an image of the irradiated zone) and (ii) an example image of a fs wave packet passing through a material bulk. Reprinted from [9] with permission from Astro Ltd

a sample polished on four sides. The Gaussian beam diameter in the focal plane of the lens in air was 1–4 μm at a level of 1/e. The light intensity on the beam axis was not very high ($\sim 10^{15}$ W cm^{-2}), being restricted by the threshold of bulk damage for the specific material. In order to prevent laser destruction of the sample surface, the distance from the linear laser caustic to the front face of the sample was sufficiently large, a few millimeters. Under the given focusing conditions ($N_a = 0.1 - 0.4$), both the distinct beam perturbation in highly non-linear material ($c : Si$–$n_2 \sim 10^{14}$ cm^2 W^{-1}) and the vanishing one in a material with relatively low Kerr non-linearity ($a : SiO_2$–$n_2 \sim 10^{16}$ cm^2 W^{-1}) were observed (see Sect. 11.3.1).

11.2.2 Interferometric Control of Laser-Induced Processes in by the Bulk of a Solid

As was mentioned in Sect. 11.1, the experimental study of transient processes in band gap materials is essential for the comprehension of radiative modification mechanisms. To visualise these processes, one of the best experimental options is an interferometric technique [10], whose major advantages can be listed as:

- As an eventual result of this approach, the change in the refractive index (Δn) and optical extinction coefficient (Δk) in the laser-irradiated region can be measured. Δn and Δk result, in turn, from every kind of matter excitation induced by light and subsequent energy dissipation. If the sensitivity of the measurements is high enough, free electrons promoted to conduction band (Sect. 11.3.1), laser-induced point defects (Sect. 11.3.3), the lattice heating (Sect. 11.3.4) and shock waves running through the bulk out of the laser-excited zone can be readily detected.

- Being an essentially optical phenomenon, the interference can be implemented in a pump-probe scheme with a delay line, which enables the extraction of time-domain data with a resolution approximately the same as the probe pulse duration ~100 fs. So Δn and Δk can be measured transiently during and after the exposure of the sample within a time window of from 0 to ~100 ns after a pump laser pulse. The variation of the delay over a wide range makes it possible to investigate both ultrafast processes such as matter ionization and relatively slow ones such as heat dissipation.

- In the experiments described here, the probe beam illuminated the excited region from the side. Such an arrangement matters a lot for the space-time analysis of the excited region, because it allows of easily tracing the light package propagation through the media. Indeed, a Gaussian beam and, consequently, a transient disturbance of the refractive index have rotational symmetry. That is the reason why the spatial distribution of Δn can be readily restored from the interference images, using the inverse Abel transform.

- The interferometric technique utilized is pretty flexible: a little smart tuning allows of measuring quite precisely the value of a permanent modification of the glass after a fs laser irradiation. In the experiments on the pulse-to-pulse modification of glass, an increase in the refractive index has been measured with an accuracy of about 10^{-5}.

The principal scheme of the arrangement is shown in Fig. 11.1. The irradiated region is imaged with a CCD camera. An interferometer is located between the projection objective and the CCD matrix. The mirrors of the interferometer are adjusted in a such a way that the angle between the beams is nearly zero, producing a broadband interference pattern on the screen. So, each beam contains a local phase perturbation caused by Δn in the irradiated zone. Since the width of the interference fringes significantly exceeds the size of the perturbed region, their shift with a change in the refractive index leads to a local change in the brightness of the images (Fig. 11.1, inset), thus making it possible to calculate Δn.

For the setup configuration used, the major difficulty arises from the relatively small focused beam radius, which is approximately a few microns. For this reason, the induced phase shift $\delta\phi$ in the probe beam is a lot smaller than in traditional interferometric schemes, where the pump and probe beams are nearly collinear [11]. To overcome this problem, image subtraction is applied. The signal from the CCD matrix is processed in real time and two interference patterns are obtained under the given irradiation conditions: with (informative) and without (background) the pump pulse action. Then one photo is matched against the other, so that they can be differenced to detect and measure any variation in the refractive index induced by the laser irradiation. Such an approach allows drastically increasing the precision of the measurements, which is limited in such a case presumably by the vibration of the interferometer mirrors. The use of a Sagnac interferometer and the accumulation of data under the same irradiation conditions with subsequent averaging have enhanced the sensitivity of the detection of phase shifts up to $\delta\phi \approx 10^{-2}$. Considering that the thickness of the investigated volume is $\sim 10\,\mu$m, the minimum detectable value of a measured variation in the refractive index can be estimated to be $\delta n = \delta\phi\lambda/d \sim 10^{-3}$.

11.3 Transient Changes in Matter

11.3.1 Kerr Non-linearity, Non-linear Ionization and Wave Packet Control

Such a sensitivity was quite sufficient to visualize all known processes induced in the medium by fs IR radiation. The example of fs pulse propagation through the bulk of fused silica ($a : SiO_2$) is shown in Fig. 11.2 (for convenience, only one interferometric image is shown). Here and throughout the chapter, a growth of the brightness corresponds to an decrease in the refractive index of the material, $\Delta n < 0$, while a drop in brightness (dark features) appears when $\Delta n > 0$.

The leading part of the laser pulse induces a growth in the refractive index as a result of the Kerr effect—a non-linear polarizability of the medium. Following the pictures at different time delays, one can observe the dark area to appear, starting from some distance away from the focal plane and then it moves at the speed of light through the sample. If the pulse intensity is high enough, the electromagnetic field induces an ionization of the covalent bonds in the glass, thus promoting electrons to the conduction band and producing holes in the valence band. As a result, the refractive index of the medium drops sharply. The carriers recombine after a little while and the corresponding lifetime was estimated to be ~ 100 fs, which is in good agreement with the characteristic carrier lifetime measured for the first time by Audebert with coauthors [12].

It is important to emphasize that the side view is an appropriate way to acquire data about pump pulse action (especially in the case of a strong optical non-linearity—see Sect. 11.3.2). First of all, the temporal and spatial parameters of a light wave-packet could be determined by tracing its image—the dark area in the interferometric picture (Figs. 11.2 and 11.3). For example, the pulse duration measured in $c : Si$ was $\tau = 250$ fs and the radius of the Gaussian beam in the focal plane inside the material was $r_g = 15\,\mu$m (at the $1/e$ level). The light intensity can be calculated as $I = \Delta n/n_2$. Note that in this approach to light pulse evaluation, there is no need to take into account any non-linear transformations of the beam (see Sect. 11.3.2) or any linear energy losses: such as reflections from the sample surface, aberrations, scattering or absorption when the light propagates through the crystal, and so forth.

The second advantage of a side view of an fs pulse propagation in a medium is that it includes, in principle, all the information needed to extract the actual time-domain dynamics of the e–h pair population. The problem is that the observed interference

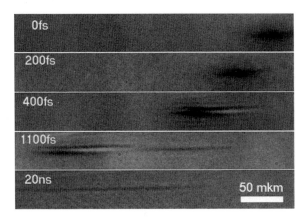

Fig. 11.2 Interference images obtained in the course of the 800 nm fs pulse propagation through a fused silica sample. The time delays are indicated in each panel. The image contrast is increased to improve visualization. Reprinted from [6] with permission from Springer

pattern is pretty puzzling. Indeed, the time of an e–h pair creation act is very small compared to the laser pulse duration. Thus, a plasma cloud starts forming in a region still occupied by the wave packet, resulting in a complex interferometric image where the sign of the refractive index change is switched over.

Moreover, the probe pulse duration equals the pump pulse duration, and consequently, is of the same order of magnitude as the plasma generation time. The experimental picture contains, therefore, a blur of plasma cloud motion along the optical axis. So, an interference pattern analysis reconstructing accurately the dynamics of Δn is still an unsolved issue.

If the carrier concentration is low and the corresponding change in n can be neglected, the Kerr effect merely causes a non-linear polarization of the medium. In this case, an interferometric picture is in fact an auto-correlation trace of the laser pulse and, as mentioned above, the actual excitation dynamics in the medium is easy to reconstruct.

If the contribution of the e–h plasma is significant, the auto-correlation integral contains a non-linear factor determined by the light absorption mechanism. To reconstruct the time-domain dynamics of Δn in such a case, one should solve a functional equation using a time dependent inverse Abel transform. Another way can be suggested, making use of some of the relevant assumptions regarding the absorption mechanism and successive simulation of the interferometric photography. It is noteworthy that the advantage of this approach is that interferometric measurements can be made with as small a time step as desired. The smallest change in the time delay is in fact only limited by the positioning accuracy of the optical system elements. The dynamics of the state of the medium in the irradiated zone can then be reconstructed with a time resolution considerably better than the pulse duration. Comparing the dynamics of the interference signal evaluated using a priori theoretical equations to

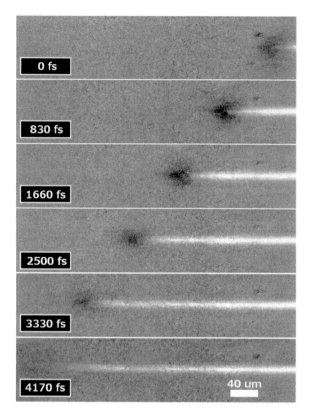

Fig. 11.3 Interference images obtained during the propagation of a 1200 nm fs pulse through a silicon crystal. The time delays are indicated in each panel. The contrast is increased to improve visualization. Reprinted from [13] with permission from Turpion Ltd

that inferred from experimental data, one can, in principle, assess the relevance of the applied models for the laser excitation of the electronic subsystem in solids.

Such an approach was realized in [13] for silicon, which has a band gap of 1.12 eV and only two photons ($\hbar\omega = 1.03$ eV) are needed to promote an electron to the conduction band. Figure 11.4 illustrates the dynamics of Δn in $c : Si$. It is on the whole consistent with the clear physical picture described above. After approximately 500 fs, when the wave packet reaches the measurement point, the refractive index and image darkness begin to increase (non-linear susceptibility); subsequently, they decrease (e–h plasma). Also presented in Fig. 11.4 are the results of numerical calculation carried out for silicon photoexcitation with two-photon absorption. The simulation is seen to agree well with the experimental data.

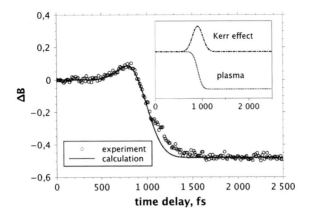

Fig. 11.4 Temporal dynamics of the relative brightness of the interference pattern in the center of the pump beam caustic (laser pulse energy 4.5 μJ). Reprinted from [13] with permission from Turpion Ltd

11.3.2 Non-linear Transformation of an Fs Beam and Radiation Losses in a Highly Non-linear Material

As pointed out previously, the subject of this chapter is a gentle modification of solids, that does not lead to cracking, void formation, etc. In these terms, the light intensity used in the experiments is not so high: $10^{13} - 10^{14}\,\mathrm{W\,cm^{-2}}$, depending on the target material. The data obtained suggest that under such conditions, the laser pulse propagation in many semiconductors (like fused silica) is quite close to that of a linear one. Despite the non-linear multiphoton absorption of radiation, the Gaussian profile of the beam is valid. Due to the relatively low magnitude of the non-linear refractive coefficient $n_2 \sim 10^{16}\,\mathrm{cm^2\,W^{-1}}$, neither the emerging plasma nor self-focusing affect substantially the laser pulse. This is quite convenient for applications: it allows controlling the material modification process and producing advanced structures in the bulk of the sample.

However, under similar conditions of irradiation, materials possessing a higher n_2 behave quite differently, which needs to be taken into consideration. Figure 11.3 demonstrates the 1200 nm fs pulse propagation in $c : Si$ that features the formation of an electron–hole (e–h) plasma filament. It starts emerging around the Gaussian beam waist position and then the plasma wave follows the self-focused light pulse until the former is completely absorbed. Due to the relatively long carrier lifetime in crystalline silicon ($\sim$10 ns), the filament resembles a long 1–2 mm and thick $\sim$10 μm plasma column.

Besides the strong filamentation, there is another factor—a non-linear absorption which must be taken into consideration because it also significantly affects the possibility of local deposition of light energy in the bulk of a material such as $c : Si$. It is usually recognized that the higher is the non-linear absorption coefficient β,

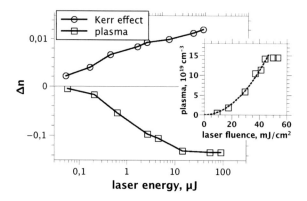

Fig. 11.5 Effect of pump pulse energy on the refractive index change in response to Kerr polarization (*up*) and the formation of an e–h plasma (*down*) in silicon

the more controllable is laser writing inside the transparent material. Indeed, in the case of small multiphoton absorption, inverse Bremsstrahlung absorption provides the plasma formation in a solid and the role of initial electrons grows with decreasing β, resulting in a quite stochastic character of breakdown process.

In the case of silicon, there is another problem. The $\beta = 0.53 * 10^9 \, \text{cm} \, \text{W}^{-1}$ for $\lambda = 1.2 \, \mu\text{m}$ [14] is so high that two-photon absorption results in a remarkable energy dissipation outside the Rayleigh range of a Gaussian beam. Recent works [13, 15] have shown that because of non-linear effects, the energy losses in silicon can reduce the laser fluence in the focal plane by a factor of more than 100 compared to linear propagation of light.

The extent of such saturation can be estimated from the data in Fig. 11.5, which show the laser-induced Δn as a function of the pulse energy for the irradiated zone. In fact, the upper curve describes the Kerr effect and the lower one corresponds to the concentration of photogenerated carriers in the lattice. The two curves correlate well with one another, are substantially non-linear, and show a well-defined logarithmic behavior, saturating with an increase in the pulse energy. Moreover, at pulse energies above 20 mJ, the e–h plasma density does not increase.

The concentration of excited carriers derived from these data is also shown in Fig. 11.5 (inset) as a function of the laser fluence in the center of the beam caustic. The plasma density N was evaluated using the Drude formula, and its highest value was $1.5 * 10^{20} \, \text{cm}^3$. The data in Fig. 11.5 suggests that the absorption probability is well approximated by the square of the light intensity up to a laser fluence of $F = 45 \, \text{mJ} \, \text{cm}^{-2}$. Such a deviation from power-law behaviour for multiphoton absorption is observed in other semiconductors (see, e.g. [16]) and, what is more, at the same plasma density, $N \sim 10^{20} \, \text{cm}^{-3}$.

11.3.3 Self-Trapped Exciton (STE) Formation

It is commonly accepted that the formation of point defects can be readily induced by a beam of high-energy particles as a result of their collisions with the lattice [1]. When the energy of the particles or photons is relatively low, any material structure modification is significantly more complicated.

The conception suggested by Frenkel of an exciton—an electron–hole pair coupled by the electrostatic Coulomb force—is one of the most important in solid state physics. Besides other things, excitons play a key role in the problem of the radiative damage to wide band-gap semiconductors, discussed here. What matters a great deal is that they are capable of converting the absorbed photon energy (1–5 eV) into energy of atomic replacement (10–20 eV) and create a lattice defect.

The phenomenon of excitons has been thoroughly investigated over the last few decades and a number of experimental and theoretical work has been performed to clarify the intrinsic mechanisms of exciton-decay assisted lattice rearrangement (see eg. [17]). At the moment, there is a general perception that e–h pairs induced in a wide band-gap material tend to create free excitons, which then decay, creating self-trapped excitons (STEs) [18]. Moreover, the emergence of localized electronic states is more probable than multiphonon recombination of free excitons. Such a scenario has been found to be preferential for many different materials: NaCl, SiO2, KBr, etc. [11].

In turn, STEs distorting a lattice locally are capable of forming point defects on lattice sites. It should be emphasized that the localization of the electronic excitation is a key factor in radiative modification. STE decay may result in a transfer of optical energy of about a few eV to a particular lattice site and induce an atom dislocation followed by the creation of an intrinsic defect. The mechanism concerned has been well studied as applied to alkali halides and fused silica [17], however there remain a lot of important questions.

It is worth applying an interferometric technique to one of them, that concerning the probability of exciton self-trapping. To estimate the free-carrier and STE concentrations, the classical Drude-Lorentz model was employed in [6]. The refractive index is given by

$$n = n_0 - \frac{N_{fr} e^2}{2m_{fr}\omega^2} + \sum_t \frac{N_{tr} e^2}{2m_{tr}(\omega_{tr}^2 - \omega^2)}, \tag{11.1}$$

where n_0 is the refractive index in the absence of the photoexcitation; N_{fr} and N_{tr} are the concentrations of free e–h pairs and the trapped carriers, respectively; t is the number of localized levels in the dielectric band gap; e is the charge of the electron; m_{fr} and m_{tr} are the free carriers mass and the effective mass of the trapped carriers; ω is the radiation frequency; and ω_{tr} is the absorption band of the defect level. This expression shows that the presence of free e–h pairs as well as the trapping of electrons at shallow defect levels (e.g., exciton levels) leads to a decrease in the

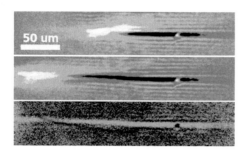

Fig. 11.6 The interferometric images of the beam waist region in diamond obtained during fs laser pulse propagation. The pulse ($E = 0.36 \mu J$) propagates from the *right side* to the *left side*. The delay (from *top* to *bottom*): 0 fs, 670 fs, 60 ps. The contrast has been increased for better visibility. Reprinted from [9] with permission from Astro Ltd

refractive index. Otherwise, the trapping of carriers at deep levels (STEs and intrinsic point defects in silica) leads to an increase in n.

The interference measurements (see Fig. 11.2) demonstrate a decrease in the refractive index related to the short-lived e–h plasma by about $\Delta n = -10^{-2}$ and a subsequent increase by $\Delta n = 10^{-3}$ due to the STEs' absorbing at 5.2 eV (240 nm). The corresponding concentrations calculated with formula (11.1) are $5 * 10^{19}$ cm^{-3} and $3 * 10^{19}$ cm^{-3}, respectively. This proves that the main part of the electronic excitation in silica indeed relaxes through localized states. The concentration of oxygen atoms whose outer electrons occupy the upper levels of the valence band in silica is about $1.7 * 10^{22}$ cm^3. Thus, the fraction of excited free e–h pairs and excitons is relatively small for the non-damaged irradiation regime. The estimated quantum efficiency of the electron–plasma excitation is about 10^{-2}.

11.3.4 Bulk Heating

Besides the fact that electronic excitation energy can be localized for the sake of STE creation, the e–h pairs are also capable of relaxing in a non-localized way. In such a case, the electronic excitation interacts with the phonon system of the solid, and this eventually results in a temperature rise. It is noteworthy that the absorbed energy could dissipate in a very puzzling way. For instance, in the case of fused silica, a certain part of the energy transfers quickly to the lattice (<1 ps) as carriers reduce their energy in the course of the cooling of free e–h pairs, the formation of excitons, and their self-trapping. But, remarkably, the rest of the energy is conserved in the STEs and is gradually released with STE decay in a few microseconds.

Such a complex process of lattice heating, as well as cooling, is the factor which significantly affects the modification of the material structure and, so, has to be comprehended in detail. However, there have been rather few publications devoted to the experimental study of fs laser induced heating of bulk material. The problem of precisely controlling the temperature in the course of the relaxation of the matter can be solved properly with the interferometric technique. Such an investigation of

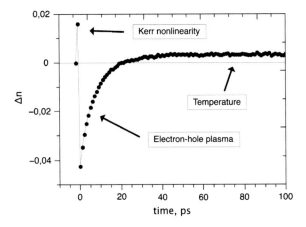

Fig. 11.7 Temporal dynamics of the refractive index of diamond on the axis of a laser beam. The laser beam intensity is $I \approx 3 \times 10^{12}\,\mathrm{W\,cm^{-2}}$. Reprinted from [9] with permission from Astro Ltd

the spreading of heat induced by a laser inside a diamond monocrystal was reported recently [9]. The diamond was chosen as a quite suitable material for this experiment because of its extremely high thermal conduction and absence of STE creation.

The sequence of images presented in Fig. 11.6 demonstrates the dynamics of Δn in diamond, which resembles that in silica and silicon (Sect. 11.3.1). The initial growth in the refractive index, induced by the Kerr effect, is followed by the appearance of a bright channel, produced by the ionization of the material. As a result, the refractive index can dramatically drop, which takes place for a high concentration of carriers. For example, as seen from Fig. 11.7, the non-linear $\Delta n \approx +0.015$ then drops to $\Delta n \approx -0.04$. After complete carrier recombination (around 20 ps), a positive change in n can be seen again. For a laser beam intensity of under $I \approx 3 \times 10^{12}\,\mathrm{W\,cm^{-2}}$, it achieves $\Delta n \approx +0.003$ (Fig. 11.7). The lower picture in Fig. 11.6 shows that the corresponding long channel with $\Delta n > 0$ exists for a long time (>40 ns) in the irradiated zone.

A slow decay of positive Δn in the channel is accompanied by the growth of its diameter over time. The exactly determined relation between them suggests that the nature of the Δn involved here is not electronic (STE) but results from the lattice heating provided by the dissipation of the carrier energy in a recombination process.

Indeed, the initial temperature distribution matches the e–h plasma distribution and can be described by a Gaussian function: $T(r, 0) = T_0 exp(-\frac{r^2}{r_{pl}^2}) + T_R$, where T_0 is the initial laser-induced temperature along the beam axis, r_{pl} is the measured radius of the plasma column, r is the distance from the beam axis, and T_R is room temperature. In the case of cylindrical symmetry, the heat equation can be written as

$$\frac{\partial T(r, t)}{\partial t} = \chi \left(\frac{\partial^2 T(r, t)}{\partial r^2} + \frac{1}{r} \frac{\partial T(r, t)}{\partial r} \right), \tag{11.2}$$

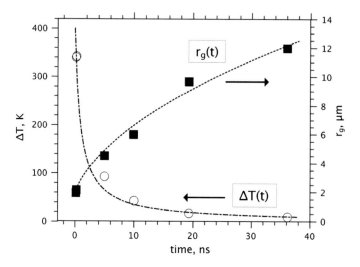

Fig. 11.8 The measured radius of the heated zone (*squares*), the maximum temperature along the pump beam axis (*circles*), calculated from the Δn data, and the corresponding results from modeling using (11.3–11.4) (*lines*). Reprinted from [9] with permission from Astro Ltd

where $\chi = 10\,\mathrm{cm^2\,s^{-1}}$ is the heat diffusivity of diamond [19]. An analytic solution shows that $T(r)$ keeps a Gaussian form over time:

$$T(r, t) = T_0 \frac{r_{pl}^2}{r_g^2(t)} exp(-\frac{r^2}{r_g^2(t)}) + T_R, \qquad (11.3)$$

where the radius r_g of the heated volume is

$$r_g(t) = \sqrt{4\chi t + r_{pl}^2} \qquad (11.4)$$

Figure 11.8 represents the experimental data for $r_g(t)$ obtained from interferometric images and its temperature rise $\Delta T(0, t)$ calculated by the formula: $\Delta T = T - T_R = \Delta n/(dn/dT)$, where for diamond, $dn/dT = 10^{-5}\,\mathrm{K^{-1}}$ [19]. The experimental points are in excellent agreement with the calculations using formulas (11.3–11.4), which proves the thermal nature of the long-lasting disturbance of the refractive index. Note that the experimental value $r_{pl} \approx 2\,\mu\mathrm{m}$ for the radius of the ionized channel is twice as small as the beam radius $r_{las} \approx 4\,\mu\mathrm{m}$ measured at the initial stages of the action of the laser pulse when Δn is determined by the Kerr effect. This difference is in accordance with the four-photon mechanism of bulk diamond ionization by intense fs laser pulses. The best agreement between the experimental points and the calculations of $\Delta T(0, t)$ was obtained for $\Delta T(0, 0) = 400\,\mathrm{K}$. This means that the non-linear absorption of fs pulses in diamond in this case resulted in lattice heating up to $T(0, 0) \approx 700\,\mathrm{K}$.

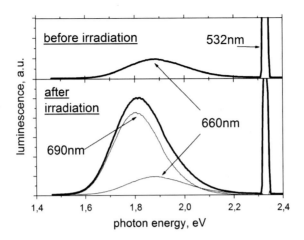

Fig. 11.9 Spectra of the optical luminescence of bismuth-doped glass (*upper part*) prior to and (*lower part*) after irradiation with femtosecond laser pulses. Thin solid lines show the spectral components for the irradiated glass. The wavelength of the excitation laser radiation is 532 nm. The scales on the two parts are identical. Reprinted from [20] with permission from Elsevier B.V.

It should be mentioned that both the ionization and the heating of the diamond at the selected intensity level had a reversible character, i.e. after the laser pulse action, no changes in the optical properties of the diamond were detected. Further increase in the beam intensity leads to bulk graphitization of diamond [7], which will be elaborated further in Sect. 11.4.3. We expect that the technique developed here will permit determining the threshold temperature for bulk diamond graphitization, T_g, which is not yet known because of the lack of experimental data.

11.4 Irreversible Modification of Material Structure

11.4.1 Formation of Radiative Defects

As mentioned above (Sect. 11.3.3), the radiative excitation triggers an STE formation mechanism whereby structural defects are produced in a perfect lattice. Such a way of lattice restructuring was well established for many wide band-gap materials with simple lattices, first of all, for fused silica and alkali halides. A deep understanding of the specific defect structure in the materials involved has allowed performing experiments which have suggested the idea of direct transformation of a STEs to Frenkel pairs. The non-bridging oxygen-hole centers (NBOHC) and oxygen vacancies (E') in the SiO_2 matrix are the kind of Frenkel defect pairs arising out of STE decay. The F-H centers in the alkali halides have the same nature.

However, in the case of a composite material, scientists have frequently been unable to explain the exact mechanism behind the formation of radiative defects. The main reason is a lack of knowledge about the actual structure of point defects. One of the many examples is the class of bismuth doped glasses, which in recent years has attracted much attention as a prospective gain material for fiber lasers [21]. The emission was found to be in the range of 1000–1700 nm [22] in alumino-phosphate, alumino-silicate, and other glasses. Since then, there have been many publications in the area of the practical implementation of lasing in the NIR range [23, 24]. Significant efforts have been made to understand the structure of the color centers and to identify the electronic transitions which are responsible for the infrared luminescence in the glass [25, 26]. Despite many investigations, the nature of the luminescent centers is still an open question and discussion is still ongoing.

Recently it was shown that femtosecond laser irradiation can be used to color bismuth-doped glass in the NIR [27] and visual [5] ranges of the spectrum. Figure 11.9 presents the spectra of optical luminescence of bismuth-doped glass before and after irradiation by fs pulses. The initial spectrum consists of a single band centered at 1.9 eV (660 nm) with a bandwidth of 0.3 eV. The intensity of the luminescence greatly increases as a result of laser-induced processes. The new band at 690 nm arises and rises rapidly as a result of the laser irradiation, while the initial band at 660 nm is stable. The observed dynamics of the luminescence coincides well with the dynamics of absorption at 460 nm (2.7 eV) and 560 nm (2.2 eV).

The model, which allows adequately describing the process of laser coloration of doped composite oxides, is a general model of photochemical reactions in solids. As stated earlier, there are a number of so-called 'precursors' in the glass matrix, i.e. metastable point defects, which have a certain probability of transforming into active centers under external perturbation. Then the photochemical reaction rate is proportional to the number of precursors which exist currently. The instantaneous concentration of defects can be found to be $N = N_0 * (1 - \exp(-\nu/\nu_0))$, where N_0 is the initial concentration of precursors and ν is the the number of laser pulses. The experimentally found value $\nu_0 \approx 2$ characterizes the rate of the photoreaction and is equal to the pulse number needed to decrease the precursor concentration to the level 1/e.

The key point clarifying the mechanisms of radiative coloration is the observed huge difference in reaction rates in the case of pure materials and those containing defects. For example, for light-induced modification of fused silica under femtosecond irradiation, the value of the constant ν_0 is more than 10^4 pulses [6, 28]. Such a big difference implies that the chemical bonds are perturbed by laser light much more strongly in the case of bismuth-doped glass.

Actually, a doping of a glass by metal ions results in the appearance of a remarkable number of electron traps, which can effectively capture free electrons that arise in the course of the ionization of the solid by laser radiation. On the one hand, such traps, being charged, create luminescent centers. On the other hand, they initiate the formation of UV defects composed of oxygen deficient and oxygen containing pairs—and very well at that. Indeed, the distance between the trapped electron and the hole is a lot more than one in the case of STEs emerging in the lattice without

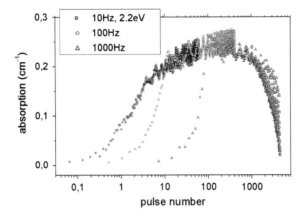

Fig. 11.10 Absorption dynamics during irradiation of bismuth-doped glass with various repetition rates of femtosecond pulses. Reprinted from [20] with permission from Elsevier B.V.

precursors. As a result, the non-compensated internal Coulomb forces are stronger, the oxygen bond is affected more deeply, and the probability of self disruption followed by the creation of a permanent defect is drastically enhanced in the molecular cluster donated electron.

It is worth noting that a similar mechanism may be very sensitive to the material heating induced by laser irradiation. Figure 11.10, illustrating this probability, presents the dependence of the absorption at 560 nm (i.e. the number of UV defects produced) on the number of laser pulses. It can be clearly seen that in the early stage of laser treatment, the rate of creation of defects diminishes with increasing pulse repetition rate. Moreover, at a pulse repetition rate of 1000 Hz, the absorption and, consequently, the number of color centers in the glass, start to decrease after reaching a maximum. The observed processes can be ascribed to the laser-induced rise of temperature, which enhances gradually the rate of decay of the color centers, which becomes at some moment higher than the rate of laser-induced coloration.

Simple estimates confirm the assumption about the significantly higher heating of the glass at a pulse repetition rate of 1000 Hz. The size of the heat affected zone during the time $t = 1/f$ (where f is the pulse repetition rate) can be estimated as $r_H \sim \sqrt{\chi t} \approx 20\,\mu$m, where χ is the temperature conductivity. The laser beam diameter $r \sim 10\,\mu$m is usually of the same order of magnitude. Then the material has not been cooling significantly in the time interval between pulses. As a result, the next pulse comes to a still heated area and the temperature gradually rises during the laser irradiation. So, at high pulse repetition rates, the laser heating can significantly influence the rates of chemical reactions and even turn them back.

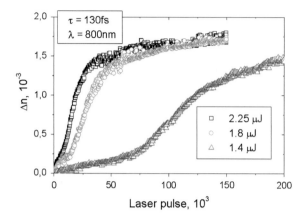

Fig. 11.11 Dynamics of the permanent modification of silica with a variation in the pulse energy at $\tau = 120\,\text{fs}$, $\lambda = 800\,\text{nm}$, and a divergence of 0.1 mrad. Reprinted from [6] with permission from Springer

11.4.2 Glass Matrix Densification

It is well established that intensive laser action on the amorphous matrix of glass can lead to irreversible variation in the refractive index n [29, 30]. The Chap. 10 of the book is devoted to variety of light-induced phenomena in chalcogenide and silicate-based glasses. We focus here on treatment with fs laser pulses witch can also result in a stable increase in n in the bulk [4, 31]. The main mechanisms which have been considered for explaining the experimental data are radiative coloration and glass matrix densification.

Indeed, point defects created by radiation cause a substantial distortion of the electron structure of the lattice cells and, hence, a permanent variation in the permittivity of the medium in the colored volume [32]. The Kramers–Kronig relation from phenomenological electrodynamics leads to a similar conclusion. Besides the formation of point defects, photo-chemical reactions can induce transformations decreasing the length of the Si–O rings and, hence, increasing the density of the material. The experiments supporting the conception of material densification were performed with Raman spectroscopy [33, 34].

It is necessary to emphasize that the principal possibility for fused silica densification is a result of the specific arrangement of its lattice. Generally speaking, silicon dioxide has a number of distinct polymorph forms in addition to an amorphous one—silica glass. The cluster structure of fused silica, as well as of silicon oxide crystalline forms, consists of a network of tetrahedral units—SiO_4. What really matters is that, in contrast to the crystals, the tetrahedra making up a glass network are able to rotate, varying their bond lengths and angles. So, laser radiation in this case is a tool for promoting the glass matrix over the continuum of its different metastable arrangements, ordered at length scales approximately the same as the Si–O bond length.

Interferometry enables us to follow immediately the change in the refractive index in the course of laser irradiation and analyze its evolution [6]. Such data is useful for a detailed understanding of the chemical reaction pathways and elementary reaction steps.

Apart from that, interferometric photography has proved the STE plasma to be localized in the vicinity of the irreversible modification region in glass (see Fig. 11.2) and enables controlling the possible development of a laser induced crack during laser irradiation.

The characteristic dependence of a permanent Δn on the pulse number contains three distinct stages (Fig. 11.11). In the initial stage, the refractive index increases linearly with the number of pulses up to a certain level ($\sim 0.5 * 10^3$). Note that this level remains unchanged for a wide range of laser fluences and apparently characterizes a transient state of the glass structure. With further irradiation, the rate of increase in the refractive index exhibits a stepwise increase by a factor of 2. Then, the rate sharply decreases when the refractive index reaches a level of $2 * 10^3$. However, the refractive index slowly increases proportionally to the number of pulses in the absence of saturation.

Thus, the instantaneous rate of modification depends on both the radiation parameters and the material characteristics at the given moment. Apparently, such a behavior implies two different processes of structural modification, with different probabilities of a single event and different rates. The most appropriate processes are the formation of radiation-induced defects, and the rotation of the Si–O tetrahedra. The latter is delayed and requires the presence of point defects (precursors) that represent weak points of the amorphous structure.

Note that the fundamental interest in laser induced processes in glass is supplemented with a technological interest. From the practical point of view, the main advantage of ultrashort pulses is the delivery of energy deep into the sample, owing to the strongly non-linear optical absorption. The application of fs pulses allows drawing waveguides inside an amorphous matrix when the sample moves through the laser focus. This feature is used in the development of laser methods for the formation of complicated 3D structures inside glasses that can operate as diffraction optical elements. The problem lies in the fact that the absolute value and sign of Δn depend on a few parameters (the radiation wavelength; the pulse energy and duration; the number of pulses; the beam divergence and polarization; and so forth). Under such conditions, on-line interferometry can be a convenient and flexible tool for monitoring permanent modifications of material in bulk during 3D laser microstructuring.

11.4.3 Allotropic Transformation in Diamond

As pointed out in Sect. 11.4.2, a glassy solid structure can be modified continuously with laser irradiation. In contrast, when exposed to intensive light, non-vitreous samples tend to get cracked inside. Only a very little portion of the matter (sapphire, silicon, etc.) concentrated on a wall of the created bulb undergoes the

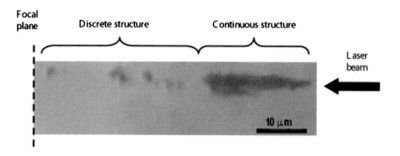

Fig. 11.12 Optical microscopy image of the laser modified region inside a diamond bulk created by multiple 120 fs pulses with energy of 320 nJ (spot size diameter 3 μm). The positions of the focal plane and the direction of laser beam propagation are shown. Reprinted from [7] with permission from Springer

crystalline-to-amorphous transition [35, 36]. Carbon, as opposed to such materials, is capable of forming many allotropes due to its varied valency. The proportion of sp_2 graphite-like to sp_3 diamond-like bonds crucially determines the physical properties of materials and their structure: crystalline (in the case of diamond and graphite) or amorphous. Since all the forms are stable or metastable under ambient conditions, allotropic transformation on external influence can be realized relatively easily.

It is well known that when exposed to intensive laser pulses, a diamond surface is graphitized. Recently, experiments have shown that fs pulses focused inside the bulk of an originally transparent diamond can also induce diamond graphitization [7]. This fact has been evidenced by the appearance of the graphite D and G peaks in Raman spectra.

When a certain threshold is exceeded, the pulse energy deposited in the bulk of the diamond target after a certain number of laser pulses initiates the emergence of a small graphitized drop. Then it grows under exposure until it reaches the diameter of the beam waist. Sometimes, several drops appear in the focal region (Fig. 11.12). The appearance of a new graphitic globule leads to the conservation of all the previously produced ones, as the highly absorbing amorphous carbon screens the other drops from the laser beam.

The graphitization process has a quite interesting feature that could be applied to the precise microstructuring of a diamond bulk. That is, the graphitized region can continuously (from pulse to pulse) spread towards the laser beam (Fig. 11.12, the right side). Putting it another way, the graphitization front moves along the laser beam—this regime has been called a laser supported 'graphitization wave' [38].

Two mechanisms for the graphitization wave in diamond have been proposed [38]. The key process is certainly the laser heating of the thin (a few tens of nanometers) layer at the diamond–graphitized phase interface. High intensity pulsed radiation heats this layer up to an extremely high temperature (more than 10^2 eV), and in such a thermal way the surrounding diamond material is graphitized. Another mechanism takes into consideration the ejection of fast electrons from the heated graphitic layer

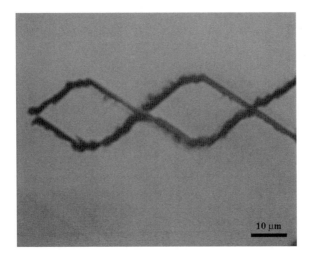

Fig. 11.13 Fragment of a laser-produced graphitic hexagonal chain. Reprinted from [37] with permission from Elsevier

into the diamond. The free electron excited by a $\hbar\omega = 1.55\,\mathrm{eV}$ photon accelerates up to $\sim 10^8\,\mathrm{cm\,s^{-1}}$. Moving in the ballistic regime, the hot electrons leave the graphite and penetrate inside the diamond to a distance of $\sim 100\,\mathrm{nm}$ during a fs pulse. Thus, the leading part of the fs pulse can assist the appearance of carriers in such a diamond layer. Afterwards, an optical breakdown can be induced there by the tail of the pulse. Note that the limited propagation distance of the external electrons in diamond restricts the thickness of the layer where an avalanche ionization is possible. The movement of the diamond sample at a constant speed along the beam outward from the laser makes possible the formation of a straight graphitized wire of practically unlimited length. Moreover, complex structures such as the chain of hexagons shown in Fig. 11.13 can be produced.

References

1. W. Primak, J. Appl. Phys. **39**(12), 5651 (1968)
2. T.A. Dellin, D.A. Tichenor, E.H. Barsis, J. Appl. Phys. **48**(3), 1131 (1977)
3. M. Rothschild, D.J. Ehrlich, D.C. Shaver, Appl. Phys. Lett. **55**(13), 1276 (1989)
4. K.M. Davis, K. Miura, N. Sugimoto, K. Hirao, Opt. Lett. **21**(21), 1729 (1996)
5. V. Kononenko, V. Pashinin, B. Galagan, S. Sverchkov, B. Denker, V. Konov, E. Dianov, Laser Phys. **21**(9), 1585 (2011)
6. V. Kononenko, V. Pashinin, M. Komlenok, V. Konov, Laser Phys. **19**(6), 1294 (2009)
7. T.V. Kononenko, M. Meier, M.S. Komlenok, S.M. Pimenov, V. Romano, V. Pashinin, V.I. Konov, Appl. Phys. A Mater. Sci. Process. **90**(4), 645 (2008)
8. S. Tzortzakis, L. Sudrie, M. Franco, B. Prade, A. Mysyrowicz, A. Couairon, L. Berge, Phys. Rev. Lett. **87**(21), 213902 (2001)

9. V.V. Kononenko, E.V. Zavedeev, M.I. Latushko, V.I. Konov, Laser Phys. Lett. **10**(3), 036003 (2013)
10. G.J. Tallents, J. Phys. D Appl. Phys. **17**(4), 721 (1984)
11. P. Martin, S. Guizard, P. Daguzan, G. Petite, P. D'Oliveira, P. Meynadier, M. Perdrix, Phys. Rev. B **55**(9), 5799 (1997)
12. P. Audebert, P. Daguzan, A. Dos Santos, J.C. Gauthier, J.P. Geindre, S. Guizard, G. Hamoniaux, K. Krastev, P. Martin, G. Petite, A. Antonetti, Phys. Rev. Lett. **73**(14), 1990 (1994)
13. V.V. Kononenko, E.V. Zavedeev, M.I. Latushko, V. Pashinin, V.I. Konov, E.M. Dianov, Quantum Electron. **42**(10), 925 (2012)
14. Q. Lin, J. Zhang, G. Piredda, R.W. Boyd, P.M. Fauchet, G.P. Agrawal, Appl. Phys. Lett. **91**(2), 021111 (2007)
15. V.V. Kononenko, V.V. Konov, E.M. Dianov, Opt. Lett. **37**(16), 3369 (2012)
16. V.V. Temnov, K. Sokolowski-Tinten, P. Zhou, A. El-Khamhawy, D. von der Linde, Phys. Rev. Lett. **97**(23), 237403 (2006)
17. K.S. Song, R.T. Williams, *Self-Trapped Excitons*, Springer Series in Solid-State Sciences (Springer, New York, 1996)
18. P.N. Saeta, B.I. Greene, Phys. Rev. Lett. **70**(23), 3588 (1993)
19. M. Prelas, G. Popovici, L. Bigelow, *Handbook of Industrial Diamonds and Diamond Films* (Marcel Dekker, New York, 1998)
20. V. Kononenko, V. Pashinin, B. Galagan, S. Sverchkov, B. Denker, V. Konov, E. Dianov, Phys. Procedia **12**(Part B), 156 (2011)
21. Y. Fujimoto, M. Nakatsuka, Appl. Phys. Lett. **82**(19), 3325 (2003)
22. M. Peng, J. Qiu, D. Chen, X. Meng, C. Zhu, Opt. Lett. **30**(18), 2433 (2005)
23. E.M. Dianov, V.V. Dvoyrin, V.M. Mashinsky, A.A. Umnikov, M.V. Yashkov, A.N. Gur'yanov, Quantum Electron. **35**(12), 1083 (2005)
24. G. Della Valle, R. Osellame, N. Chiodo, S. Taccheo, G. Cerullo, P. Laporta, A. Killi, U. Morgner, M. Lederer, D. Kopf, Opt. Express **13**(16), 5976 (2005)
25. V.G. Truong, L. Bigot, A. Lerouge, M. Douay, I. Razdobreev, Appl. Phys. Lett. **92**(4), 041908 (2008)
26. B.I. Denker, B.I. Galagan, V.V. Osiko, I.L. Shulman, S.E. Sverchkov, E.M. Dianov, Appl. Phys. B **98**(2–3), 455 (2010)
27. M. Peng, Q. Zhao, J. Qiu, L. Wondraczek, J. Am. Ceram. Soc. **92**(2), 542 (2009)
28. N. Kuzuu, Y. Komatsu, M. Murahara, Phys. Rev. B **44**(17), 9265 (1991)
29. T.P. Seward III, C. Smith, N.F. Borrelli, D. Allan, J. Non-Cryst. Solids **222**, 407 (1997)
30. R.E. Schenker, W.G. Oldham, J. Appl. Phys. **82**(3), 1065 (1997)
31. E.N. Glezer, M. Milosavljevic, L. Huang, R.J. Finlay, T.H. Her, J.P. Callan, E. Mazur, Opt. Lett. **21**(24), 2023 (1996)
32. A.M. Streltsov, N.F. Borrelli, J. Opt. Soc. Am. B **19**(10), 2496 (2002)
33. C.B. Schaffer, J.F. Garcia, E. Mazur, Appl. Phys. A Mater. Sci. Process. **76**(3), 351 (2003)
34. C.W. Ponader, J.F. Schroeder, A.M. Streltsov, J. Appl. Phys. **103**(6), 063516 (2008)
35. S. Juodkazis, K. Nishimura, H. Misawa, T. Ebisui, R. Waki, S. Matsuo, T. Okada, Adv. Mater. **18**(11), 1361 (2006)
36. A.H. Nejadmalayeri, P.R. Herman, J. Burghoff, M. Will, S. Nolte, A. Tünnermann, Opt. Lett. **30**(9), 964 (2005)
37. T.V. Kononenko, V.I. Konov, S.M. Pimenov, N.M. Rossukanyi, A.I. Rukovishnikov, V. Romano, Diam. Relat. Mater. **20**(2), 264 (2011)
38. T.V. Kononenko, M.S. Komlenok, V.P. Pashinin, S.M. Pimenov, V.I. Konov, M. Neff, V. Romano, W. Lüthy, Diam. Relat. Mater. **18**(2–3), 196 (2009)

Chapter 12
A Decade of Advances in Femtosecond Laser Fabrication of Polymers: Mechanisms and Applications

Mangirdas Malinauskas and Saulius Juodkazis

Abstract We overview principles and developments of three-dimensional (3D) direct laser writing in polymers. Challenges to reach efficient structuring with sub-100 nm spatial resolution are presented. Research into the structuring by ultrashort laser pulses has seen an immense growth over the last decade due to its flexibility, easy handling and variety of applications. Here, a discussion regarding the mechanisms of the linear and nonlinear light absorption at tight focusing conditions and typical writing parameters are provided. The traditional and novel polymers together with their photosensitization and sample developing strategies are reviewed. Sub-1 ps pulses are capable to create cross-linkable species by direct absorption and bond breaking at $\sim$TW/cm^2 irradiance. Confined thermal and linear absorption via avalanche ionization is an efficient use of light energy for localized polymerization. This is a unique feature of ultrashort laser. Applications in microoptics, photonics, microfluidics and cell scaffolds are presented. Directions of up-scaling the fabrication throughput for industrial demands are introduced. 3D laser writing is becoming a part of wider field of additive manufacturing techniques which is innovating for creation of microdevices.

M. Malinauskas (✉)
Physics Faculty, Department of Quantum Electronics, Vilnius University,
Saulėtekio Avenue 10, LT-10223 Vilnius, Lithuania
e-mail: mangirdas.malinauskas@ff.vu.lt

S. Juodkazis
Centre for Micro-Photonics, Faculty of Engineering and Industrial Sciences,
Swinburne University of Technology, Hawthorn, VIC 3122, Australia

Department of Semiconductor Physics, Vilnius University, Saulėtekio Avenue 10,
LT-10223 Vilnius, Lithuania
e-mail: sjuodkazis@swin.edu.au

V. P. Veiko and V. I. Konov (eds.), *Fundamentals of Laser-Assisted Micro- and Nanotechnologies*, Springer Series in Materials Science 195, DOI: 10.1007/978-3-319-05987-7_12, © Springer International Publishing Switzerland 2014

272 M. Malinauskas and S. Juodkazis

12.1 3D Direct Laser Writing in Polymers at Nanoscale

Direct femtosecond Laser Writing (DLW) by photopolymerization is a technology enabling the construction of mm-size structures with sub-100 nm spatial resolution. When the beam of an ultra-fast laser is tightly focused into the volume of a transparent, photosensitive material, the cross-linking process can be initiated by non-linear absorption within tightly confined focal volume (or smaller) [1–4]. By moving the beam focus three-dimensionally inside the material, 3D structures can be fabricated in a pinpoint writing. The technique has been implemented with a variety of acrylate, epoxy, organic-inorganic hybrid as well as hydrogel materials. Up to date several components and microdevices have been fabricated such as photonic crystal templates [5–7], micromechanical and microfluidic devices [8–10] as well as microoptical devices [11–13].

The DLW technology is unique in the sense that it allows the fabrication of computer-aided-design (CAD), fully 3D structures with resolution beyond the diffraction limit [14, 15]. No other competing technology currently can achieve this. Classic 3D prototyping techniques such as UV laser microstereolithography [16, 17], 3D ink jet printing [18, 19] and laser sintering [20, 21] can also produce fully 3D structures, however, they cannot provide resolution better than a few microns. On the other hand, lithographic techniques with superior resolution, such as electron beam [22, 23] or atomic force lithography microscopy [24, 25], cannot produce anything more complicated than high-aspect ratio 2D structures and the fabrication throughput is relatively very low, equipment requires clean room facilities. The progress of DLW technology is driven by a demand to produce increasingly smaller features and more complex functional structures. Current trends in micro/nano-electronics require 3D integration due to an increasing complexity and restrictions of size/mass for more efficient operation. New materials and procedures have been introduced to improve spatial resolution and feature precision, reduce costs, and add new functionalities. Employing DLW complex 3D structures of meso-scale dimensions (0.1–100 μm) are fabricated exploiting the nonlinear character (in terms of light intensity, I) of light-matter interaction, i.e., the resulting photo-modification of material scaling as $\propto I^N$ with $N > 1$ being the order of nonlinearity. The nonlinear material response is required to realize volumetric sectioning and modification inside the bulk of the polymer. The very same principle is employed in two-photon (or multi-photon) microscopy, where layer by layer raster scanning is used to obtain 3D images [26]. It should be noted that the threshold-like response of the material to the light intensity, e.g., cross-linking or polymer chain scission, is by definition a nonlinear photochemical reaction by itself.

Here, we summarize the principles of 3D structuring by DLW based polymerization and outline current and potential future applications. A stress for this fast growing field of science and technology is given for its simplicity and distinct character to produce 3D structures over a large nano-/micro-/macro scale of dimensions via a variety of intra-coupled mechanisms specific to the choice of wavelength, pulse duration, and material.[1]

12.1.1 Tight Focusing of Ultrashort Pulses: Multi-Photon and Avalanche Ionization Polymerization

In direct laser writing a focused laser beam is used. The spatial bell-shaped Gaussian profile fits well the electrical field strength, E [V/m] , energy En [J], fluence F [J/cm^2], irradiance I [J/(s cm^2)] (intensity $I = \frac{1}{2}\varepsilon_0 c n_0 |\mathbf{E}|^2$),[2] and dose $D = I \times t_{ex}$ [J] (t_{ex} is the exposure time) distributions. The generic Gaussian distribution at focus of all those parameters is described by A:

$$A(r) = A_0 \exp(-2r^2/w_0^2), \tag{12.1}$$

where w_0 is the waist of the beam, A_0 is the amplitude of either E (En, I, F, D), r is the radius (Fig. 12.1a). For the Gaussian beam the fluence is given by $F_0 = 2E_0/(\pi w_0^2)$. Even at very tight focusing with an oil-immersion objective lenses (usually numerical aperture $NA > 1.25$) when vectorial Debye focusing should be considered, the focal intensity is always close to a Gaussian envelope. When a laser beam of Gaussian-like intensity cross section is clipped by the aperture of the objective lens, there is a departure form the conditions of the Gaussian beam/pulse focusing [27]. However, the resultant focal intensity distribution still can be considered close-to-Gaussian.

The fluence and dose are usually relevant parameters for consideration of modifications induced by the absorbed laser light energy at the focal region of a sample. When the material modification is induced by nonlinear optical absorption, the intensity becomes an important parameter since the modification by an N-photon process scales as $\propto I^N$. In order to reveal scaling mechanism and order of the process, it is customary to plot the lateral cross section of modification (e.g., ablation or polymerized line width) versus the intensity, I, or En, F, D. Very often, a linear representation is chosen similar to the one shown in Fig. 12.1b, where the radius of modification. e.g., polymerization, is presented for the different energy E_n (can be I, F, D). The plots are obtained from (12.1) by setting different threshold energy, $E_{th} = E_0 \exp(-2r^2/w_0^2)$ and solving it for radius r:

$$r = w_0\sqrt{(\ln(E_0) - \ln(E_{th}))/2}. \tag{12.2}$$

---

[1] Other frequently encountered terms which describe the mechanism and become generic are: TPP/2PP two-photon polymerization, MPP—multi-photon polymerization.

[2] Here ε_0 is the dielectric permittivity of vacuum, c is the speed of light, n_0 is the refractive index of the ambient.

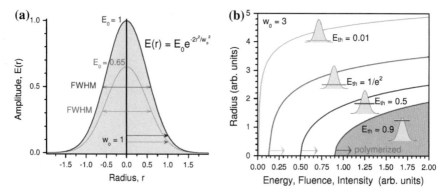

Fig. 12.1 Spatial radial distribution of E-field (intensity, dose, etc.) at the focus of (**a**) Gaussian beam/pulse in a normalized presentation: $E(r) = E_0 \exp(-2r^2/w_0^2)$, here the waist (radius) of the beam is $w_0 = 1$ (dimensionless for clarity). Two bell-shaped profiles with the same waist $w_0 = 1$ and FWHM only for different amplitude values $E_0 = 1$ and 0.65. **b** The radius of the Gaussian envelop which is above a certain threshold $E_{th}(r) = E_0 \exp(-2r^2/w_0^2)$ as a function of the incident energy (intensity, dose), E_0. The arrow depicts the span of E_0 values at which polymerization (or other modification) is expected (a shadowed region) in direct write applications when the modification threshold is $E_{th} = 0.9$

Figure 12.1b reveals the usual practical difficulty to judge upon the mechanism on a modification process—the scaling $\propto I^N$—since typically the experimental data are not spanning over at least one order of magnitude in terms of the pulse intensity I (or F, E_n, D, etc.). Close to the threshold when $E_0 \approx E_{th}$ a very steep dependence of $r \propto E_0$ is expected but it becomes saturated very quickly for larger E_0. It is very easy to claim that the mechanism of modification is nonlinear as it follows prediction E^N or I^N. For the accurate definition of an absorption mechanism, a log-log plot or a linearized presentations, e.g. (12.2), is plotted as $r^2 \propto \ln(E_0)$; then slope, γ, reveals the scaling exponent (an order of the N-photon process) and the threshold value is determined from the intersection with the abscise axis. This is a standard technique employed in analysis of laser ablation craters.

There is almost no practical restrictions to write tall 3D structures inside liquid or gel state resins, however, there are just several solid resists which can be used to make thick-enough films without structural defects, e.g., peeling off and cracking. For definition of 3D resolution, resolution bridges are recorded in thicker film polymers [28] since arbitrary small features can be recorded on the surfaces of transparent substrates on which the resist is deposited. Thiel et al. introduced a scheme of the novel dip-in 3D-DLW approach based on immersion of the objective into the optical refractive index matching photoresist. This way the sample height is no longer fundamentally limited as the photoresist serves both the immersion oil and polymerizable material [29]. In addition, this helps to avoid depth-dependent optical aberrations; otherwise a specially designed objective and/or beam shaping should be used [30].

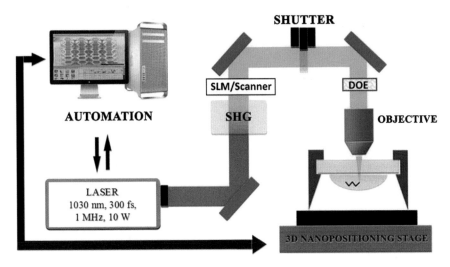

Fig. 12.2 Laser 3D micro/nano-fabrication setup: ultrafast pulsed beam is guided and focused inside the photosensitive material. A structure is written in the volume following the trace of the sample translation 3D trajectory or beam deflection. Beam control options: (1) space light modulator (SLM) and/or scanner or a specialized optical diffraction element (DOE) for industrial applications

12.1.2 Laser 3D Nanopolymerization Setup

In DLW micro/nano-structuring a precise positioning system optimized for sample translation (or a beam deflection) in 3D space is a necessity. With only a few exceptions, fs-laser oscillators or amplified laser systems are used as irradiation sources. The most common parameters are in the range of: repetition rate ~ 1 kHz to 80 MHz, average optical power ~ 200 mW to 20 W, pulse duration -20 fs to 10 ps, wavelength 515 to 1030 nm. The diagram of a typical experimental setup is shown in Fig. 12.2. It is noteworthy, that pulses obtained by frequency doubling or parametric up-/down-conversion can have an advantage in laser polymerization by over ultra-short pulses since the nanosecond-long pedestal of fs-pulses (a background of a pumping laser source) is fully suppressed by the optical nonlinear generation of the wavelength-shifted pulses. This is important in laser structuring with presence of free charges seeded by ultra-short high peak intensity pulses since an ensuing avalanche ionization (also discussed in Sect. 13.2.1 by Bityurin) by the following laser pulse pedestal continue ionization on nanosecond scale. This jeopardize the control and precision of energy delivery which can be best controlled by ultra-short femtosecond laser pulses [31].

The expanded fs-laser beam is guided through the objective focusing it into the volume of the photoresin. Additional beam control and shaping by spatial light modulator (SLM), digital mirror array (DMD), galvano scanners [5, 32, 33], etc., can be implemented on the beam delivery path. The sample is mounted on xyz positioning stages. By translating the sample three-dimensionally, the position of laser focus moves accordingly inside the photo-polymer, allowing the 3D writing of complex structures, which can be recovered after the immersion of the sample in a solvent and the removal of the unexposed resist. Alternatively (or in combination with the stages) galvanometric scanner can be used to direct the beam to the desired location. As the setup can also serve as a microscope—a camera can be integrated in the system to enable online process monitoring. The control of such systems usually is automated using custom made or commercial software for 3D DLW applications. Accordingly, 3D structures can be designed directly or imported from stereolitography CAD (STL) files. In industrial-scale fabrication,—multiple laser beams can be employed by using diffractive optical elements or spatial light modulators (Fig. 12.2) [32, 34–36].

12.2 Instrumentation

12.2.1 Laser Sources and Surface Tracking

At tight beam focusing conditions, when the focal spot size of the fs laser beam is comparable to the laser wavelength $\sim\lambda$, multiphoton polymerization can occur using only a few mW of the output power of the of a Ti:sapphire or other fs-laser MHz oscillator. With optimization of the excitation wavelength for linear and nonlinear absorption, it is possible to expose and leave irreversible modification in the resin with a single shot from a fs-laser ($\sim$1 nJ/pulse). In principle, this allows structures to be produced as fast as the repetition rate of the laser. Thus the rapid writing of 3D structures is limited only by the ability to scan from voxel-to-voxel (the volume element).[3]

It is also possible to employ a SLM or a multiple mirror array to fabricate different structures at different locations simultaneously. Such an approach was already demonstrated for fabrication of identical structures at hundreds of positions simultaneously and has the potential for industrial application [32, 35]. For periodic large area structures, two other techniques can be employed for rapid production: multi-beam interference lithography ([15], and references therein) and multiple beam generation using a multi-lens array (MLA) [34, 37]. Despite the possibility to fabricate over the larger area simultaneously, holographic interference has the disadvantage that the structures have to be periodic. MLA also cannot provide such flexibility as single beam pinpoint structuring, though it increases the fabrication speed by the number of lenses on the array. Here, we focus on fabrication of structures written witha a single focused beam as it is the most flexible way to make micro-devices. With fast positioning stages it is possible to fabricate large area structures (up to 1 × 1 cm²) area structures very fast (within one hour or less). Though an amplified fs-laser

---

[3] In practice, using high speed translation stages and fabricating complex shaped structures the limiting factor becomes not the linear scanning velocity itself but the rate of how fast the scanning direction can be changed. In other words acceleration and deceleration, especially at turning points.

system is not a necessity for a pinpoint structuring, it provides a basis for future system improvements towards industrial-scale production. When discussing industrial applications, a Yb:KGW laser system working at few 100 kHz repetition rate is enough to ensure relevant voxel overlapping even scanning at up to $\sim$10 cm/s speed [38, 39]. Conventional amplified Ti:Sapphire laser systems performing at 1–10 kHz repetition rate have $\sim$100$^{\times}$ slower production speed.

Typical sample positioning systems are high precision piezo-stages which are combined with galvano-mirror scanners and linear stages (for large range/step and repeat fabrication). Piezo stages are the most accurate, $\sim$1 nm in positioning precission, but their speed is limited to $\sim$100 μm/s and working field is also restricted to typically <500 μm. A mirror scanning system using galvanometric scanners is more suitable for making microstructures with intricate geometry more rapidly (as it virtually has no mechanical inertia), but the structure size is restricted by the objective field-of-view (40 $\times$ 40 μm^2 in case of 100$^{\times}$ magnifying objective). When joining write fields by translation, stitching errors are inevitable. Hence, for fabricating over larger area at 100 nm-order precision, linear stages are commonly preffered.

For nano-structuring a precise sample positioning is required. When a CMOS or CCD camera is used for the online monitoring of the stucture fabrication, the correct focal position can be easily determined. While scanning along the X- or Y-axis, a deviation of Z-axis sample position can occur, if the sample plane is not perpendicular to the beam. This means that the polymerized structure might be built at an angle, be "*buried*" into the substrate, or not connected to the substrate at all [40]. This problem becomes becomes more important when the lateral structure size increases. To solve this problem, a Z-axis pre-compensation can be integrated into micro/nano-fabrication software to keep track of the Z-axis. A simple way to realize this is based on a passive autofocus by determining three exact focal positions at three different lateral locations and applying the plane equation. Such basic method could ensure the focal position with 1–2 μm precision for a standard flat sample. A manual pre-compensation, an autofocussing or a machine vision system can be implemented in order to track the surface of the substrate and ensure optimal structuring [40–42].

12.2.2 Spatial Resolution and Feature Precision

Though the spatial resolution is defined as minimal distance distinguished between two separate features, the surface roughness as well as surface height modification (profile) depends mainly on the voxel overlap (δ) and tends to minimize, when the axial or lateral scanning steps become smaller than a critical value [43, 44]. Figure 12.3 shows the geometry of a focal spot, a schematic of the voxel overlap in axial and lateral directions and real polymerized structures. The degree of voxel overlap can be expressed as the product of the voxel displacement in axial and lateral directions according to the equation

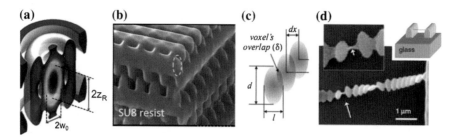

Fig. 12.3 a Ideal focus intensity cross section produced by objective lens of numerical aperture $NA = 1.35$; the outside contour is set at 1 % intensity [45]. **b** A photonic crystal woodpile structure in SU-8 with produced by focusing with $NA = 1.35$ lens; ratio of axial-to-lateral cross sections of the polymerized features is $\sim$3 [47]. **c** Voxel overlapping dependence on scanning direction and pulse repetition rate. **d** Resolution bridge in SU-8 at the limit of overlap; single polymerized voxels are discernable [28]

$$\delta = \frac{d - dz}{d} \times \frac{l - dx}{l} \qquad (12.3)$$

where δ is the degree of the voxel overlap, d and l are the axial and longitudinal dimensions of the voxel, respectively, and dz and dx are the axial and lateral scanning steps, respectively. Thus it allows the production of nanogratings over a freeform microstructures in a single step [12].

The focal spot size w_0 (diameter) is defined by the numerical aperture, NA, of the focusing lens and wavelength, λ as $w_0 = 1.22\,\lambda/NA$ (Fig. 12.3a). The axial extent of the focus is defined as double the Rayleigh length $2z_R = 2\pi w_0^2/\lambda$. At aberration-free ideal focusing with numerical apertures $NA = 1.2 - 1.4$ the axial-to-lateral cross section of the beam is $3 - 2.5$ and is always observed in polymerized structures (Fig. 12.3c, d).

When making planar structures, the DLW technique can be employed to fabricate microtructures with nanometer features in a single step following a high overlap and voxel truncation procedure. For instance, a flat pad is fabricated on the top of the cover glass slide with the 90 % voxel's overlap to ensure high surface smoothness (RMS roughness value of <10 nm can be obtained [40, 43]). Then 2D nanolines are fabricated on the top of the polymerized pad by additional scans of the laser beam with the beam focal position shifted in the Z-axis according to the top of substrate. Such a procedure can lead to the fabrication of 2D nanolines varying in height from 10 nm to more than 40 nm [46]. Thus, the multiple scanning of ascending focus enables the manufacturing of nanolines with dimensions comparable in resolution to the ones written by electron beam lithography. This is possible because the minimal reproducible voxel size depends on its threshold core dimensions [48], yet part of it can be extruded out of the pre-polymerized structure. However, higher resolution or feature precision is restricted by material properties. As the extraction of DLW-made structures requires wet developing, the evaporation of the solvent can distort or destroy the created structures [49], unless critical point drying is used [50]. On the other hand, such drawback can serve as a method to create nanofeatures by self-formation. An indirect (non-localized) polymerization effect has been studied and its efficiency depends on exposure conditions [51], distance [52] and laser scanning direction [44].

12.2.3 Fabrication Strategies, Throughput and Scaling

The issue of fabrication efficiency was raised since the first demonstrations of the DLW technology [53]. Structuring with ultra high spatial resolution brings the disadvantage of low manufacturing throughput. To calculate time needed for the fabrication of a specific structure one can use a simple estimation:

$$t = \frac{xyzF}{Rv} \tag{12.4}$$

where t is fabrication time, x, y, and z are width, length, and height of the structure, R is the structuring resolution depending on the NA of the used objective and applied laser power P, v is the sample scanning speed, and F is the fill factor (the ratio of polymerized versus non-polymerized volume of the whole structure). It is obvious that the increase of structure volume and/or the fill factor dramatically increases the fabrication time. By changing to appropriate optics and using high sample scanning velocity, it is possible to increase the volume and filling ratio in shorter time.

In order to speed up the fabrication process, only the outer shells of bulky structures can be made using DLW lithography, and their interior can be polymerized using a post-fabrication UV treatment. The diameter d and height l of single photopolymerized volume pixels (voxels) is controlled by modifying the laser power, scanning speed, and focusing optics [1, 39]. This allowed the fabrication of enclosure shells mechanically strong enough and the formation of unexposed regions inside the optical elements for further processing steps. The procedure of shell fabrication, development, and post-treatment with UV light enables the fabrication time reduction by approximately one-two orders of magnitude depending on pattern complexity [12]. This approach has been already demonstrated for producing 3D microstructures of complex geometry and components for microfluidics [54]. The time required to fabricate a compact, hemi-spherical lens of radius, R, by volumetric raster scanning with pitch, d, (the same pitch in lateral and axial directions is considered) at scan speed, v, is: $t_v = \frac{V}{d^2 v}$ where the volume of hemi-sphere $V = \frac{1}{2} \times 4\pi R^3$; the pitch is at least two times smaller than the cross section of the voxel actual fabrication. One the contrary, for only the shell fabrication the required time is $t_s = \frac{S}{dv}$ with the surface of hemi-sphere $S = \frac{1}{2} \times \frac{4}{3}\pi R^2$. Hence, the factor of an efficiency increase is $f_c \equiv \frac{t_v}{t_s} = 3\frac{R}{d}$. As the radius of a lens increases, or the voxel size decreases (hence, the pitch, d, too), the fabrication time becomes impractical for the volumetric raster fabrication. In the case of the micro-optical components conventionaly 10–1000 μm in their crossection, an improvement of approximately $(1-3) \times 10^3$ times is achieved using the shell-fabrication approach. The actual exposure time required for the fabrication of a $R = 50$ μm radius lens of focal length $\sim$100 μm at volume filling by a raster scan with 200 nm overlap between voxels is 14.7 h, reducing to approximately 2.6 min by employing the proposed outer shell-exposure [12, 55, 101].

12.2.4 Wet Developing

12.3 Materials for Laser Nanopolymerization

A material suitable for 3D volume selective structuring by nonlinear light-matter interaction must contain at least two basic components: a polymerizable material, which will become the structure backbone, and a photo-initiator, which will absorb the laser light and provide the active species that will cause this polymerization. To date, a large combination of polymeric materials and photo-initiator combinations have been used for this purpose. These are mostly negative photoresists such as acrylate materials (and their mixtures) [2, 56], the epoxy-based photoresist SU-8 [15, 57], and hybrid sol-gel materials [58–60]. There are also a few examples of positive resists being used, such as S1813 [61–63]. Various microtructures from photosensitive protein-resin can be also produced [64–66]. Light induced swelling and densification have been proposed for laser writing of 3D structures inside the bulk of the polymer [67]. In the following sections, we will show how these photo-initiators and negative polymeric materials are used for applications in 3D laser polymerization. Latelly, employing fs pulses, direct 3D structuring in non-photosensitized polymers was reported [68, 69].

12.3.1 Structuring with and Without Photo-Initiators

During photo-polymerization a liquid or gel state monomer is converted into a solid state polymer; this transformation is induced by light or/and temperature delivered by light. In conventional, one-photon lithography a light sensitive molecule, the photo-initiator, absorbs the incident radiation and produces an active species which causes a chain reaction of photopolymerization. In case of non-linear absorbtion induced polymerization a low linear absorbtion is still present [67]. Recent studies have shown that pure monomer materials without the presence of the photoinitiator can be reproducibly stuctured. And thought the fabrication window is decreased, the reachable resolution is significantly increased [38, 69, 70]. This can be achieved via avalanche (impact) ionization mechanism which is dominating in the localised light introduction in the dielectric material at a close to breakdown conditions [38, 68]. Additionally, confined heating can ensure more reproduceble structuring when high rep rates >200 kHz [71]. A simple method to distinguish accumulation effects by using varying repetition rate bursts was introduced by Baldacchini [72]. Finally, in composite organic-inorganic materials polymerization can be induced via photoinduced reduction created free radicals as well [73].

12.3.2 Organic Photopolymers

The first materials used for D@3D laser nanopolymerization (also discussed in Sect. 13.3.2 by Bityurin) were optical glues which are acrylate based photopolymers [74, 76]. They polymerize under UV i-line Hg-lamp illumination at 365 nm or eximer laser exposure at 308 nm. These materials have several properties which make them attractive for two-photon polymerization applications: a wide variety of the full composites or their monomers are commercially available (ususally at low cost and long shelf-life); they are transparent at visible and near-infra-red wavelengths, and can be therefore be processed by NIR and green ultra-fast lasers; they can be developed in common, non-aggressive solvents such as ethanol, isopropanol or acetone; they can be polymerized fast and with low shrinkage; and, after polymerization they are mechanically and chemically stable.

Due to their versatile chemistry, acrylate photopolymers have been used in their pure form but also doped with other materials to add them functionality. They have been doped with TiO_2 nanoparticles to increase their refractive index for photonic crystal applications [77]; with CdS nanoparticles for the fabrication of light-emitting 3D structures [78, 79]; with metal-binding materials to cover them with metals using electroless plating [80, 81]; with chitosan to make them suitable for bioapplications [82]; with polymer MEH-PPV (poly(2-methoxy-5-(2′-ethylhexyloxy)-1,4-phenylenevinylene) to make them electroluminescent [83].

12.3.3 SU-8

SU-8 is a negative, epoxy-based photo-resist commonly used for the fabrication of high-aspect ratio structures, using standard contact lithography or LIGA [84]. Its absorption maximum is at 365 nm. When exposed to the right light, SU-8's long molecular chains cross-link, causing the material to become insoluble to most common solvents. For this to happen, the SU-8 samples require post-exposure heat treatment. The possibility to structure SU-8 by two-photon polymerization was first demonstrated by Witzall et al., who employed single shots from Ti:Sapphire femtosecond amplifier to fabricate small dots [85]. More complex 2D and 3D structures were built by Belfield et al. and by Kuebler et al., respectively, who worked on the development of novel, efficient photo-initiators for cationic polymerization [86, 87]. Since then, SU-8 has been used by a large of the research teams involved in two-photon fabrication have used SU-8 [28, 88–91]. SU-8 is thermally stable, transparent in the visible and highly resistant to solvents, acids and bases. These properties make it suitable for permanent use application; by itself, SU-8 has been employed for the fabrication of photonic [92] and microfluidic [54, 93, 94] structures,while it has also been covered with metal for metamaterial fabrication [95, 96].

12.3.4 Novel Hybrid Materials

Over the last several years, 3D laser writing research has focused on photosensitive sol-gel hybrid materials. Especially silicate-only based photopolymers have proved to be a very popular choice, as they can be commercially available and they combine properties of silicate glasses such as hardness, chemical and thermal stability, and optical transparency with the laser processing at low temperatures of organic polymers; properties impossible to achieve with just inorganic or polymeric materials [60, 97–107]. The most widely used silicate material is the photopolymer ORMOCER; commercially available from Microresist Technologies, Germany, and it has been used for a variety of mostly photonic applications, like the optically active polymer microdisc [108].

ORMOCER and other silicate-only based hybrid materials have provided the possibility to fabricate high resolution 3D structures, they do not allow, however, the optimization and "fine-tuning" of the materials properties for specific applications. The versatile chemistry of sol-gel composites allows the co-polymerization of more than one metal alkoxides; this has been shown to enhance the material's mechanical stability and allow the modification of its optical properties. There are a few examples of composite photosensitive sol-gels used in 3D laser polymerization applications [9, 109–113]; recently, Ovsianikov et al. showed that under specific fluence conditions and material combinations can be structured into complex 3D structures without shrinkage [107, 114]. In addition to metal alkoxides, hybrid materials chemistry provides the possibility of the inclusion of functional groups, such as nonlinear optical molecules [115], quantum dots [116], fluorescent dyes [117] and metal binding materials [118]. Recently, a zirconium silicate (ORMOSIL) doped with a monomer containing amine moieties was used for the fabrication by DLW of 3D photonic crystals with optical bandgaps [119].

12.3.5 Biopolymers and Elastomers

One of the applications proposed for DLW is fabrication of scaffolds for tissue engineering, and drug delivery. To this end, there has been some limited research into the synthesis and structuring of biodegradable polymers, hydrogels and proteins. Seidlits et al. [120] reported the 3D structuring of a hydrogel in 2007. The same team used a picosecond pulse green wavelength laser to structure pure protein (bovine serum albumin, BSA) [121]. BSA and also avidin using flavin mononucleotide were crosslinked by Turunen et al. [65]. Claeyssens et al. synthesized and structured a novel, photostructurable composite based on polycaprolactone [122], while the same researchers later moved to a polylactide-based polymer, which they used to for the fabrication of 3D structures for the alignment of neural cells [123]. Ovsianikov et al. made structures using acrylated poly(ethylene glycol) [124]. There has also been some research by FORTH on the functionalization of the surface of 3D structures by biomolecules. They made 3D structures using hybrid materials, and functionalized them with biotin [125, 126] and amyloid peptides [105], targeting applications such as sensing and tissue engineering. Usage of elastomeric [68] and elastomeric degradable biomaterials are desired for soft tissue, such as vascular, engineering [127]. Recent studies have shown that most of DLW structurable materials are biocompatible in vivo [128].

12.4 Applications

12.4.1 Micro-Optics: Integrated and Multi-Functional Components

Uniqueness of DLW for micro- and nano-photonics is due to inherent possibility to shape and form optical elements without requirement of polishing, which is not scalable to the size of tens-of-micrometers and limits the standard optical polishing and lapping techniques available at larger scale. It does allow not only minimization of standard optical elements [11] or arrays of them [129] but also enables creation of hybrid refractive/diffractive components combining both of the functions [62, 130]. Polymerized surfaces as smooth as 10–20 nm (min–max) roughness over optical wavelength scale $\sim$0.5 μm is achievable [12]. Free-space writing allows to form micro-optics on optical fiber tips, to combine diffractive and refractive functions [12] which might prove essential for handling light on sub-1 μm spatial scale in opto-micro-fluidic applications. Moreover, free-form micro-optical components such as axicon lenses [131] and optical vortex generators [132] as well as hybridization of both [13] are fabricated and performs at high quality. There is no competing technology in terms of simplicity for fabrication of micro-optics. It still to be seen the aging and stability of new materials in those applications where high light intensities/powers are used. Yet initial laser induced damage tests shows their optical resistance to be comparable to that of commonly used dielectric coatings (order of 10 J/cm^2) [46]. Additionaly, it should be noted, that polymerized optical elements such as gratings and lenses can be transferred into glasses and crystaline transparent materials via plasma assisted etching [134]. Furthermore, such components enables direct integration of miniaturized optical elements [12, 135], and are currently active fields of research (Fig. 12.4).

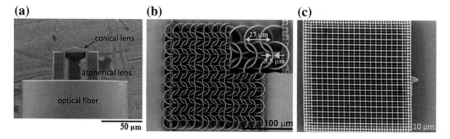

Fig. 12.4 Functional microdevices fabricated via DLW: **a** A bifunctional aspheric-conical microoptical component integrated on the tip of the single-mode optical fiber [133]; **b** A flexible "*chain-mail*" scaffold consisting of separate intertwined rings [148]; **c** A photonic crystal template produced without the presence of the photoinitiator [69]

12.4.2 2D and 3D Photonics

Photonic crystal templates and other 3D nano-micro-structures can be recorded by free scanning inside the photo-polymer via direct write. Since the focal spot has an axial elongation which is 2–3 times larger as compared with the axial cross section (the actual number depends on numerical aperture of the objective lens), it is possible to use this feature, e.g., an optical anisotropy can be introduced. A photonic crystal template, which has different transmission coefficients for the right and left circular polarizations along the spirals as well as for the linear polarizations along the horizontal and vertical directions [136]. As traveling light encounters different amount of polymer matrix aligned in respect of the incident polarization, the transmission and reflection shows polarization dependent values. 3D spirals of a photonic crystal (PhC) can be oriented vertically [90, 137] or horizontally [138] in respect to the substrate. Due to shape asymmetry of the focal spot, lateral scanning in horizontal writing of spirals allows to reduce up to twice the spiral pitch. Such low refractive index templates of SU-8 can be inverted into silicon [139] by well established and high fidelity procedures [140]. Employing positive tone photoresist and metal infiltration its possible to create 3D golden nanostructures working as optical cloaks [142]. Alternatively, medium contrast structures can be used for as wavelength selective flat lensing components [143]. Recentle, a miniature chiral beamsplitter for circularly polarized light based on gyroid photonic crystals was reported [144]. A special galvo-dithering technique was employed to produce such structure.

12.4.3 Biomedicine and Microfluidics

Mechanical properties of 3D structured polymer materials can be controlled by their volume fraction. It was demonstrated that woodpile and similar structures, which are promising for 3D bio-scaffolds, behaves as foams under mechanical loading [141].

The elastic response depends on the ratio between the spatially averaged mass density $\langle \rho \rangle$ of the structure and the density of monolithic polymer ρ_s (SU-8 in [141]); note, the latter value can be dependent on the nonuniform crosslinking due to the Gaussian intensity distribution which brings extra measure of mechanical control of single polymerized structures. The structural modulus generally scales with the ratio $\rho_r = \langle \rho \rangle / \rho_s$ as $E \sim E_s \times \rho_r^2$, were E_s is the Young modulus of the monolithic polymer. Thus, the elastic response can be controlled via the relative display of the polymer filling fraction. This is important for a bio-/medical implants where matching of mechanical properties is crucial for longevity and wear of the artificial part.

Among biomedical microdevices two main directions could be distinguished—the passive scaffolding serving as supporting constructs or micro-mechanic and/or micro-fluidic functional devices. Regarding the first type of compartments for cell growth it is worth mentioning that a variety of cell studies can be performed, such as cell migration [145], stem cell growth [146], composite compartments with varied cell adhession properties and pore size/shape investigations [147, 148]. Second class, an active elements, could consist of movable parts [9], microneedles for precise drug delivery [149] or cell sorters [10]. Despite their functions all of the microobjects can be produced with controlled pore size, general porosity and overal size [150]. They can be manufactured out of biostable [151, 152] or biodegradable [124, 153] as well as consist of composite parts [154, 155]. It is noteworthy, that up to some extent such biomedical sample structures can be easily coated, replicated or transfered to another materials [156] keeping its high structural quality [148, 157].

An area where DLW has seen enormous growth recently is its application for scaffolds in tissue engineering. This is a discipline which applies the principles of engineering and life sciences toward the development of biological substitutes that restore, maintain, or improve tissue function or a whole organ [158]. An important part of the development of artificial tissue is the choice of scaffold, as it can influence the attachment, migration, and proliferation of cells. 3D cell cultures offer a realistic environment where the functional properties of cells can be observed and manipulated [159–161]. Although producing scaffolds using laser-based, user controlled manufacturing techniques such as DLW or classic stereolithography is time consuming and therefore costly, recent studies have shown that they can provide tissue engineering solutions for aligned and complex tissue growth. An important advantage is that a controlled topological environment for cell growth can be achieved. In addition, ordered 3D scaffolds are ideally suited for exploring the relationship between 3D topology and cell growth. A recent review by Narayan and Goering [162] describes in great detail the use of lasers in biomedical applications, including tissue engineering. In this field, most groups have used permanent scaffolds for hard tissue engineering or investigating cell growth [82, 146, 147, 151, 154]. More recently, there has been a lot of active research into the synthesis of new, biocompatible and biodegradable materials [65, 122, 163].

As calculated in [150] DLW employing femtosecond lasers allows one to microstructure using light exposure doses within the therapeutic range (1–10 J/cm^2). Having the light density needed for the irreversible photomodification confined within the focal volume one can think of artificial scaffold creation during in vitro tests [164], and due to already high fabrication throughput even during in vivo surgery [165]. This would enable dynamic experiments of cell biology or creation of custom shaped artificial scaffolds in a real time incorporating live cells [166].

The flexible direct laser polymerization approach for the production of custom shaped and pore size as well as porosity polymer scaffolds has reached the level of fabrication throughput requisite for practical applications [4]. True 3D microstructured scaffolds of size up to 1 cm^3 and pore sizes in the range of 10–100 μm with controlled porosity can be manufacturing overnight, thus be created for a individual patient once required. Precise control of the form ant filling factor of the artificial scaffold enables one to control it's mechanical properties. For the direct laser polymerization technology this is the step out of the laboratory to the market. Using high pulse repetition femtosecond lasers operating at the visible wavelengths the exposure dose needed for the localized photopolymerization is not exceeding the values tolerable for performing in in vitro [164] and in vivo encapsulation [165]. This opens a unique way to photostructure the scaffolds in real time during the cell biology or surgery experiments. Further experiments are targeted towards using elastic or biodegradable and addionally biologically functionalized materials for the creation of artificial scaffolds with desired properties [68]. Next to these mentioned application targetted for soft-tissue regeneration, movable parts [9] and hard-tissue constructs [167] or surface functionalization for cell adhession optimization is also in the scope of interest [168]. One more path way is to use three-dimensional nanofabrication based on simultaneous additive and material removal processing by non-linear laser matter interaction [169]. Similar approach combining two regimes of laser impact can be applied to produce a 3D scaffold and transport live cells into it using laser induced forward transfer method [170]. This approach enables to print sorted cells selectively in space into already prefabricated compartments.

12.5 Future Outlook

Thanks to the efforts of synergetic work of many research groups worldwide, the capabilities of the DLW fabrication technique is pushed forwards to the state being applicable for industrial demands. While such issues as improvement of fabrication spatial resolution and incorporation of new materials continue to attract considerable attention, this technology is already at a mature enough stage that the focus is evolving towards applications. Construction and developments of laser 3D fabrication systems is targeted as a laboratory prototype of machinery for rapid and routine micro/nanostructuring over large areas (volumes) as well as custom small scale production. This has already drawn a great interest for micro-optical, photonic, biological, microfluidic and optofluidic applications. The latter growing interest forces further progress in further improving DLW technology. Successive introduction of novel ultrafast laser excitation sources, study on photopolymerization mechanisms and material response at nanoscale revealed important issues in the DLW nanostructuring. Avalanche ionization and self-polymerization are of critical importance in order to manufacture high spatial resolution and not distorted functional micro/nanodevices. Lastly, though addressed in this chapter, possibility to photostructure various materials on different substrates, capapability to integrate simple elements into integrated microsystems, as well as incorporation of several functions in a single component, are still of a great importance.

Despite recent advances in the field of direct laser micro-/nano-fabrication, it is still a great challenge for material scientists to offer suitable materials for true 3D photostructuring. Though there is a way to introduce shape pre-compensation in order to minimize structural distortions [171, 172], yet it seems more like the optimization of the materials could be ultimate solution [38, 58]. As in has been presented in this review, DLW technique based on femtosecond laser offers opportunity to fabricate 3D functional nano/microdevices for various specific applications. Further improvements of this technique could be realized applying Optical Parametric Amplifiers (OPAs) and adjusting laser wavelength to the materials two-photon or more photons absorption. It would enable easy laser-to-material tuning which would optimize the multi-photon polymerization efficiency, thus providing higher resolution. Furthermore integration of STED method to DLW can offer unmatched axial resolution which would mean the dimensions of voxel to be far less 100 nm and would open a way to nano-realm using VIS/NIR light (see [173, 174] and a Sect. 13.3.2.1 by Bityurin). An excess of available power in amplified laser systems could be utilized by applying multi lens array [34, 37]. Alternatively advances in beam shaping could also be applied here, like Bessel beam [175], DOE or holographic mask [36] could be used to modify light intensity distribution for exposure. SLM [35] or DMD [33] could be attractive due to the dynamic modulation of beam profile. This would enable DLW nanostructuring in 3D with increased throughput up to few hundred times. Recently demonstrated simultaneous manipulation and fabrication in aqueos media offers new posibilites to create composite microstructures for micro scale bioengineering [176].

DLW produced structures will be more often used as a templates for further steps as selective metallization, infiltration and double material inversion techniques are being developed rapidly. Besides to that, it will make an advance in new extensively explored fields of nanophotonics and plasmonic metamaterials. Additionally, it will enable fabrication of true 3D electric circuits, thus transferring planar microchips to the third dimension, which has already been demonstrating to operate in optical regime [177, 178]. Posibility to exploit tuning the mechanical properties via nanostructure by manufacturing auxetic materials is also highly expected in the near future [29]. Authors strongly believe that the technology will experience a breaktrough as the junction of nano-macro scale will be reached, offering new features in a practicaly applicable scale. One os such examples could be hierarchical structures having high mechanical rigidity and desired surface adhesion [179].

In summary, 3D laser micro-/nano-fabrication has become a matured and versatile technology over last 10–15 years with one of its embodiments—the 3D laser polymerization—being in a forefront of actively interconnected and evolving fields: nano-/micro-optics, nano-/micro-mechanics, opto-fluidics, biomedical scaffold and tissue engineering as briefly reviewed in this chapter. While decade ago it was hampered with perception that only femtosecond laser pulses have to be used, the most recent developments in new photo-materials and better understanding of the underlying principles of light-matter interaction shows that simpler setups with less expensive lasers delivers excellent resolution at high throughput. A wider uptake of laser fabrication technologies into industrial applications is a next expected direction. The genesis of this field strongly advanced with availability of ultra-short pulsed lasers and engineering trials to structure available photopolymers, proceeded next with development of new materials which opened new fields into bio-/medical and micro-optical applications.

Acknowledgments MM appreciates Dr. A. Ovsianikov, Dr. C. Reinhardt and Prof. B. Chichkov for introducing and engaging into the direct laser polymer structuring, Dr. M. Farsari and Dr. E. Brasselet for continuous collaboration. Part of the research was funded by a grant MIKROŠVIESA (No. MIP-12241) from the Research Council of Lithuania.
SJ acknowledges support from the Australian Research Council DP120102980 and VP1-3.1-MM-01-V-02-001 from the Research Council of Lithuania grants.

References

1. H.-B. Sun, S. Kawata, in *Two-Photon Photopolymerization and 3D Lithographic Microfabrication*, ed. by N. Fatkullin, NMR. 3D Analysis. Photopolymerization (Springer, Berlin/Heidelberg, 2004). http://link.springer.com/chapter/10.1007/b94405
2. C. LaFratta et al., Angew. Chem.-Int. Edit. **46**, 6238 (2007)
3. A. Ovsianikov, B. Chichkov, in *Two-Photon Polymerization—High Resolution 3D Laser Technology and its Applications*, Nanoelectronics and Photonics, vol (Springer, 2008). http://link.springer.com/chapter/10.1007%2F978-0-387-76499-3_12
4. M. Malinauskas, M. Farsari, A. Piskarskas, S. Juodkazis, Phys. Rep. **533**, 1 (2013)
5. J. Serbin, A. Ovsianikov, B. Chichkov, Opt. Express **12**, 5221 (2004)
6. M. Deubel et al., Opt. Lett. **31**, 805 (2006)
7. J. Trull et al., Phys. Rev. A **84**, 033812 (2011)
8. P. Galajda, P. Ormos, J. Opt. B-Quantum S. O. **4**, 78 (2002)
9. C. Schizas et al., Int. J. Adv. Manuf. Technol. **48**, 435 (2010)
10. L. Amato et al., Lab Chip **12**, 1135 (2012)
11. R. Guo et al., Opt. Express **14**, 810 (2006)
12. M. Malinauskas et al., J. Opt. **12**, 124010 (2010)
13. A. Žukauskas, M. Malinauskas, E. Brasselet, Appl. Phys. Lett. **103**, 181122 (2013)
14. S.-H. Park, D.-Y. Yang, K.-S. Lee, Laser Photon. Rev. **3**, 1 (2009)
15. S. Juodkazis, V. Mizeikis, H. Misawa, J. Appl. Phys. **106**, 051101 (2009)
16. M. Farsari et al., Opt. Lett. **24**, 549 (1999)
17. M. Farsari et al., J. Mat. Process. Tech. **107**, 167 (2000)
18. P. Calvert, Chem. Mat. **13**, 3299 (2001)
19. E. Tekin, P. Smith, U. Schubert, Soft Matt. **04**, 703 (2008)
20. I. Gibson, D. Shi, Rapid Prototyp. J. **3**, 129 (1997)
21. I. Yadroitsev et al., Appl. Surf. Sci. **255**, 55235527 (2009)
22. C. Vieu et al., Appl. Surf. Sci. **164**, 111 (2000)
23. A. Grigorescu, C. Hagen, Nanotechnology **20**, 292001 (2009)
24. A. Majumdar et al., Appl. Phys. Lett. **61**, 2293 (1992)
25. S. Lyuksyutov et al., Nat. Mat. **2**, 468 (2003)
26. W. Denk, J.H. Strickler, W.W. Webb, Science **248**, 73 (1990)
27. C. Mauclair et al., Opt. Lett. **36**, 325 (2011)

28. S. Juodkazis et al., Nanotechnology **16**, 846 (2005)
29. T. Buckmann et al., Adv. Mater. **24**, 2710 (2012)
30. F. Burmeister, U.D. Zeitner, S. Nolte, A. Tunnermann, Opt. Express **20**, 7994 (2012)
31. S. Juodkazis et al., Opt. Express **17**, 15308 (2009)
32. H. Lin, B. Jia, M. Gu, Opt. Lett. **36**, 406 (2011)
33. R. Nielson, B. Kaehr, J.B. Shear, Small **5**, 120 (2009)
34. J.-I. Kato et al., Appl. Phys. Lett. **86**, 044102 (2005)
35. S. Gittard et al., Biomed. Opt. Express **2**, 3167 (2011)
36. E. Stankevicius et al., J. Micromech. Microeng. **22**, 065022 (2012)
37. S. Matsuo, S. Juodkazis, H. Misawa, Appl. Phys. A **80**, 683 (2004)
38. M. Malinauskas et al., Opt. Express **18**, 10209 (2010)
39. M. Malinauskas et al., Lith. J. Phys. **5**, 201 (2010)
40. M. Malinauskas, et al., Proc. SPIE **7204**, 72040C1 (2009)
41. B.J. Jung et al., Opt. Express **19**, 22659 (2011)
42. T. Kildusis, T. Kazakevicius, Laser Focus World (2011), pp 29–36. http://www.laserfocusworld.com/articles/print/volume-47/issue-11/columns/software-computing/laser-micromachining-software-attains-research-friendly-status.html
43. H. Sun, S. Kawata, Appl. Phys. Lett. **86**, 071122 (2005)
44. M. Malinauskas et al., Lith. J. Phys. **50**, 135 (2010)
45. T. Kondo et al., Thin Solid Films **453–4**, 550 (2004)
46. A. Žukauskas et al., SPIE **8428**, 84280K (2012)
47. V. Mizeikis, S. Matsuo, S. Juodkazis, H. Misawa, in *Femtosecond Laser Microfabrication of Photonic Crystals*, Three-Dimensional Laser Microfabrication: Fundamentals and Applications, vol (Wiley, Weinheim, 2006), p 2001. http://eu.wiley.com/WileyCDA/WileyTitle/productCd-3527608400,subjectCd-PH40.html
48. A. Pikulin, N. Bityurin, Tech. Phys. **57**, 697 (2012)
49. S.-H. Park et al., Microelectron. Eng. **85**, 432 (2008)
50. S. Maruo, T. Hasegawa, N. Yoshimura, Opt. Express **17**, 20945 (2009)
51. S. Park, T. Lim, D.-Y. Yanga, N. Cho, K.-S. Lee, Appl. Phys. Lett. **89**, 173133 (2006)
52. D. Tan et al., Appl. Phys. Lett. **90**, 071106 (2007)
53. R. Borisov et al., Laser Phys. **8**, 1105 (1998)
54. D. Wu et al., Lab Chip **9**, 2391 (2009)
55. D.-Y. Yang et al., Appl. Phys. Lett. **90**, 013113 (2007)
56. M. Farsari et al., J. Photochem. Photobiol. A **181**, 132 (2006)
57. W. Teh et al., Appl. Phys. Lett. **84**, 4095 (2004)
58. M. Farsari, M. Vamvakaki, B. Chichkov, J. Opt. **12**, 124001 (2010)
59. A. Radke et al., Adv. Mater. **23**, 3018 (2011)
60. M. Malinauskas et al., Opt. Lasers Eng. **50**, 1785 (2012)
61. W. Zhou et al., Science **296**, 1106 (2002)
62. B.N. Chichkov et al., Proc. MRS **850**, 251 (2005)
63. J. Gansel et al., Science **325**, 1513 (2009)
64. B. Kaehr et al., Anal. Chem. **78**, 3198 (2006)
65. S. Turunen et al., Biofabrication **3**, 045002 (2011)
66. H.-B. Sun, S. Matsuo, H. Misawa, Appl. Phys. Lett. **74**, 786 (1999)
67. R. Infuehr et al., Appl. Surf. Sci. **254**, 836 (2007)
68. S. Rekštytė, M. Malinauskas, S. Juodkazis, Opt. Express **21**, 17028 (2013)
69. R. Buividas, et al., Opt. Mat. Express **3**(10), 1674–1686 (2013). http://dx.doi.org/10.1364/OME.3.001674
70. M.T. Do et al., Opt. Express **21**, 26244 (2013)
71. P. Danilevicius, M. Malinauskas, S. Juodkazis, Opt. Express **19**, 5602 (2011)
72. T. Baldacchini, S. Snider, R. Zadoyan, Opt. Express **20**, 29890 (2013)
73. E. Kabouraki et al., Nano Lett. **13**, 3831 (2004)
74. S. Maruo, O. Nakamura, S. Kawata, Opt. Lett. **22**, 132 (1997)
75. S. Maruo, K. Ikuta, Appl. Phys. Lett. **76**, 2656 (2000)

76. T. Baldacchini et al., J. Appl. Phys. **95**, 6072 (2004)
77. S. Maruo, O. Nakamura, S. Kawata, Thin Solid Films **453–4**, 518 (2004)
78. Z.B. Sun et al., Adv. Mater. **20**, 914 (2008)
79. Z. Sun et al., Nanotechnology **19**, 035611 (2008)
80. F. Formanek et al., Appl. Phys. Lett. **88**, 3 (2006)
81. N. Takeyasu, T. Tanaka, S. Kawata, Appl. Phys. A **90**, 205 (2008)
82. D. Correa et al., JNN **9**, 5845 (2009)
83. C. Mendonca et al., Appl. Phys. Lett. **95**, 113309 (2009)
84. V. Saile, et al., *LIGA and Its Applications* (Wiley, 2009). http://eu.wiley.com/WileyCDA/
    WileyTitle/productCd-3527316981.html
85. G. Witzgall et al., Opt. Lett. **23**, 1745 (1998)
86. K. Belfield et al., J. Phys. Org. Chem. **13**, 837 (2000)
87. S. Kuebler et al., J. Photoch. Photobio. A **158**, 163 (2003). http://eu.wiley.com/WileyCDA/
    WileyTitle/productCd-3527316981.html
88. M. Deubel et al., Appl. Phys. Lett. **85**, 1895 (2004)
89. V. Mizeikis, K. Seet, S. Juodkazis, H. Misawa, Opt. Lett. **29**, 2061 (2004)
90. K. Seet, V. Mizeikis, S. Matsuo, S. Juodkazis, H. Misawa, Adv. Mater. **17**, 541 (2005)
91. K.K. Seet, V. Mizeikis, S. Juodkazis, H. Misawa, J. Non-Crystal, Solids **352**, 2390 (2006)
92. M. Deubel et al., Nat. Mater. **3**, 444 (2004)
93. G. Kumi, C. Yanez, K. Belfield, J. Fourkas, Lab Chip **10**, 1057 (2010)
94. M. Stoneman, M. Fox, C. Zeng, V. Raicu, Lab Chip **9**, 819 (2009)
95. Y. Chen, A. Tal, D. Torrance, S. Kuebler, Adv. Funct. Mater. **16**, 1739 (2006)
96. M. Rill et al., Nat. Mater. **7**, 543 (2008)
97. M. Straub, L. Nguyen, A. Fazlic, M. Gu, Opt. Mater. **27**, 359 (2004)
98. R. Houbertz et al., Thin Solid Films **442**, 194 (2003)
99. R. Houbertz et al., Adv. Eng. Mater. **5**, 551 (2003)
100. Y. Jun, P. Nagpal, D. Norris, Adv. Mater. **20**, 60610 (2008)
101. J. Serbin et al., Opt. Lett. **28**, 301 (2003)
102. J. Li et al., Adv. Mater. **19**, 3276 (2007)
103. J. Li, B. Jia, M. Gu, Opt. Express **16**, 20073 (2009)
104. T. Woggon, T. Kleiner, M. Punke, U. Lemmer, Opt. Express **17**, 2500 (2009)
105. V. Dinca et al., Nano Lett. **8**, 538 (2008)
106. I. Sakellari et al., Appl. Phys. A **100**, 359 (2010)
107. A. Ovsianikov et al., ACS Nano **2**, 2257 (2008)
108. T. Grossmann et al., Opt. Express **19**, 11451 (2011)
109. B. Bhuian, R. Winfield, S. O'Brien, G. Crean, Appl. Surf. Sci. **252**, 4845 (2006)
110. R. Winfield, B. Bhuian, S. O'Brien, G. Crean, Appl. Phys. Lett. **90**, 111115 (2007)
111. R. Winfield, B. Bhuian, S. O'Brien, G. Crean, Appl. Surf. Sci. **253**, 8086 (2007)
112. M. Farsari, B. Chichkov, Nat. Photon. **3**, 450 (2009)
113. M. Malinauskas et al., Metamaterials **5**, 135 (2011)
114. A. Ovsianikov et al., Opt. Express **17**, 2143 (2009)
115. M. Farsari et al., Appl. Phys. A-Mater. **93**, 11 (2008)
116. B. Jia, D. Buso, J. van Embden, J. Li, M. Gu, Adv. Mater. **22**, 2463 (2010)
117. A. Zukauskas, et al. Lith. J. Phys. **50**, 55 (2010)
118. K. Terzaki et al., Opt. Mater. Express **1**, 586 (2011)
119. N. Vasilantonakis et al., Adv. Mater. **24**, 1101 (2012)
120. S. Seidlits, C. Schmidt, J. Shear, Adv. Funct. Mater. **19**, 3543 (2009)
121. E. Ritschdorff, J. Shear, Anal. Chem. **82**, 8733 (2010)
122. F. Claeyssens et al., Langmuir **25**, 3219 (2009)
123. V. Melissinaki et al., Biofabrication **3**, 045005 (2011)
124. A. Ovsianikov et al., Acta Biomater. **7**, 967 (2011)
125. T. Drakakis et al., Appl. Phys. Lett. **89**, 144108 (2006)
126. M. Farsari et al., Appl. Surf. Sci. **253**, 8115 (2007)
127. S. Baudis et al., Biomed. Mater. **6**, 055003 (2011)

128. M. Malinauskas et al., Appl. Phys. A **108**, 751 (2012)
129. D. Wu et al., Appl. Phys. Lett. **97**, 031109 (2010)
130. M. Malinauskas et al., Eur. Phys. J. Appl. Phys. **58**, 20501 (2012)
131. A. Zukauskas et al., Appl. Opt. **51**, 4995 (2012)
132. E. Brasselet et al., Appl. Phys. Lett. **97**, 211108 (2010)
133. A. Žukauskas, et al., JLMN **9**(1), 68 (2014)
134. M. Mizoshiri, H. Nishiyama, J. Nishii, Y. Hirata, Appl. Phys. A **98**, 171 (2010)
135. G. Cojoc et al., Microelectron. Eng. **87**, 876 (2010)
136. M. Thiel, J. Ott, A. Radke, J. Kaschke, M. Wegener, Opt. Lett. **38**, 4252 (2013)
137. K. Seet, V. Mizeikis, S. Juodkazis, H. Misawa, Appl. Phys. A **82**, 683 (2005)
138. K. Seet, V. Mizeikis, S. Juodkazis, H. Misawa, Appl. Phys. Lett. **88**, 221101 (2006)
139. K.K. Seet et al., IEEE J. Sel. Top. Quant. **14**, 1064 (2008)
140. N. Tetreault et al., Adv. Mat. **18**, 457 (2006)
141. S. Juodkazis et al., Appl. Phys. Lett. **91**, 241904 (2007)
142. T. Ergin et al., Science **328**, 337 (2010)
143. L. Maigyte et al., Opt. Lett. **38**, 2376 (2010)
144. P.D. Turner et al., Nat. Photon **7**, 801 (2010)
145. P. Tayalia et al., Adv. Mat. **20**, 4494 (2008)
146. M. Malinauskas et al., Lith. J. Phys. **50**, 75 (2010)
147. S. Psycharakis et al., Biomed. Mater. **6**, 045008 (2011)
148. P. Danilevicius et al., Opt. Laser. Technol. **45**, 518 (2013)
149. S. Gittard et al., Expert Opin. Drug Deliv. **7**, 513 (2010)
150. P. Danilevicius et al., J. Biomed. Opt. **17**, 081405 (2012)
151. A. Ovsianikov et al., J. Tissue. Eng. Regen. Med. **1**, 443 (2007)
152. A. Ovsianikov et al., Appl. Surf. Sci. **253**, 6603 (2007)
153. A. Ovsianikov et al., Biomacromolecules **12**, 851 (2011)
154. F. Klein et al., Adv. Mater. **23**, 1341 (2011)
155. S. Rekštytė, et al., JLMN **9**(1), 25 (2014)
156. S. Gittard et al., Biofabrication **1**, 041001 (2009)
157. A. Koroleva et al., J. Opt. **12**, 124009 (2010)
158. R. Langer, J. Vacanti, Science **260**, 920 (1993)
159. S. Zhang, F. Gelain, X. Zhao, Semin. Cancer Biol. **15**, 413 (2005)
160. A. Abbott, Nature **424**, 870 (2003)
161. E. Stratakis et al., Prog. Quant. Electron. **33**, 127 (2009)
162. R. Narayan, P. Goering, MRS Bulletin **36**, 973 (2011)
163. S. Engelhardt et al., Biofabrication **3**, 025003 (2011)
164. K. Aarcaute, B. Mann, R. Wicker, Annals. Biomed. Eng. **34**, 1429 (2006)
165. J. Torgersen et al., J. Biomed. Opt. **17**, 105008 (2012)
166. A. Ovsianikov, et al., Langmuir **30**(13), 3787 (2014)
167. A. Ovsianikov, et al., Expert Rev. Med. Devices **9**(6), 613 (2012)
168. R. Wittig, E. Waller, G. von Freymann, R. Steiner, J. Laser Appl. **24**, 042011 (2012)
169. W. Xiong et al., Light Sci. Appl. **1**, e6 (2012)
170. A. Ovsianikov et al., Biofabrication **2**, 014104 (2010)
171. H. Sun et al., Appl. Phys. Lett. **85**, 3708 (2004)
172. R. Houbertz, Appl. Surf. Sci. **247**, 504 (2005)
173. L. Li, R. Gattass, E. Gershgoren, H. Hwang, J. Fourkas, Science **324**, 910 (2009)
174. J. Fischer, M. Wegener, Opt. Mater. Express **1**, 614 (2011)
175. M. Duocastella, C. Arnold, Laser Photon. Rev. **6**, 607 (2012)
176. F. Dawood, S. Qin, L. Li, E. Lina, J. Fourkas, Chem. Sci. **3**, 2449 (2012)
177. M. Schroder et al., J. Europ. Opt. Soc. Rap. Public. **7**, 12027 (2012)
178. C.-W. Lee, S. Pagliara, U. Keyser, J. Baumberg, Appl. Phys. Lett. **100**, 171102 (2012)
179. M. Rohrig, M. Thiel, M. Worgull, H. Holscher, Small **8**, 2918 (2012)

Chapter 13
Laser Nanostructuring of Polymers

Nikita M. Bityurin

Abstract We discuss laser nanostructuring of polymer surfaces by means of interference and colloidal particle lens array approaches. We focus on laser swelling as a mechanism of surface structuring. 3D laser nanopolymerization, laser induced formation of metal nanoclusters within a polymer matrix as well as laser bubbling are also considered.

13.1 Introduction

Laser surface nanostructuring is an effective tool for large area processing. In this respect, we consider the laser interference approach and application of the colloidal/contact particle lens array technique. The latter is studied in more detail taking into account the effect of the neighboring spheres on their focusing ability. The advantages of using bi-chromatic femtosecond laser beams are discussed. Laser irradiation results in formation of either ablation craters or bumps. The features of the mechanisms of creation of such structures on nanoscales are discussed. Attention is paid primarily to laser swelling—the least studied but very promising effect, which is used first of all for the surface nanostructuring of polymers and glasses. The relaxation mechanism of laser swelling of polymers describing this phenomenon as a volume relaxation of glassy materials within the glass transition region is considered.

Laser nanostructuring within the material bulk can be achieved by a tightly focused laser beam using both the nonlinear light absorption and the strongly nonlinear material response. Some theorems concerning 3D photochemical information recording are formulated. In more detail, the 3D nanostructuring is considered in terms of

N. M. Bityurin (✉)
Institute of Applied Physics, RAS 46, Ul'yanov str,
Nizhny Novgorod 603950, Russian Federation
e-mail: bit@appl.sci-nnov.ru

V. P. Veiko and V. I. Konov (eds.), *Fundamentals of Laser-Assisted Micro- and Nanotechnologies*, Springer Series in Materials Science 195, DOI: 10.1007/978-3-319-05987-7_13, © Springer International Publishing Switzerland 2014

nanopolymerization. The fluctuation limit for voxel size is determined and a way for the spatial resolution improvement by means of quencher diffusion is proposed.

Nanostructured materials can also be produced even by laser exposure without tight focusing if irradiation leads to the development of instability resulting in the formation of nanoclusters or nanoinhomogeneities in an initially homogeneous medium. Examples of such an instability are the photo-induced formation of metal nanoclusters within dielectric matrices and the laser-induced bubbling. We consider the synthesis and the properties, including the nonlinear optical ones, of photo-induced nanocomposites based on a polymer matrix with photo-induced gold nanoclusters. The nanocluster growth model based on the theory of first-order phase transition is presented. We review the existing experimental data on laser bubbling of polymers. Two models for laser bubbling of polymers based on the droplet explosion mechanism and the cavitation mechanism are considered.

13.2 Surface Nanostructuring

13.2.1 Colloidal Particle Lens Arrays and Interference Lithography

Nanostructuring of materials has a significant effect on their physical and chemical properties, which underlies many advanced nanotechnologies. Owing to recent advances in laser systems, laser radiation is one of the most effective tools for modifying materials. At the same time, fabrication of nanostructures using the most readily available visible and near-UV lasers encounters the problem of overcoming the diffraction limit in focusing the laser beam. To produce nanostructures by laser pulses, the laser radiation field should be highly localized, using near-field optics in particular. This can be achieved by a variety of techniques [1], including those that employ field enhancement beneath an atomic force microscope tip [2], near-field optical microscope probes, various near-field masks [3], and interference lithography [4]. A promising area of research is the so-called laser particle nano-lithography [3], namely, fabrication of submicron- and nanometer-scale structures via laser radiation field localization using transparent micro- and nanoobjects placed on the surface of the material. Here, colloidal particle lens array (CPLA) proved to be an efficient near-field focusing device for laser nanoprocessing of materials [5–11] (Fig. 13.1). It should be noted that surface nanostructuring can be performed by different ways without laser radiation. For instance, e-beam technologies offer opportunities for creating nanofeatures with quite sophisticated structure [12]. Thus, in order to be competitive, laser technology should, at least in principle, provide an opportunity for creating nanostructures simultaneously on large surface areas. From this point of view, the latter two approaches, namely, interference lithography and CPLA-mediated laser structuring, are very promising.

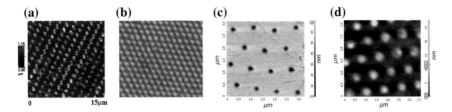

Fig. 13.1 a The structure on polyimide (swelling) obtained by four-beam interference technique: an AFM image (XeCl laser, 308 nm, fluence = 60 mJ/cm^2). (See paper by Verevkin et al. [13]). **b** Polystyrene spheres 1 μm in diameter on a polymethylmethacrylate (PMMA) substrate. **c** Typical AFM picture of structures on the surface of PMMA (*left*) and vitreous glass (*right*) substrates. CPLA mediated irradiation by femtosecond pulses from a TiSa laser (see paper [14] and the text below)

Comparing these two approaches reveals their advantages and disadvantages. It is worth noting that interference lithography of vacuum UV laser beams is considered in paper [15] as a future of laser lithography enabling a spatial resolution of up to 11 nm. At the same time, while femtosecond laser pulses are employed for the material surface nanostructuring, the CPLA approach has evident preference over the interference technique from the point of view of the ability of simultaneous modification of large areas. There are several papers devoted to the femtosecond interference approach for laser nanostructuring of material surfaces [16]. Nevertheless, it is evident that this technique has some constraints from this point of view. Indeed, the longitudinal spatial length of a 50 femtosecond pulse is about d = 15 μm. This significantly limits the number of features contained in the interference structures obtained by this technique. Below, we mainly focus on the CPLA mediated surface nanostructuring by means of a femtosecond laser pulse.

High focusing ability of microspheres together with high nonlinearity of the material response to irradiation by femtosecond pulses provides an opportunity for fine-structuring of the material surfaces. On the contrary, multiple re-scattering of the laser light within an array of spheres [17–20] can diminish the advantages of the considered setup. It is shown that electrodynamic interaction between the spheres can reduce the field enhancement, elongate the focus volume drastically, thus changing the aspect ratio of the laser-irradiated zone, and shift the maximum of the laser field towards the inner part of the sphere. All of these effects can have a significant impact on the material response and thereby on the nanostructuring process (see recent paper [20] for detail).

Recently, efforts to provide a more efficient modification of materials by optimization of the femtosecond pulse shape (typically via spectral phase modulation) have been reported [21–23]. The idea is that the more powerful front part of the pulse efficiently promotes seed electrons by multi-photon ionization, while the following tail could be effective in the impact ionization process. However, the second harmonic can also be a powerful tool for generation of seed electrons [24]. Below, we consider several reasons why conversion of some part of the beam energy to the second harmonic can be useful for the CPLA-mediated nanostructure formation.

For a laser intensity level of more than 10^{12} W/cm^2 and a short pulse duration of about 50 fs, any modification of a material that is linearly transparent at the wavelength of the irradiation is caused by ionization (or electron excitation from the valence to the conduction band). For such a short pulse length, either multiple photon or tunnel ionization mechanism dominates, depending on the value of the adiabatic (Keldysh) parameter, γ_A [25]. If $\gamma_A \gg 1$, then the direct transition from the ground state to the conduction band can be interpreted as a multi-photon transition with rate given by $W_\omega \propto I_\omega^K$. For the typical cases described below, the value of $\gamma_A = 1$ corresponds to the intensity $I \approx 10^{14}$ W/cm^2. Since the order of transition $K_{2\omega} = K_\omega/2$, the transition rate for the same laser intensities not exceeding $I \approx 10^{14}$ W/cm^2 is larger at SH than at FF. Moreover, *conversion of several percent of the initial pulse energy into the second harmonic can significantly increase the multi-photon ionization rate* (see [24] for detail). Combined effect of FF and SH, as is shown in [24], can be more effective than that of FF and SH applied separately because the seed electrons generated by SH can be multiplied by the impact ionization process promoted by FF. Thus, when some part of the laser energy is converted into SH, one can expect a lower modification threshold.

When the laser light is focused by means of a spherical microlens, the strong field maximum is formed beneath the sphere. The smaller wavelength of SH compared to FF allows the microlens to focus the light into a smaller spot. Indeed, FDTD calculation of the field intensity distribution on glass surfaces beneath the polystyrene sphere (typical experimental conditions) shows that irradiation at SH results in about a factor of 1.9 smaller focal spot compared to FF (800 nm). As was mentioned above, the multi-photon absorption order at FF is twice that at SH. However, it can easily be shown that even with allowance for the different orders of the multi-photon absorption processes, SH is advantageous over FF for the localized material modification.

It was discussed above that when CPLAs are employed for the laser beam focusing, coupling of the spherical modes of the constituent microspheres results in multiple cross scattering of light within the array. However, this deteriorative effect may be different at the FF and SH wavelengths. For both the SH and the FF, the presence of the neighbors results in a smaller enhancement and a weaker localization of the field beneath the central sphere. Our calculations show that *the effects of the neighboring spheres on both the laser field enhancement and the localization by a spherical microlens is less for SH than for FF.* It is interesting to note that the focal volume elongation effect is much more pronounced for FF than for SH.

Thus, calculations show that the CPLA focusing systems provide a better field localization at SH than at FF. As was argued above, conversion of only a small fraction of the beam energy into SH can significantly lower the modification threshold. This also means that *even if the major part of the energy remains in FF, the localization of the modification process would be governed by SH for near-threshold fluences.* Thus, more localized structures can be obtained.

Experimentally, PS spheres about 1 μm in diameter were deposited both on PMMA and glass plate substrates. In our experiments [14], we used the Titanium Sapphire laser system Spitfire-Pro (Spectra Physics Co.) in a single-shot regime.

The pulse duration was 50 fs, the energy of a single pulse was 1.7 mJ, the central wavelength was 800 nm, and the beam diameter was 7 mm. A flat-convex lens with a focal length of 15 cm was used for the beam focusing. We studied the formation of periodic pit and hillock (hole and bump) nanostructures in different irradiation regimes. The samples were irradiated by single femtosecond pulses of fundamental frequency (FF), of the second harmonic (SH), or by bi-chromatic FF+SH pulses. A thin (100 μ) BBO crystal (oee or II type) was used for the SH generation with a maximum integral efficiency of 5 %. Crystal orientation was varied for the efficiency (phase matching) adjustment. The crystal was placed *after* the lens to avoid space separation of the FF and SH pulses [26]. A blue glass filter was used for the SH selection. The fluence was changed by moving the sample along the axis of the focused beam, keeping it far from the air breakdown area. In what follows, when speaking about the fluence of a bi-color pulse, we assume the fluence of the FF alone before matching the BBO orientation.

When the fluence of the laser pulse is increased above a certain threshold level (threshold intensity is about 5×10^{11} W/cm^2 for SH+FF and 10^{12} W/cm^2 for FF), the spheres are eliminated from the substrate within the irradiated area. This process is similar to the cleaning phenomenon (see [27]). When increasing the fluence up to fifteen percent from the cleaning threshold, we obtain well-defined structures. They are ablation craters on PMMA substrates and swelling bumps on glass substrates (see Fig. 13.1c, d). For glass substrates, the elimination threshold, and correspondingly the fluence of the structure formation, are almost twofold lower in the case of irradiation by an FF+SH combination compared to the sole FF. For PMMA substrates, this difference proves to be even more pronounced. In both cases, the fluence of the structure formation for the FF+SH irradiation is significantly smaller than the fluence of cleaning for irradiation by the FF alone. The addition of SH results in more localized structures both on glass and PMMA substrates, allowing one to reach an ablation-pit radius of about 100 nm. When filtering out the FF radiation and keeping the SH alone, the structures appear only if the sample is shifted closer to the focus of the beam relative to the position in which the FF+SH structuring occurs. The shift provides an SH fluence higher than that in the FF+SH beam taking into account the attenuation by the filter. This means that FF radiation in a bi-color beam significantly contributes to the modification process.

Conversion of a part of the energy of the fundamental frequency into the second harmonic *beyond* the focusing lens precludes spatial separation of the SH and FF signals. If the BBO crystal were located before the lens, the FF pulse would go first and the SH pulse behind. Our calculations of temporal intensity distributions of the FF and of the SH within a bi-chromatic pulse at the output of a BBO crystal taking into account the dispersion of group velocity shows that in our case both pulses (FF and SH) propagate together. It is clear, however, that for a more efficient use of the bi-chromatic pulses, the SH pulse should be the leading one.

Thus, it is shown that conversion of about 5 % of the laser pulse energy into the second harmonic provides a decrease in the modification threshold and a change in the morphology of obtained structures. The results indicate a higher sensitivity of the materials to SH and also suggest that for the near-threshold fluences, the localization of the modification process is governed by SH, resulting in production of finer structures.

13.2.2 Nanoablation and Nanoswelling

Laser ablation, a technique for polymer surface processing via layer-by-layer material removal by laser pulses, has been extensively studied in the past 25 years. The main trends in the development of this technique were reviewed elsewhere [28]. In particular, a model was constructed for the laser ablation of strongly absorbing polymers by nano and femtosecond laser pulses [29, 30]. Note, however, that there is no appropriate model for the ablation of weakly absorbing polymers and that specific features of the ablation of such polymers with femtosecond laser pulses have not been fully examined.

Convex surface structures can be produced through laser swelling with no material removal. Laser exposure of polymers and polymer-like materials below the ablation threshold produces a bump (hump) (see Fig. 13.1a, d and [31–35]). This effect may be due to both expansion of the irradiated material and substance redistribution over the surface through hydrodynamic effects. The former process is usually referred to as laser swelling, whereas the formation of bumps on an initially smooth surface exposed to laser radiation is often referred to as bumping. In producing nanofeatures, swelling is preferable because the response of the material to irradiation is then more local. Swelling is also of interest because it leads to the formation of regions with increased free volume. The kinetics of chemical reactions in polymer matrices is sensitive to free volume. Therefore, laser swelling can be used to produce surface nanostructures with enhanced reactivity. Given that chemical reactions in "nanoreactors" have a number of specific features, it is reasonable to anticipate that studies of laser-induced surface nanostructuring through swelling will give interesting, unexpected results. Porous convex structures can be selectively doped with luminescent dyes.

Sometimes (see, e.g., [36, 37]) swelling is due to the formation of micro- and nanocavities in the irradiated region during the stress relief. We consider this phenomenon in a special section below. At the same time, surface nanoswelling near the swelling threshold seems to obey a different mechanism.

Relaxation model for the laser swelling of polymers has been formulated in [38]. Laser heating of a polymer material to a temperature greater than the glass transition temperature converts the material to a rubber state. In this state, the Young modulus significantly decreases while the thermal-expansion coefficient significantly increases, reaching about 10^{-3} K^{-1}.

This transition requires the rearrangement of some parts of polymer chains containing 5–20 monomer parts. This process is a cooperative one and consists of simultaneous movement of the whole group of segments. In polymers, this process is related to α-relaxation. The relaxation time strongly depends on temperature.

The heating occurs during the laser pulse, while the cooling is provided by heat diffusion. Due to the finite time of relaxation, the change in volume follows the change in temperature with some delay. Typically, this delay means that sharp heating of the material during a short laser pulse will continue to increase the material volume for some time after the pulse is over. During cooling of the material after the end of the pulse, the volume at some time can reach an equilibrium value for the temperature at that time. When the cooling goes further, the volume will decrease with delay and when the cooling proceeds up to the room temperature there is a chance that the relaxation being slower and slower cannot compensate for the change in temperature because the relaxation time strongly increases with decreasing temperature. This results in the creation of a permanent hump, or the residual swelling.

If u_r is the relaxing volume change in the rubber state, then the simplest equation yielding the evolution of u_r can be written as

$$\dot{u}_r = (u_{r0}(T) - u_r)/\tau(T). \qquad (13.1)$$

Equation (13.1) is similar to the Kelvin-Voigt equation for a viscous elastic body when the outer stress is given and the deformation relaxes to the steady state. In this case, the temperature provides a steady value of $u_{r0} = \alpha_1(T - T_1)$ to which the actual volume change relaxes. Here, α_1 is the thermal-expansion coefficient in the rubber state, and we use T_1 instead of T_g to allow for some uncertainty in determining of the glass transition temperature.

It can be shown that experimental data on the swelling of dye-doped PMMA exposed to nanosecond frequency-doubled Nd : YAG laser pulses [39] and swelling dynamics in undoped PMMA exposed to 248-nm KrF excimer laser radiation [40] fit well the relaxation model. When comparing to the model with the experimental data, we employed the approximation [41] for the dependence of relaxation time on temperature with $T_2 = $ const:

$$\tau = \tau_0 \exp(T^*/(T - T_2)). \qquad (13.2)$$

Thus, near the ablation threshold the swelling seems to have a relaxation nature.

When considering swelling on nano and microscales [42], one cannot use the above point-like model because of the fundamental non-uniformity of the laser heating and the influence of the outer, not heated neighboring parts of the material.

The simplest generalization of the relaxation model could be a hydrodynamic model which, taking into account that the shear modulus above the glass transition point is three orders of magnitude smaller than the bulk modulus, considers swelling in terms of heating and cooling of a viscous compressible liquid with strongly temperature-dependent shear and bulk viscosities. In the limit of the homogeneous heating corresponding to the point-like model (13.1) one easily obtains Kelvin-Voigt equation (13.1) with the relaxation time $\tau = (\xi + 4\eta/3)/K$, where ξ and η are the bulk and shear viscosities, respectively, and K is the bulk modulus.

In the case where the laser heating is significantly localized as it is with the nanoswelling, the so-called "beaker" approximation [43] is used. Here, the simultaneous heating of the beaker, a cylinder domain of radius R and length L, within the material bulk contacting the material surface to a temperature above the glass transition one is considered. During the cooling it is assumed that the temperature is equal within the cylinder and the material outside the beaker is kept in glass state. Approximate consideration of this simplified model for the height of the hump h << R reduces the height evolution problem to the Kelvin-Voigt equation with the temperature-dependent relaxation time

$$\tau = \left(\xi + \eta \left(4/3 + 2L^2/R^2\right)\right)/2K(1+\beta) \text{ and } \beta = (8\alpha_s L)/KR^2,$$

where α_s is the surface tension. The above expressions indicate important dependence of the swelling kinetics on different kinds of viscosity, dimensions of the heated region, R and L, as well as on the surface tension.

It should be noted however, that the complete hydrodynamic theory for laser swelling should have a solution (either analytical or numerical) of the full hydrodynamic equation together with the heat diffusion equation in deformed medium allowing for the temperature dependence of the surface tension and the outer stress due to the thermo acoustic response of the medium to sharp laser heating. This theory is now in progress.

A fundamental problem of the theory of laser swelling is the absence of both real experimental and analytical data on bulk viscosity and its dependence on temperature and pressure, while the temperature dependence of shear viscosity on temperature is known to follow relation (13.2) or similar relation (see, e.g., [44] for a review).

13.3 3D Nano-Structuring

Contrary to the surface nano patterning, lasers have no real competitors in 3D nano and micro-structuring within the material bulk. Thus, the successive recording modes such as direct laser writing (DLW) are acceptable. Below, we consider the two main ways for 3D structuring. The first is 3D bitwise information recording, and the second is formation of connected nanostructures by 3D laser nanopolymerization. In both cases, we deal with the spatial resolution problem. It should be noted that the peak of scientific activity within the field of 3D laser information recording was in the beginning and the middle of the 2000s, whereas the 3D nanopolymerization is a rapidly developing and topical technology at the moment. Nevertheless, the approaches developed for information recording are interesting and apply for nanopolymerization as well, especially since femtosecond laser polymerization is one of the possible ways of information recording although now it is used mainly for the creation of photonic structures.

13.3.1 3D Photochemical Information Recording

3D optical memory provides increased density of information storage compared
to the conventional 2D systems where bitwise information is recorded on the surfaces of the disk. There are different approaches to recording and reading of bitwise
information within the bulk of material [45–50]. In what follows, we consider bitwise
information recording provided by the local absorption of an appropriate amount of
photons. Thereby, the speed of photon absorption does not matter. An example is the
well-known fluorescent memory [46, 47] in which the absorbed photons are used to
provide the photochemical reaction with a product that is fluorescent when excited
at the wavelength of reading. The widely discussed recording with photochromic
molecules [48] is also of this type. Here, the photochemical isomerization is used to
provide information reading through changes in the local optical properties of the
materials. Since this type of recording is often associated with the photochemical
reaction, we call it photochemical recording.

Usually, the 3D bitwise recording is provided by sequential (parallel) focusing
of the laser beam (beams) in the position (positions) of recorded bit (bits). Namely,
"unity" is recorded if a specific bit position is irradiated and "zero" is recorded
otherwise. However, the problem is that when unity is recorded in a given position, spurious cross-talk writing occurs in the neighboring positions. The limitations
imposed by this circumstance are discussed in this section for both single- and two-
photon absorption (see [51] for detail).

We now consider how to maximize the recording density of a 3D array of voxels
(bits of information) using focused laser beams. This problem was solved for a
3D photochemical bitwise laser-assisted information recording [51], but it is also
important, e.g., for the laser polymerization considered below. In [51], a quantity
was introduced to characterize the spurious recording level, namely, the ratio of the
number of photons absorbed at a given point when unity is written into all other
points to the number of photons that would be absorbed at a given point when unity
is written into it. We introduce the concept of a permissible cross-talk recording
level η_p. The laser beams are assumed to be Gaussian. Analysis of single-photon
recording indicates that an increase in the number of layers with a permissible value
of η_p should be accompanied by an increase in the distance between the bits within the
layers in such a way that the information density calculated per square centimeter of
the disk does not increase. This means that the transition from 2D to 3D information
recording with single-photon absorption does not increase the recording density per
unit surface area. Therefore, 3D single-photon information recording is ineffective.
At the same time, analysis of two-photon information recording indicates that for a
given permissible spurious exposure, there is an optimal configuration of recording
points which maximizes the volume recording density. In particular, for a numerical
aperture of the objective NA $= 1$, the refractive index of the medium $n = 1.5$,
wavelength $\lambda = 800\,nm$, and permissible spurious recording level $\eta_p = 0.1$, the
volume recording density ρ_{inf} may reach 3×10^{13} bit cm^{-3}. This means that a disk
12 cm in diameter and 0.5 mm in thickness may contain about 20 TB of information.

At $\eta_p = 0.1$, the separation between the layers will be $0.75\,\mu$m. Thus, the disk will have 670 layers. The permissible spurious exposure level is determined by the reading procedure. At NA $= 1$ and $\eta_p = 0.7$, such a disk may contain 65 TB of information.

13.3.2 3D Nanopolymerization

13.3.2.1 Fluctuation Limit

The most versatile technique for producing 3D nanostructures is sequential processing of a polymerizable medium using a well-focused femtosecond laser beam (for more detail, see in Chap. 12). Femtosecond lasers make it possible to effectively use two-photon absorption for polymerization initiation. Since there is a nonlinear absorption and gelation threshold, the process takes place only in a small region around the beam spot. The forming blob of polymer gel, commonly referred to as a voxel, may be less than 100 nm in size [52, 53]. Translating the sample relative to the beam by a precision positioning system, one can create 3D patterns composed of individual voxels (see [54]). Even though laser polymerisation has been the subject of intense experimental studies, very little work has been directed towards theoretical modeling of the process.

In laser nanopolymerization, the threshold for the response of a substance to laser exposure is determined by a percolation-like transition associated with the formation of gel, namely, a 3D network of macromolecules. In the case of homogeneous polymerization, the percolation transition has a threshold in terms of the monomer conversion to polymer. Recent work [55] has shown, however, that the minimum size of the nanofeatures produced by laser nanopolymerization is limited by random inhomogeneities in the network of polymer molecules (Fig. 13.2). When the threshold is exceeded only slightly, the resulting voxel is of a purely fluctuation nature. The properties of such voxels were studied in [55]. It has been shown that different samples of fluctuation voxels are dissimilar. The centroid and size of voxels fluctuate from implementation to implementation, and the number of voxels may differ from unity. In other words, irradiation results are irreproducible under such conditions. To make up reproducible voxels, these should have a non-fluctuating core, as is shown in Fig. 13.2c. A generalized analytical formula for various spatial distributions of the monomer conversion to polymer is derived in [55]. This formula can be used to estimate the minimum radius of a reliably produced voxel with negligible fluctuations. To this end, the results of the existing gradient percolation theory (percolation for a given spatial distribution of the percolation parameter) were generalized to a wider range of spatial distributions compared to those considered in the literature. The formula was verified using Monte Carlo simulation.

Analysis shows that despite the threshold-like response of the media to the laser irradiation fluctuation nature of the percolation transition precludes reducing the size of a voxel to much below the size of the laser beam limited by diffraction. Additional

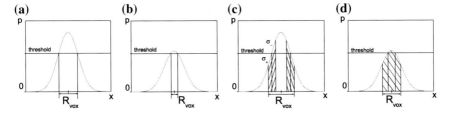

Fig. 13.2 Schematic illustrating the fabrication of a small voxel by laser exposure using a threshold response of the material: **a, b** ideal case in which any small increase in intensity (or in appropriate integral characteristics of the exposure) above the threshold produces a voxel, which can thus be made as small as desired; **c, d** percolation transition as a threshold process in which σ_+ and σ_- fluctuation zones form at the voxel boundary; **d** the resulting voxel is of a purely fluctuation nature when the threshold is exceeded only slightly (R_{vox} is the voxel radius)

tools should be used to lower this limit. One of the ways is to employ a two-beam technique of direct laser writing DLW STED [56]. Another approach is addressed below.

13.3.2.2 Diffusion-Assisted Direct Writing

Quencher or inhibitor of polymerization is used to introduce one more threshold in the polymerization process. It is believed that this additional threshold helps improve the spatial resolution. A quencher molecule interacts with a free radical one, thereby preventing the polymerization process to proceed. Due to this reaction, the quencher molecule is consumed. Thus, the laser light should produce some amount of radicals through the initiation reaction in order to consume almost all the quenchers to start the polymerization process at the particular point. It is an old idea of the additional threshold [57]. A new idea is to employ the diffusion of the quencher [58]. While the diffusion of small radicals as well as macromolecules can be detrimental to spatial resolution [59], the diffusion of the quencher, which inhibits the polymerization, can have a positive effect.

One of the prominent positive effects of the quencher diffusion is recovery of the initial quencher spatial concentration upon irradiation. When writing two adjacent lines, the quencher in the interstice is consumed. When the lines are written at a sub-diffraction distance to each other, the quencher in the interstice is totally consumed and the polymerization is started as though it was a single feature, as is seen in Fig. 13.3b. However, when the quencher recovers to its initial concentration before writing the next neighboring line, such limitation of the spatial resolution relaxes (Fig. 13.3c).

Moreover, quencher diffusion also helps improve the spatial resolution in terms of the size of an elementary voxel. To understand this quencher diffusion effect, we consider a basic photo-polymerization model. The evolution of spatial distributions of the number densities of the quencher molecules (Q), free radicals (R), and free monomer molecules (M) can be modeled by the following set of equations:

$$\frac{\partial R}{\partial t} = S\left(\vec{r},t\right) - k_{tQ}QR, \quad \frac{\partial Q}{\partial t} = D_Q \Delta Q - k_{tQ}QR, \tag{13.3}$$

$$-\frac{\partial M}{\partial t} = k_p RM, p = \frac{M_0 - M}{M_0} \tag{13.4}$$

Free radicals are generated as a result of the absorption of laser light by a photo-initiator. For two-photon absorption, the source term is proportional to the square of the local field intensity I, pulse duration t_p, and pulse repetition rate R_p: $S \propto I^2(\vec{r},t)\,t_p R_p$. As a result of the quenching reaction, both the quencher and the radical are consumed with the reaction rate K_{tQ}. The diffusion coefficient of the quencher is D_Q and Δ is the Laplacian. The polymerization rate is given by (13.4) and is proportional to the monomer number density M, the radical number density, and the propagation constant k_p. The conversion p indicates the degree of polymerization (here, M_0 is the starting number density of the monomer).

Consider the set of equations (13.3) in more detail. If the line scan is much slower than the diffusion of the quencher, one can consider long-lasting irradiation with a source that is constant in time. In order to understand the main effect, we consider the one-dimensional problem in half-space (coordinate $x > 0$) with $S = S_0$ for $x > 0$ and with the boundary condition $Q = Q_0$ at $x = 0$. The stationary solution of set (13.3) with the above boundary condition satisfies the equation $D_Q \partial^2 Q/\partial x^2 = S_0$ and yields $Q = Q_0 (x/x_c - 1)^2$ for $x < x_c$, and $Q = 0$ for $x > x_c$. Here, $x_c = \sqrt{2D_Q Q_0/S_0}$ (see Fig. 13.4).

This means that the quencher is localized within the scale x_c which depends on the laser intensity. The larger the intensity, the larger S and the smaller the scale of quencher localization x_c. It should be noted that for $x > x_c$ the stationary solution for radical concentration R is no longer valid and the process in this domain occurs as if there is no quencher at all.

A similar consideration was employed in [60–62] to clarify the UV oxidation phenomenon for polymers such as PVC and PMMA. Here, oxygen effectively reacting with the radicals plays the role of a quencher. The above model explained the discovered kinetic features of the process by the formation of an oxidized domain just beneath the irradiated surface and a non-oxidized one within the material bulk. In the latter domain, the photo-activated processes occur as if there is no oxygen at all. It is important to note that the length of the oxidized domain depends on the light intensity.

Coming back to the polymerization, we consider the size of the irradiated domain of finite length L and allow the quencher to diffuse on both sides (see Fig. 13.4b). In this case, as is seen in Fig. 13.4b, a domain of size $d = L - 2x_c$ where the quenching is not effective, will form in the middle of the irradiated zone. The length d decreases with decreasing irradiation intensity. For a sufficiently small intensity, d could be significantly smaller than L. While L is typically comparable to the beam size, d can be made much smaller.

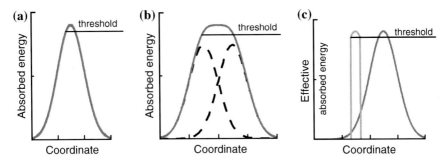

Fig. 13.3   a Formation of a single sub-diffraction sized feature is possible due to the polymerization threshold. **b** When the quencher diffusion is not effective, the formation of two features at a sub-diffraction distance is limited by tails of the distributions of the energy absorbed during scans (*blue dotted lines*). The total absorbed energy (*black line*) exceeds the threshold not only where the nanofeatures are expected to form, but also in the interstice. **c** Formation of the second feature at a sub-diffraction distance from the first one assisted by the quencher diffusion. The energy absorbed during the first scan causes both the consumption of the quencher and the formation of the polymer feature. Since the quencher is diffusion-regenerated between scans, the only effect of the irradiation that remains is the formation of the polymer feature (*green line*). This allows creation of the second feature at a sub-diffraction distance

The irradiation time, conversely, does not affect the size of the quencher-free domain. An increase in the irradiation time (or a decrease in the scan speed in the case of direct laser writing DLW) only causes an increase in the polymerization degree (conversion) within the quencher-free domain (Fig. 13.4d). On the contrary, in the threshold model (without the quencher diffusion), the maximum polymerization degree of a nanofeature can be increased by applying a higher irradiation dose either by increasing the beam intensity or the irradiation time (Fig. 13.4c). However, the higher dose inevitably results in increased size of the polymer feature. By employing the diffusion of the quencher, one can handle both the size and the maximum conversion of the polymer feature separately. The formation of sharp spatial distributions of the conversion is important for achieving a high spatial resolution, a high mechanical stability, and resistance to fluctuations discussed above.

Use of the approach based on the quencher diffusion allowed the authors of [58] to fabricate woodpile structures with an interlayer period of 400 nm, which is comparable to what has been achieved by the two-beam DLW-STED technique currently regarded as state-of-the-art.

13.3.3 Nano-Stucturing Through Instabilities

Above we considered laser nano- and micro-structuring of materials that was provided by tightly focused laser beams. However, nanostructured materials can also be produced even by laser exposure without tight focusing if irradiation leads to

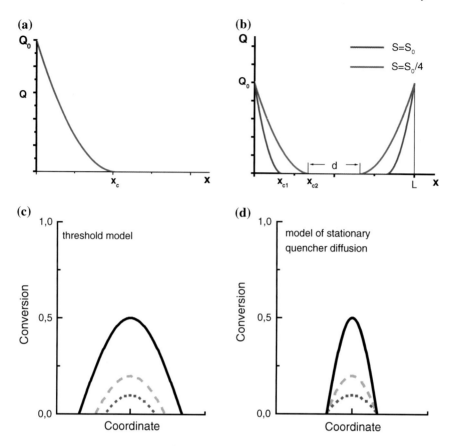

Fig. 13.4 a Concentration of the quencher is localized within the scale $x_c = \sqrt{2D_Q Q_0 / S_0}$. **b** In the case of the irradiated domain of finite length L, there is a quencher-free domain of size d dependent on the irradiation intensity. By decreasing the laser intensity, d can be decreased to sizes much smaller than L. **c** Schematic of the conversion profiles in threshold polymerization regime for different irradiation doses. **d** Schematic of the conversion profiles in the model of stationary quencher diffusion for different irradiation times and fixed irradiation intensity

the development of an instability [63] resulting in the formation of nanoclusters or nanoinhomogeneities in an initially homogeneous medium. The examples of such instabilities are laser- induced bubbling and photo-induced formation of metal nanoclusters within dielectric matrices. The development of such inhomogeneities typically has a significant effect on the optical properties of such materials, which may be of considerable practical interest. In this context, it is important to produce materials capable of such nanostructuring under laser irradiation.

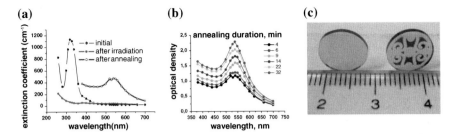

Fig. 13.5 Effects of the UV irradiation and heat treatment on the attenuation spectrum of a composite: **a** UV irradiation eliminates the peak at 320 nm due to HAuCl4. The subsequent heat treatment gives rise to a peak due to the formation of gold nanoparticles in the polymer; **b** effect of heat treatment at 75 °C on the attenuation spectrum of a 50-mm thick PMMA film (peak-attenuation wavelength $\lambda = 540$ nm). **c** Bulk PMMA sample synthesized by polymerization. The *left* sample is a freshly prepared polymer. The *right* sample is UV-irradiated by a XeCl laser through a figured steel stencil over 20 min and subsequently annealed over 3 min at 160 °C (for details see [71]). Color parts are the domains containing Au nanoparticles

13.3.3.1 Photo-Induced Nanocomposites

In this chapter, we consider the photo-induced formation of metal nanoclusters, i.e., nanocomposites. The most widespread chemical method for the preparation of nanoparticles is the reduction of metal compounds in solution in the presence of various stabilizers [64]. Another approach to the fabrication of polymer-matrix nanocomposites was developed in [65]. Here, the nanoparticles are not implanted into the polymer, but produced directly in a polymer matrix by reducing dopants. The key feature of this approach is that the polymer performs a number of functions, acting, on the one hand, as a matrix for nanoparticles and, on the other hand, as a stabilizer intended to prevent nanoparticle aggregation and ensure a uniform nanoparticle distribution throughout the polymer and temporal stability of the nanocomposite. In these studies, films 20–200 μm in thickness were produced by casting and spin-coating with PMMA solutions. The atomic gold precursor in the film was HAuCl4, which was added to a polymer solution. The formation of gold nanoparticles was initiated by UV irradiation of the HAuCl4-containing PMMA films.

As the UV source, we used a DRP-400 high/medium pressure mercury lamp or a XeCl excimer laser. The nucleation and growth of gold nanoparticles were followed using attenuation coefficient measurements in the UV and visible spectral regions. Figure 13.5a illustrates the effect of UV irradiation on the attenuation coefficient of the composites. The spectrum of the non-irradiated sample has a prominent peak at 320 nm, which corresponds to the peak-absorption wavelength of HAuCl4. It is clear from Fig. 13.5a that UV exposure leads to HAuCl4 photolysis, as is evidenced by the disappearance of the 320-nm peak. At the same time, there is no absorption in the visible range after this step (initiation of the nanoparticle formation). Next, the irradiation was ceased and the sample was placed in a thermostat maintained at a certain temperature. The formation of nanoparticles was inferred from changes in the attenuation spectrum in the plasmon resonance region of gold nanoparticles. The time variation of the attenuation spectrum was studied in a wide temperature range (20–80 °C). Typical spectra are presented in Fig. 13.5b. As is seen in Fig. 13.5b, thermostating gives rise to an attenuation peak in the plasmon resonance region of gold nanoparticles. Based on the Mie theory [66] and modern models for the size-dependent permittivity of metal particles [67, 68], one can derive the attenuation spectrum of spherical particles, neglecting particle-particle interactions. Note that the shape of the attenuation spectrum is size-dependent. Namely, as the particle radius increases, the attenuation peak grows both in magnitude and relative to the UV absorption, becomes sharper, and shifts towards the longer wavelengths. For the larger particle radii, the peak begins to broaden, and another feature corresponding to the next mode of the plasmon resonance emerges. Analysis of the evolution of the attenuation spectrum during annealing demonstrates that the shape of the spectrum varies insignificantly. The same refers to the peak position and width. Since the evolution of the spectrum is associated with the formation and growth of gold nanoparticles in the film, this behavior of the spectrum suggests that annealing increases the number density of nanoparticles, whereas the particle size distribution remains essentially unchanged. We propose the following model for the photo-induced formation of gold nanoparticles in polymer films [69]. Absorption of an UV photon by a precursor molecule initiates a sequence of chemical reactions, leading to the reduction of gold atoms and the formation of a supersaturated solid solution of gold atoms in the polymer matrix. Subsequent decomposition of the solid solution gives gold nanoparticles, which act as nuclei of a new phase, metal gold. Raising the annealing temperature markedly accelerates gold diffusion and, accordingly, the formation of nanoparticles. Therefore, the formation of gold nanoparticles in this model can be described in terms of the theory of first-order phase transitions. However, attempts to describe the kinetics of gold nanoparticle formation using Zeldovich and Lifshitz–Slezov theories [70] have been unsuccessful because these theories predict a shift of the particle size distribution to larger sizes over time, i.e., an increase in average particle size, whereas in our experiments the particle number density increased with the time, but their size distribution remained unchanged. The above experimental data can be understood if the stabilizing effect of the matrix is taken into account [69]. In a rate equation similar to the Fokker—Planck equation in the Zeldovich theory for the time variation of the particle size distribution, the stabilization effect is represented by an extra relaxation term which corresponds to the transition of a growing nanoparticle to a stabilized, inactive state with a particular transition frequency. The stabilization leads to a steady-state particle size distribution and a monotonic increase in the total number of nanoparticles with a fixed size distribution. Taking this into account allows one to adequately describe the measured spectra and particle growth data and determine model parameters, such as the parameter related to the particle lifetime in the active state and the particle flux through the critical point in particle size space.

As was pointed out above, the synthesis of metal nanoparticles in transparent dielectric matrices considerably changes both the linear and nonlinear optical properties of the material. That the nanocomposites described above result from photo-induced processes allows one to produce nanostructured regions of arbitrary shape and to control the properties and size distribution of the resulting clusters by adjusting the irradiation conditions. Such nanocomposites can be produced using UV lamps, but the use of laser radiation is critical for the fabrication of complex architectures within bulk materials for photonic applications.

The above UV-induced changes in the optical properties of studied materials permit one to prepare diffraction gratings [71]. Good efficiency of diffraction transformation was shown with visible and IR laser radiation. Diffraction gratings made of polymer materials containing gold nanoparticles demonstrated the best efficiency for an IR laser beam (wavelength 1550 nm) at a level of 1.8 %, which corresponds to a refractive-index change of 4×10^{-5}.

To examine the UV-induced nonlinear optical properties, a highly sensitive method based on a spectrally resolved two-beam coupling technique was used for detection of electronic optical nonlinearity in thin polymeric films in the infrared spectral range. The experimental setup [71, 72] has an erbium-doped fiber laser generating 100 fs pulses with high repetition rate at a wavelength of 1570 nm as the light source. Cross phase modulation during interaction of two beams with high and low intensity in the focal area leads to large relative spectrum changes at the edge of the laser spectrum band. Thus, spectral analysis allows getting information about the nonlinear optical properties of studied media. The nonlinearity activation time was measured using a time delay between the pump and probe femtosecond laser pulses; its common value for a fast electronic nonlinear response does not exceed the pulse duration (100 fs). High values of the UV-induced nonlinear refractive index (n_2) and two-photon absorption coefficients due to UV-induced gold nanoparticles were obtained in studied materials. The nonlinear optical coefficients were found for the samples based on PMMA with gold nanoparticles making up about 5 % in mass. The nonlinear refractive index was -1.3×10^{-13} cm^2/W, which is a factor of 300 larger than the nonlinearity of quartz. The two-photon absorption for these samples was 8×10^{-9} cm/W.

Thus, we demonstrate the initially homogeneous materials that become nanostructured due to UV laser irradiation. Metal nanoparticles are formed within the irradiated domains. This results in a dramatic change in the linear and nonlinear optical properties of these materials, thus showing good prospects for their application in photonics.

One example of optical nanocomposite devices is the so-called random lasers [73–76], in which feedback is due to scattering by nanoparticles. Popov et al. [77] reported random lasing in gold nanoparticles embedded in a polymer matrix (PMMA) containing a laser dye. But those particles were inserted into the matrix. The approach described above can in principle be used to produce random lasers in UV-irradiated regions of the material. At the same time, the above photo generation of gold nanoclusters in a polymer matrix encounters several problems which limit the use of this technology in photonics. One problem is the large number of small particles formed, which make a significant contribution to the optical loss. One way to alleviate this problem is to increase the free volume in the matrix, which might cause the smallest particles to aggregate. Yakimovich et al. [78] used methyl methacrylate/ 2-ethylhexyl acrylate co-polymers as matrices. Ethylhexyl acrylate is known to increase the free volume. The results demonstrate that, all other factors being the same, the attenuation peak of the nanoparticles in such matrices is shifted towards the longer wavelengths, indicating an increase in average nanoparticle size. In general, the larger the nanoparticle, the greater the ratio of its scattering and absorption coefficients, which is also essential for a number of applications.

Another problem is that the above procedure gives films, whereas many applications, including random laser fabrication, require bulk materials rather than films. These bulk materials can be synthesized by means of polymerization. The problem here is that [79] the synthesis of bulk PMMA containing $HAuCl_4$ is very difficult, despite the good solubility of this precursor in MMA. This compound converts to a metal precipitate in the course of polymerization even if the concentration of $HAuCl_4$ is less than 0.5×10^{-3} mol/L. Polymerization starts only when the reduction of the precursor has completely finished. This problem was overcome in our group by Agareva et al. [71, 80]. The proper choice of the precursor, initiator, and regime of polymerization yielded bulk PMMA samples containing a wide range of precursor concentrations. It was demonstrated that the UV irradiation of synthesized bulk PMMA followed by annealing allowed gold nanoparticles to form within only UV-irradiated areas of the polymer matrix. The result of the XeCl laser irradiation of a bulk polymer performed through a figured steel stencil followed by annealing at $160\,^\circ$C for 3 min is presented in Fig. 13.5c.

13.3.3.2 Laser Bubbling

Another example of structuring through an instability is the bubbling phenomenon. Under the effect of laser radiation just near the ablation threshold sometimes one can observe numerous cavities within the irradiated polymer. The cavity (bubble, void) formed within the laser irradiated materials can be due to cavitation bubbles [81] created during the rarefaction wave propagation (originated from the reflection of a pressure wave from the free surface of the sample). The creation of cavitation bubbles relies on the glass transition temperature (T_g) of the polymer and excess of the laser heated material temperature over T_g, as well as on the value of the maximum tensile stress provided by the laser pulse.

A detailed approach for the KrF excimer laser polymer bubbling is published in [36]. According to [36], the threshold of laser bubbling is related to a homogeneous nucleation threshold followed from the Zeldovich formula. However, at the moment, this theory is not developed enough to compare its results with the experimental data. However, a more comprehensive analysis shows that other steps of the cavity creation, namely, the cavity growth due to tensile stress during propagation of the rarefaction wave followed by its collapse or relaxation upon passage also are very important and crucial.

In [37], the phenomenon of bubbling within the Paraloid B72 polymer samples prepared by a casting technique due to single pulse irradiation by a KrF laser for the sub-ablation fluences was clarified experimentally. A methodology relying on the observation of morphological alterations in the bulk material (Paraloid B72) by using third-harmonic generation is developed. This non-destructive procedure permits detailed and accurate imaging of the structurally laser-modified zone extent in the vicinity of the irradiated area. Visualization and quantitative determination of the contour of the *laser-induced swelling/ bulk material interface* are carried out and the data on the position of this interface for the laser fluence are reported.

In order to address these data theoretically, an alternative to the above cavitation bubble generation mechanisms, which can be considered for polymer films fabricated by the casting method, is suggested.

Consider a liquid droplet of radius r within the polymer matrix. This droplet can be originated from solvent residuals due to casting. Upon laser irradiation, both the matrix and the droplet are heated, and the droplet reaches its boiling temperature. If the pressure of the (liquid) droplet saturated vapor is equal to the surface tension pressure, then the cavity will expand. The evaporating droplet will provide enough gaseous molecules to support the bubble growth. If the growth proceeds up to the complete evaporation of the liquids, then the pressure will change from the saturated pressure to the pressure of an ideal gas with a fixed number of gas molecules within the bubble. Due to the subsequent cooling of the matrix, the pressure of the vapor inside the bubble decreases, and the bubble starts to collapse once the surface tension pressure overcomes the vapor pressure. Strong dependence of the viscosity on temperature can prevent the elimination of a bubble and stabilize it with some final size. According to this model, the distance from the surface at which the bubbles remain is the position of the rear border of the bubbling zone z_{rear} measured experimentally. The growth and collapse of a bubble was addressed using the Rayleigh-Plesset equation [82] for the case of small bubbles in a high-viscous liquid (in which inertial terms are neglected). In this simplified model, it was assumed that the gas temperature inside the bubble is equal to the temperature of the surrounding material. The temperature evolution was described by a simple heat diffusion equation, $\partial T / \partial t = a \partial^2 T / \partial z^2$, with the initial conditions $T(z, 0) = T_{room} + (\alpha \phi / c_P \rho) e^{-\alpha z}$ and boundary condition $\partial T / \partial z|_{z=0} = 0$. Here, ϕ is the laser fluence and α is an effective absorption coefficient. It follows from the solution of this equation that for each point $z > 0$ the temperature initially increases, then approaches the maximum value $T_{max}(z)$, and finally decreases. Analysis of the above model of bubble dynamics shows that for different fluences the position of the rear border of the bubbling zone z_{rear}, corresponds to a fixed value of the maximum temperature $T_{max}(z_{rear}, \phi) \approx T^*$ with $T^* =$ const. For the considered experimental data, this constant was determined to be 394 K. It was shown that $z_{rear}(\phi)$ determined from the above equation fits the reported experimental data with $\alpha_{eff} = 1000 \, cm^{-1}$.

References

1. L. Novotny, B. Hetch, *Principles of Nano-Optics* (Cambridge University Press, Cambridge, 2008)
2. Y.F. Lu., B. Hu, Z. H. Mai, W. J. Wang, W. K. Chim, T. C. Chong, Jpn. J. Appl. Phys. **40**, 4395 (2001)

3. T.C. Chong, M.H. Hong, L.P. Shi, Las. Phot. Rev. **4**, 123 (2010)
4. T.M. Bloomstein, M.F. Marchant, S. Deneault, D.E. Hardy, M. Rothschild, Opt. Express **14**, 6434 (2006)
5. Y.F. Lu, W.D. Song, Y.W. Zheng, B.S. Luk'yanchuk, JETP Lett. **72**, 457 (2000)
6. H. Hasegawa, T. Ikawa, M. Tsuchimori, O. Watanabe, Y. Kawata, Macromolecules **34**, 7471 (2001)
7. H.J. Muenzer, M. Mosbacher, M. Bertsch, J. Boneburg, J. Micros. **202**, 129 (2001)
8. B.S. Luk'yanchuk, Z.B. Wang, W.D. Song, M.H. Hong, Appl. Phys. A **79**, 747 (2004)
9. G. Langer, D. Brodoceanu, D. Bäuerle, Appl. Phys. Lett. **89**, 261104 (2006)
10. W. Wu, A. Katsnelson, O.G. Memis, H. Mohseni, Nanotechnology **18**, 485302 (2007)
11. A. Khan, Z.B. Wang, M.A. Sheikh, D.J. Whitehead, L. Li, Appl. Surf. Sci. **258**, 774 (2011)
12. A. del Campo, E. Arzt, Chem. Rev. **108**, 911 (2008)
13. Y. Verevkin, N.G. Bronnikova, V.V. Korolikhin, Y.Y. Gushchina V.N. Petryakov, D.O. Filatov, N.M. Bityurin, A.V. Kruglov, V.V. Levichev, Tech. Phys. **48**(6), 757 (2003)
14. N. Bityurin, A. Afanasiev, V. Bredikhin, A. Alexandrov, N. Agareva, A. Pikulin, I. Ilyakov, B. Shishkin, R. Akhmedzhanov, Opt. Express **21**, 21485 (2013)
15. M. Rothschild, Opt. Photonics News **21**(6), 26–31 (2010)
16. Y. Nakata, K. Murakawa, K. Sonoda, K. Momoo, N. Miyanaga, T. Hiromoto. Appl. Phys. A **112**, 191 (2013)
17. A. Pikulin, N. Bityurin, G. Langer, D. Brodoceanu, D. Bäuerle, Appl. Phys. Lett. **91**, 191106 (2007)
18. Z.B. Wang, W. Guo, B. Luk' yanchuk, D.J. Whitehead, L. Li, Z. Liu, J. Laser Micro-Nanoengin. **3**, 14 (2008)
19. N. Arnold, Appl. Phys. A **92**, 1005 (2008)
20. A. Pikulin, A. Afanasiev, N. Agareva, A. Alexandrov, V. Bredikhin, N. Bityurin, Opt. Express **20**, 9052 (2012)
21. L. Englert, B. Rethfeld, L. Haag, M. Wollenhaupt, C. Sarpe-Tudoran, T. Baumert, Opt. Express **15**, 17855 (2007)
22. C. Mauclair, M. Zamfirescu, J.P. Colombier, G. Cheng, K. Mishchik, E. Audouard, R. Stoian, Opt. Express **20**, 12997 (2012)
23. L. Englert, M. Wollenhaupt, C. Sarpe, D. Otto, T. Baumert, J. Laser Appl. **24**, 042002 (2012)
24. N. Bityurin, A. Kuznetsov, J. Appl. Phys. **93**, 1567 (2003)
25. L.V. Keldysh, Sov. Phys. JETP **20**, 1307 (1965)
26. R.A. Akhmedzhanov, I.E. Ilyakov, V.A. Mironov, E.V. Suvorov, D.A. Fadeev, JETP **109**, 370 (2009)
27. B. Luk'anchuk (ed.), *Laser Cleaning* (World Scientific, Singapore, 2002)
28. N. Bityurin, Ann. Rep. Prog. Chem. C **101**, 216 (2005)
29. N. Bityurin, A. Malyshev, J. Appl. Phys. **92**(1), 605 (2002)
30. N. Bityurin, B.S. Luk'yanchuk, M.H. Hong, T.C. Chong, Chem. Rev. **103**, 519 (2003)
31. H. Fukumura, N. Mibuka, S. Eura, H. Masuhara, Appl. Phys. A **53**, 255 (1991)
32. F. Beinhorn, J. Ihlemann, K. Luther, J. Troe, Appl. Phys. A **68**, 709 (1999)
33. H.M. Phillips, R. Sauerbrey, Opt. Eng. **32**, 2424 (1993)
34. M. Himmelbauer, E. Arenholz, D. Bäuerle, K. Schilcher, Appl. Phys. A **63**, 337 (1996)
35. M. Himmelbauer, N. Arnold, N. Bityurin, E. Arenholz, D. Bäuerle, Appl. Phys. A **64**, 451 (1997)
36. S. Lazare, I. Elaboudi, M. Castillejo, A. Sionkowska, Appl. Phys. A **101**, 215 (2010)
37. A. Selimis, G.J. Tserevelakis, S. Kogou, P. Pouli, G. Filippidis, N. Sapogova, N. Bityurin, C. Fotakis, Opt. Express **20**, 3990 (2012)
38. N. Bityurin, Appl. Surf. Sci. **255**, 9851 (2009)
39. A.Yu. Malyshev, N.A. Agareva, O.A. Mal'shakova, N.M. Bityurin, Opt. Zh. **74**, 80 (2007)
40. T. Masubuchi, H. Furutani, H. Fukumura, H. Masuhara, J. Phys. Chem. B **105**, 2518 (2001)
41. J.D. Ferry, *Viscoelastic Properties of Polymers*, 3rd edn. (Willey, N.Y., 1980)
42. E. McLeod, C.B. Arnold, Nat. Nanotechnol. **3**, 413–417 (2008)

43. N. Bityurin, in *Abstracts of 11 international conference on laser ablation*, COLA 2011, Playa del Carmen, Mexico, p. 29 (2011)
44. G. Floudas, M. Paluch, A. Grzybowski, K.L. Ngai, *Molecular Dynamics of Glass-Forming Systems* (Springer, Berlin Heidelberg, 2011)
45. D. Parthenopoulos, P. Rentzepis, J. Appl. Phys. **68**(11), 5814 (1990)
46. M.M. Wang, S.C. Esener, F.B. McCormick, I. Cokgor, A.S. Dvornikov, P.M. Rentzepis, Opt. Lett. **22**(8), 558 (1997)
47. A.S. Dvornikov, P.M. Rentzepis, Opt. Commun. **136**, 1 (1997)
48. S. Kawata, Y. Kawata, Chem. Rev. **100**, 1777 (2000)
49. E.N. Glezer et al., Opt. Lett. **21**, 2023 (1996)
50. M. Sugiyama et al., Appl. Phys. Lett. **79**, 1528 (2001)
51. N. Bityurin, B.S. Luk'yanchuk, M.H. Hong, T.C. Chong, Opt. Lett. **29**, 2055 (2004)
52. S.H. Park, T. W. Lim, D.-Y.Yang, N.C. Cho, K.-S. Lee, Appl. Phys. Lett. **89**, 173133 (2006)
53. K. Takada, H.-B. Sun, S. Kawata, Appl. Phys. Lett. **86**, 071122 (2005)
54. M. Farsari, B.N. Chichkov, Nat. Photonics **3**, 450 (2009)
55. A. Pikulin, N. Bityurin, Phys. Rev. B. **82**, 085406 (2010)
56. J. Fischer, M. Wegener, Opt. Mater. Express **1**, 614 (2011)
57. S. Kavata, H.-B. Sun, T. Tanaka, K. Takada, Nature **412**, 697 (2001)
58. I. Sakellari, E. Kabouraki, D. Gray, V. Purlys, C. Fotakis, A. Pikulin, N. Bityurin, M. Vamvaraki, M. Farsari, ACS Nano **6**(3), 2302 (2012)
59. A. Pikulin, N. Bityurin, Phys. Rev. B **75**, 195430 (2007)
60. N.M. Bityurin, V.N. Genkin, V.P. Lebedev, L.V. Nikitin, V.V. Sokolov, L.D. Strelkova, G.T. Fedoseeva, Vysokomolec. Soed. **25A**(1), 80 (1983)
61. A.P. Alexandrov, A.A. Babin, N.M. Bityurin, S.V. Muraviov, F.I. Feldchtein, Techn. Phys. Lett. **21**(4), 249 (1995)
62. N. Bityurin, S. Muraviov, A. Alexandrov, A. Malyshev, Appl. Surf. Sci. **110**, 270 (1997)
63. N. Bityurin, Quant. Electron. **40**(11), 955 (2010)
64. M.-C. Daniel, D. Austruc, Chem. Rev. **104**, 293 (2004)
65. A. Alexandrov, L. Smirnova, N. Yakimovich, N. Sapogova, L. Soustov, A. Kirsanov, N. Bityurin, Appl. Surf. Sci. **248**(1–4), 181 (2005)
66. M. Born, E. Wolf, *Principles of Optics* (Pergamon, Oxford, 1969)
67. U. Kreibig, M. Vollmer, *Optical Properties of Metal Clusters* (Springer, New York, 1995)
68. P.B. Johnson, R.W. Christy, Phys. Rev. B **6**, 4370 (1972)
69. N. Sapogova, N. Bityurin, Appl. Surf. Sci. **255**, 9613 (2009)
70. E.M. Lifshitz, L.P. Pitaevskii, *Physical Kinetics* (Pergamon, Oxford, 1981)
71. N. Bityurin, A. Alexandrov, A. Afanasiev, N. Agareva, A. Pikulin, N. Sapogova, L. Soustov, E. Salomatina, E. Gorshkova, N. Tsverova, L. Smirnova, Appl. Phys. A **112**, 135 (2013)
72. A.V. Afanas'ev, A.P. Aleksandrov, N.A. Agareva, N.V. Sapogova, N.M. Bityurin, A.E. Mochalova, L.A. Smirnova, J. Opt. Technol. **78**, 537 (2011)
73. V.S. Letokhov, Zh. Eksp. Teor. Fiz. **5**, 262 (1967)
74. X.H. Wu, A. Yamilov, H. Noh, H. Cao, E.W. Seelig, R.P.H. Chang, J. Opt. Soc. Am. B **21**, 159 (2004)
75. D. Anglos, A. Stassinopoulos, R.N. Das, G. Zacharakis, M. Psyllaki, R. Jakubiak, R.A. Vaia, E.P. Giannelis, S.H. Anastasiadis, J. Opt. Soc. Am. B **21**, 208 (2004)
76. E.V. Chelnokov, N. Bityurin, I. Ozerov, W. Marine, Appl. Phys. Lett. **89**, 171119 (2006)
77. O. Popov, A. Zilbershtein, D. Davidov, Appl. Phys. Lett. **89**, 191116 (2006)
78. N.O. Yakimovich, N.V. Sapogova, L.A. Smirnova, A.P. Aleksandrov, T.A. Gracheva, A.V. Kirsanov, N.M. Bityurin, Rus. J. Phys. Chem. B. **2**(1), 128 (2008)
79. J. Yang, T. Hasell, W. Wang, S.M. Howdle, Eur. Polym. J. **44**, 1331 (2008)
80. N.A. Agareva, A.P. Aleksandrov, L.A. Smirnova, N.M. Bityurin, Perspekt. Mater. **1**, 5 (2009)
81. G. Paltauf, P.E. Dyer, Chem. Rev. **103**, 487 (2003)
82. J.P. Frank, J.-M. Mishel, *Fundamentals of Cavitation* (Kluwer Ac. Publ, Dordrecht, 2004)

Index

Made in the USA
Columbia, SC
13 September 2019